FRACTIONAL
QUANTUM
MECHANICS

FRACTIONAL QUANTUM MECHANICS

Nick Laskin
TopQuark Inc., Canada

 **World Scientific**

NEW JERSEY · LONDON · SINGAPORE · BEIJING · SHANGHAI · HONG KONG · TAIPEI · CHENNAI · TOKYO

Published by

World Scientific Publishing Co. Pte. Ltd.

5 Toh Tuck Link, Singapore 596224

USA office: 27 Warren Street, Suite 401-402, Hackensack, NJ 07601

UK office: 57 Shelton Street, Covent Garden, London WC2H 9HE

Library of Congress Cataloging-in-Publication Data

Names: Laskin, Nick, author.

Title: Fractional quantum mechanics / by Nick Laskin (TopQuark Inc., Canada).

Description: Singapore ; Hackensack, NJ : World Scientific, [2018] |
   Includes bibliographical references and index.

Identifiers: LCCN 2017051779| ISBN 9789813223790 (hardcover ; alk. paper) |
   ISBN 9813223790 (hardcover ; alk. paper)

Subjects: LCSH: Quantum theory--Mathematics. | Fractals. | Path integrals. | Schrödinger equation.

Classification: LCC QC174.17.M35 L37 2018 | DDC 530.1201/514742--dc23

LC record available at https://lccn.loc.gov/2017051779

British Library Cataloguing-in-Publication Data

A catalogue record for this book is available from the British Library.

For any available supplementary material, please visit
http://www.worldscientific.com/worldscibooks/10.1142/10541#t=suppl

Printed in Singapore

Contents

Preface

Published in 1965, the brilliant book by R. P. Feynman and A. R. Hibbs, "Quantum Mechanics and Path Integrals" (McGraw-Hill, New York, 1965) presented an unusually new at the time quantum-mechanical concept - the path integral approach. This approach was an alternative to the Schrödinger wave equation and Heisenberg matrix dynamics. The discovery of the quantum mechanical path integral approach was originated by the attempt of P. A. M. Dirac to find what corresponds in quantum theory to the Lagrangian method of classical mechanics. Dirac searched for the role which classical mechanics fundamentals like the Lagrangian and 'least-action' principle play in quantum mechanics. Based on Dirac's findings, Feynman developed his idea of *"the sum over paths"* and called it *"a path integral"*. Since then, the path integral approach and the perturbation technique based on it, Feynman diagrams, became powerful tools in quantum mechanics and quantum field theory, solid state and quantum liquid theory, equilibrium and non-equilibrium statistical physics, theory of turbulence and chaotic phenomena, theory of random processes and polymer physics, mathematics, chemistry, economic studies.

Soon after Feynman's remarkable invention, M. Kac observed that replacing time t with imaginary time $-i\tau$ in Feynman's path integral brings us Wiener's path integral, which is the integral over Brownian motion paths. Wiener's integral is a well-defined object from the stand point of integration in functional spaces. Feynman's path integral is a rather heuristic object well adjusted to implement perturbation expansions and to develop nonperturbative techniques in quantum theory. This circumstance is the reason to use the term *"Brownian-like paths"* when talking about Feynman's path integral, and use the term *"Brownian motion paths"* when talking about Wiener's path integral. The trajectories of Brownian motion are in fact the first example of fractional physical objects. Hence, the Feynman path integral approach to quantum mechanics based on Brownian-like paths can be considered the first successful attempt to apply the fractality concept to quantum physics.

In the past two decades, it has been realized that the understanding of complex quantum and classical physics phenomena requires the implementation of the Lévy flights random process instead of Brownian motion. What is the motivation behind the involvement of Lévy flights into consideration? It is well known that the position of a diffusive particle evolves as a square root in time, $x(t) \sim t^{1/2}$. The 1/2 law or "square root law" is an attribute of Brownian motion. However, for many complex quantum and classical physics phenomena this temporal diffusive behavior has not been observed. Instead, the more general evolution law $x(t) \sim t^{1/\alpha}$ with $0 < \alpha \leq 2$, where α is the Lévy index, has been found. Thus, the well-known diffusion law $x(t) \sim t^{1/2}$ is included as a special case at $\alpha = 2$. The adequate mathematical model to describe $1/\alpha$ scaling is known as the Lévy flights. The scaling law $1/\alpha$ assigned to the Lévy flights has been empirically observed in quantum phenomena like the laser cooling of atoms, ion dynamics in optical lattice, quantum anomalous transport, and in the measurement of the momentum of cold cesium atoms in a periodically pulsed standing wave of light. Besides this, Lévy flights are widely used to model a variety of classical physics phenomena such as kinetics and transport in classical complex systems, anomalous diffusion, chaotic dynamics, plasma physics, dynamics of economic indexes, biology, physiology and social science.

The Brownian-like quantum dynamics results in the Feynman path integral. So too, the Feynman path integral method ought to be extended to approach complex quantum dynamic phenomena where Lévy-like flight manifests at the quantum level. To extend the Feynman approach the integration has to be expanded from Brownian-like to Lévy-like paths. The basic outcome of the path integral over Lévy flight paths is a new fundamental quantum equation – the fractional Schrödinger equation. In other words, if the Feynman path integral approach to quantum mechanics allows one to derive the Schrödinger equation, then the path integral over Lévy flight paths leads to a fractional Schrödinger equation. This is a manifestation of a new non-Gaussian physical paradigm, based on deep relationships between the structure of fundamental physics equations and fractal dimensions of "underlying" quantum paths. The fractional Schrödinger equation includes the space derivative of fractional order α instead of the second order space derivative in the well-known Schrödinger equation. The Lévy parameter α becomes a new fundamental parameter. The fractional Schrödinger equation is the fractional differential equation in accordance with modern terminology. This is the main point for the term, *"fractional Schrödinger equation"*, and the more general term, *"fractional quantum mechanics"*.

To date, there is no book on fractional quantum mechanics and its applications. This book gives a systematic, in-depth, fully covered and self-consistent presentation on the subject. In addition to providing a deep insight into fractional quantum mechanics fundamentals, the book advances applications of fractional calculus to solve fractional quantum mechanics physical problems. The book pioneers quantum mechanical applications of the α-stable Lévy probability distribution and the H-function, invented by C. Fox in the '60s. It has occurred that the H-function, never before used in quantum mechanics, is a well-suited mathematical tool to solve the fractional Schrödinger equation for quantum physical problems.

The book discovers and explores new fractional quantum mechanics equations. Remarkably, these equations turn into the well-known quantum mechanics equations in the particular case when the Lévy index $\alpha = 2$, and Lévy flights become Brownian motion.

For advanced readers, young researchers and graduate students the book serves as the first monograph and the first handbook on the topic, covering fundamentals and applications of fractional quantum mechanics, time fractional quantum mechanics, fractional statistical mechanics and fractional classical mechanics.

Chapter 1

What is Fractional Quantum Mechanics?

1.1 Path integral

Non-relativistic quantum mechanics can be formulated in the frameworks of three different approaches: the Heisenberg matrix algebra [1]-[3] the Schrödinger equation [4]-[6] and the Feynman path integral [7]-[9]. These three, apparently dissimilar approaches, are mathematically equivalent.

The discovery of the quantum mechanical path integral approach was originated in the early '30s by the attempt of Dirac [10] to find a relationship between classical and quantum mechanics in terms of the classical mechanics 'least-action' principle. In essence, Dirac searched for the role that classical mechanics fundamentals like the Lagrangian and the 'least-action' principle play in quantum mechanics. While he considered the Lagrangian approach to classical mechanics more fundamental than the Hamiltonian one, at the time it seemed to have no important role in quantum mechanics. Dirac speculated on how this situation might be rectified, and concluded that the propagator in quantum mechanics is "analogous" to $\exp(iS/\hbar)$, where S is the classical mechanical action and $\hbar$ is Planck's constant.

Based on Dirac's findings Feynman developed the "integral over all paths", the path integral. How it came to him was well described by Feynman in his Nobel lecture [11]. *I went to a beer party in the Nassau Tavern in Princeton. There was a gentleman, newly arrived from Europe (Herbert Jehle) who came and sat next to me. Europeans are much more serious than we are in America because they think a good place to discuss intellectual matters is a beer party. So he sat by me and asked, "What are you doing" and so on, and I said, "I'm drinking beer." Then I realized that he wanted to know what work I was doing and I told him I was struggling with this problem, and I simply turned to him and said "Listen, do you know any*

1

way of doing quantum mechanics starting with action – where the action integral comes into the quantum mechanics?" "No," he said, "but Dirac has a paper in which the Lagrangian, at least, comes into quantum mechanics. I will show it to you tomorrow." Next day we went to the Princeton Library (they have little rooms on the side to discuss things) and he showed me this paper. Dirac's paper [10] claimed that a mathematical tool which governs the time development of a quantal system was "analogous"to the classical Lagrangian.

Professor Jehle showed me this; I read it; he explained it to me, and I said, "What does he mean, they are analogous; what does that mean, analogous? What is the use of that?" He said, "You Americans! You always want to find a use for everything!" I said that I thought that Dirac must mean that they were equal. "No," he explained, "he doesn't mean they are equal." "Well," I said, "let's see what happens if we make them equal." So, I simply put them equal, taking the simplest example, but soon found that I had to put a constant of proportionality A in, suitably adjusted. When I substituted and just calculated things out by Taylor-series expansion, out came the Schrödinger equation. So I turned to Professor Jehle, not really understanding, and said, "Well you see Professor Dirac meant that they were proportional." Professor Jehle's eyes were bugging out – he had taken out a little notebook and was rapidly copying it down from the blackboard and said, "No, no, this is an important discovery."

Feynman's path integral approach to quantum mechanics [12] brings deep insights into the relationship between quantum and classical mechanics and became a new efficient mathematical tool in quantum theory and mathematical physics. Feynman's thesis advisor, J. A. Wheeler wrote in Physics Today **42**, 24 (1989), *"Feynman has found a beautiful picture to understand the probability amplitude for a dynamical system to go from one specified configuration at one time to another specified configuration at a later time. He treats on a footing of absolute equality every conceivable history that leads from the initial state to the final one, no matter how crazy the motion in between. The contributions of these histories differ not at all in amplitude, only in phase. And the phase is nothing but the classical action integral, apart from the Dirac factor, $\hbar$. This prescription reproduces all of standard quantum theory. How could one ever want a simpler way to see what quantum theory is all about!"*

The path integral approach due to its new overall space-time perspective gives an intuitive way of viewing quantum dynamics and understanding the classical limit of quantum mechanics.

Today Feynman's path integrals are widely used in quantum gauge field theory and the theory of nonperturbative phenomena [13]-[17], physics of dense strongly interacting matter (quark-gluon plasma) [18], statistical physics [19], physics of polymers, theory of phase transitions and critical phenomena [20], atomic physics and high-energy radiation phenomena, quantum optics [21], [22], theory of stochastic processes and its numerous applications [23]-[26], option pricing and modeling of financial markets [20].

1.2 Fractals and fractional calculus

The Feynman path integral is in fact the integration over Brownian-like quantum paths. The Brownian path is an example of a fractal. Following Mandelbrot [27], it is said that a fractal is a self-similar object, that is the whole object looks like any of its parts (for instance, see [27], [28]). The wording *"when the sum of independent identically distributed random quantities has the same probability distribution as each random quantity distributed"* can be considered as an expression of self-similarity - *"when does the whole object look like any of its parts"*. The trajectories of Brownian motion and Lévy flight are self-similar curves. In other words, the trajectories of Brownian motion and Lévy flight can be considered as random fractals. The fractal dimension of a Lévy flight trajectory can be expressed in terms of the Lévy index α which serves as a measure of self-similarity. In fact, the Lévy index α is equal to the Hausdorff–Besicovitch dimension of trajectory of random motion [29]. That is, the fractal dimension of a trajectory of Brownian motion is 2, while the fractal dimension of a Lévy flight path is α, $0 < \alpha \leq 2$.

If Brownian diffusion is described by the well-known diffusion equation with scaling $x(t) \sim t^{1/2}$, then the diffusion governed by the Lévy flight process is described by a fractional diffusion equation with scaling $x(t) \sim t^{1/\alpha}$, $0 < \alpha \leq 2$. The fractional diffusion equation is a well-suited mathematical model to study anomalous diffusion displaying the scaling $x(t) \sim t^{1/\alpha}$.

The standard diffusion equation includes a second-order spatial derivative. The fractional diffusion equation includes a fractional α-order spatial derivative. Thus, the appropriate mathematical tool to study diffusion and diffusion-like processes with $1/\alpha$ scaling is a fractional diffusion equation, which belongs to the mathematical field called fractional calculus [30]-[32].

Fractional calculus is concerned with the generalization of differentiation and integration to non-integer (fractional) orders. The subject has a long history. In a letter to L'Hospital in 1695, Leibniz [33] raised the following question: *"Can the meaning of derivatives with integer order be generalized to derivatives with non-integer orders?"* L'Hospital was somewhat curious about the question and replied to Leibniz: *"What if the order will be 1/2?"* From this question, the study of fractional calculus was born. In a letter dated September 30, 1695 [33], [34] Leibniz responded to the question, *"$d^{1/2}x$ will be equal to $x\sqrt{dx} : x$. This is an apparent paradox from which, one day, useful consequences will be drawn."* Over the centuries many brilliant mathematicians, among them Liouville, Riemann, Weyl, Fourier, Abel, Lacroix, Leibniz, Grunwald, Letnikov and Riesz, have built up a large body of mathematical knowledge on fractional integrals and derivatives. Although fractional calculus is a natural generalization of calculus, and although its mathematical history is equally long, it has just recently become the attractive efficient tool in physics and mathematics.

The important fields of application of fractional calculus in physics are the phenomena of anomalous diffusion, fractional quantum mechanics, quantum and classical dynamics with long-range interactions, long memory relaxation processes with non-exponential time decay, long tail phenomena observed in network communication systems and in the behavior of financial markets. Fractional calculus is an efficient tool to study chaotic dynamic systems and their transport properties. In particular, it helps to obtain some quantitative and qualitative results based on the known features of fractional derivatives.

In mathematics, considerable attention has been recently focused on the study of problems involving fractional spaces and nonlocal equations, where nonlocality is treated in terms of pseudo-differential operators - fractional differential operators, for example, the fractional Laplace operator. Fractional differential operators and the fractional Sobolev spaces attract the attention of many researchers, both from a pure mathematical stand point and from a stand point of various applications, since these operators naturally arise in many different contexts, such as, anomalous diffusion, fractional kinetics, fractional quantum mechanics, fractional statistical mechanics, financial mathematics, signal processing and network communication systems, theory of equilibrium and non-equilibrium phenomena in the systems with long-range interaction, theory and applications of fractional stochastic processes, etc. The bibliography on the topic is so extensive that it is a real challenge to come up with a reasonable list of references. Instead, we direct the readers to papers [31], [32], [35], [36] where an introduction to the subject and an extensive bibliography can be found.

A review of recent developments in the field of fractional calculus and its applications has been presented in [31], [32].

Among the many monographs and textbooks on fractional calculus fundamentals and their wide range of applications, books [30], [37]-[43], which target the application of fractional calculus to a variety of important natural science problems.

1.3   Lévy flights

The understanding of quantum and classical physics phenomena governed by long-range space processes has required the implementation of the Lévy flights random process [44]. It is well known that displacement of a particle from some initial point follows the well-known square root law, $x(t) \sim t^{1/2}$. The square root law is an attribute of the well-known Brownian motion model for diffusion. However, for complex quantum and classical physics phenomena this temporal diffusive behavior has not been observed. Instead, the more general diffusion law $x(t) \sim t^{1/\alpha}$ with $0 < \alpha \leq 2$, where α is the Lévy index, has been found. The mathematical model to describe $1/\alpha$ scaling is known as Lévy flights. What is the reason for the term *"Lévy flights"*? The cloud of Lévy particles spreads faster than the cloud of Brownian particles. The space scale $x(t)$ of the Lévy cloud as a function of time t follows the law $x(t) \sim t^{1/\alpha}$. It is obvious that the lower bound is reached for the Brownian cloud, $\alpha = 2$. That is when $\alpha = 2$ Lévy flights turn into Brownian motion. The typical Lévy flight path looks like a set of diffusion islands connected by long-step straight line flights, the length of which depends on the value of the index α. The 2D Brownian motion is displayed in Fig. 2, and the typical 2D trajectory of Lévy flight is shown in Fig. 3.

The Lévy stochastic process is widely used to model a variety of complex dynamic and kinetic processes, such as anomalous diffusion and kinetics [45], [44], turbulence [46], chaotic dynamics and transport [47], [48], plasma physics [49], dynamics of economic indexes [50], option pricing [20], biology and physiology [51], [52].

The phenomenon of Lévy flights became an object of intensive study in quantum physics. The scaling $1/\alpha$ law assigned to the Lévy flights has been empirically observed in the laser cooling of atoms [53]-[55], in ions dynamics in optical lattice [56], in multiple scattering of light in hot vapours of rubidium atoms [57], in multiple scattering of light in hot vapors of rubidium atoms in the measurement of the momentum of cold cesium atoms in a periodically pulsed standing wave of light [58], in anomalous transport phenomenon in the quantum kicked systems [59], [60].

Lévy flights are a natural generalization of Brownian motion. In mid '30s, Lévy [61] and Khintchine [62] posed the question: When does the sum of N independent identically distributed random quantities $X = X_1 + X_2 + ... + X_N$ have the same probability distribution $p_N(X)$ (up to scale factor) as the individual steps $p_i(X_i)$, $i = 1, ..., N$? The traditional answer is that each $p_i(X_i)$ should be a Gaussian distribution, because of a central limit theorem [63]. In other words, a sum of N Gaussians is again a Gaussian, but with N times the variance of the original one. Lévy and Khintchine found a way to generalize the central limit theorem. They discovered a class of non-Gaussian α-stable (stable under summation) distributions. So, from the stand point of the probability theory the α-stable probability law [63]-[66] is a generalization of the well-known Gaussian law. In other words, the α-stable distribution with index α results from the sum of independent identically distributed variables with probability density $p_\alpha(x) \sim x^{-1-\alpha}$, $x \to \infty$ in the same way that the Gaussian distribution results from sum of independent identically distributed variables with probability density $p_G(x) \sim \exp\{-x^2/2\}$. Each α-stable distribution has a stability index α, called the Lévy index $0 < \alpha \leq 2$. The Lévy index defines the boundary of convergence of statistical moments of fractional order. That is, a fractional statistical μ-order moment of a stable law is finite only if its order μ is strictly smaller than its Lévy index α (i.e. $\mu < \alpha$). Every moment of higher order (including $\mu = \alpha$) is infinite or, as often said, divergent. The only exception is the normal distribution which corresponds to the particular stable law with the Lévy index $\alpha = 2$ and it has an exceptional property that all its statistical moments are finite.

1.4 Schrödinger equation

The *Schrödinger equation* is a fundamental equation of quantum mechanics. Discovered by Schrödinger in 1926 [4], it was historically the second fundamental formulation of the quantum mechanics. The first formulation was the *Heisenberg matrix mechanics* developed by Heisenberg, Born and Jordan in 1925.

Invented by Feynman in 1948 the *Feynman path integral* [7] is an alternative to Heisenberg matrix mechanics [1], [2] and Schrödinger wave equation [4]-[6]. The Feynman path integral is in fact the functional integral over

Brownian-like quantum paths. Therefore, if we go from Brownian-like to Lévy-like quantum paths we come to the *Laskin path integral* discovered by Laskin in 2000 [67]. The Laskin path integral is the generalization of the Feynman path integral. If one applies the Wick rotation $(t \to -i\tau)$ to the Feynman path integral then the Feynman path integral becomes the *Wiener path integral*, invented by Wiener in 1923 [68]. The Wiener path integral is an efficient tool to study standard diffusion phenomena. If one applies the Wick rotation to the Laskin path integral, then the Laskin path integral becomes the generalization of the Wiener path integral. Laskin path integral with implemented Wick rotation is an efficient tool to study anomalous diffusion phenomena [35], [44], [45]. In other words, the well-known mapping between standard quantum mechanics and standard diffusion has been extended by means of the Laskin path integral to the mapping between fractional quantum mechanics and anomalous diffusion. It allows us to apply the path integral technique to study physical phenomena related to anomalous diffusion.

The fractional Schrödinger equation discovered by Laskin [67] is a fundamental equation of fractional quantum mechanics. The fractional Schrödinger equation is a manifestation of the Lévy-like quantum path scaling $\Delta x \sim (\Delta t)^{1/\alpha}$, while the Schrödinger equation can be considered as a manifestation of the Brownian-like quantum path scaling $\Delta x \sim (\Delta t)^{1/2}$ with Δx and Δt being space and time increments of a quantum path.

Therefore, fractional quantum mechanics can be thought of as the quantum mechanics where the *underlying* Brownian-like quantum paths are substituted with the Lévy-like paths. In fractional quantum mechanics the underlying quantum path has fractal dimension α and exhibits Lévy-like scaling $\Delta x \sim (\Delta t)^{1/\alpha}$, while in standard quantum mechanics the underlying quantum path has fractal dimension 2 and exhibits Brownian-like scaling $\Delta x \sim (\Delta t)^{1/2}$. When $\alpha = 2$, fractional quantum mechanics becomes standard quantum mechanics, and fundamental equations of fractional quantum mechanics are transformed into the well-known fundamental equations of standard quantum mechanics.

When we work with the path integral over Lévy flight quantum paths we derive the fractional generalization of the Schrödinger equation - fractional Schrödinger equation [67]. The fractional Schrödinger equation includes the space derivative of fractional order α (α is the Lévy index), instead of the second order ($\alpha = 2$) space derivative in the standard Schrödinger equation. Thus, the fractional generalization of the Schrödinger equation is the fractional differential equation in accordance with modern well-established terminology (see, for example, [30], [31], [37]-[41], [43], [48], [69], [70]). This is one of the reasons to coin the new term *fractional Schrödinger equation* and the more general term *fractional quantum mechanics*. Due to the link between the path integral over the Lévy flights and the fractional Schrödinger equation we can apply the path integral tool to study the fractional Schrödinger equation. And vice versa, we can use the fractional Schrödinger equation to solve the path integrals over Lévy flights.

1.5 Physical applications of fractional quantum mechanics

Despite the involvement of fractional Riesz derivative into the fractional Schrödinger equation, it has occurred that exact solutions to many canonical quantum mechanics problems can be obtained. From a stand point of quantum mechanical fundamentals the following canonical quantum mechanics problems are solved analytically in the framework of fractional quantum mechanics:
 - A free particle;
 - A particle in δ-potential;
 - A particle in double δ-potential;
 - Infinite potential well;
 - Finite square potential well;
 - Quantum particle in a box with δ-potential;
 - Quadrupole triple δ-potential;
 - Linear potential field;
 - One-dimensional Coulomb potential;
 - Penetration through δ-potential barrier;
 - Penetration through rectangular finite potential barrier;
 - Tunneling in fractional quantum mechanics;
 - Quantum kernel for a particle in the infinite potential well.

Fractional Bohr atom model has been introduced and analyzed in the framework of fractional quantum mechanics. Energy spectrum of fractional Bohr atom and equation for fractional Bohr radius were found.

A new quantum model - *quantum fractional oscillator* has been introduced and studied in semiclassical approximation. Quantum fractional oscillator is a generalization of the well-known quantum oscillator model of the standard quantum mechanics. In other words, in fractional quantum mechanics the fractional oscillator plays the same role as the quantum oscillator does in the standard quantum mechanics.

From a stand point of empirical confirmation of fractional quantum mechanics, a few interesting applications of fractional Schrödinger equation and fractional calculus have been recently developed aiming to come up with experimental setups to observe fractional quantum mechanical phenomena. Stickler [71] launched a solid state physics realization of fractional quantum mechanics by introducing a 1D infinite range tight-binding chain, which he referred to as the 1D Lévy crystal characterized by the Lévy parameter of order $\alpha \in (1, 2]$. The dispersion relation and the density of states of this many-particle system were studied for arbitrary $\alpha \in (1, 2]$. It has been shown that in the limit of small wave numbers all interesting properties of the fractional Schrödinger equation are recovered, while for $\alpha \to 2$ the well-established nearest-neighbor one-dimensional tight-binding chain arises.

Another solid state physics implementation of fractional quantum mechanics is the mean-field model for polariton condensates [72]. Polaritons are quasiparticles consisting of excitons and cavity photons within semiconductor micro-cavities. Polaritons obey Bose–Einstein statistics and, thus, can condense at certain physical conditions into a single particle mode. Excitons are coupled pairs of electrons and holes of oppositely charged spin-half particles in a semiconductor held together by the Coulomb force between them. Excitons interact with light fields and can form integer spin polariton quasiparticles in the strong coupling regime that are confined to the micro-cavity [73]. The initiation of quantum mean-field model for polariton condensates has been inspired by the well established solid state physics concept of velocity dependent mass of a particle due to the interaction with environment. It has been shown in [72] that the kinetic energy of the polariton condensate deviates significantly from the well-known parabolic law. Linear waves analysis and numerical simulations show significant impact velocity dependency of polariton mass on dynamic behavior of polariton condensate when compared to the classic regime with velocity independent polariton mass. The effect of dependency of polariton mass on velocity serves as a test of the feasibility of fractional quantum mechanics to study the dynamics of polariton condensate.

In the field of quantum liquid physics, an interesting application of the fractional Schrödinger equation and Heisenberg dynamic equations for fractional quantum mechanic operators of coordinate and velocity was developed by Tayurskii and Lysogorskiy [74]. They built a fractional ·two-fluid hydrodynamic model to study the motion of superfluid helium in nanoporous media. As a physical application of the developed fractional two-fluid hydrodynamic model, new nonlinear equations for pressure-temperature and pure temperature oscillations in superfluid helium were obtained in [74]. It was shown that at low temperatures the pressure-temperature coupling constant has linear dependence on temperature, which can be tested experimentally to verify two-fluid fractional model.

A nuclear physics realization of fractional quantum mechanics was initiated by Herrmann [75], who developed the fractional symmetric rigid rotor model to study the low energy excitations of ground state band spectra of even-even nuclei. A fractional extension of the rotation group $SO(n)$ has been developed to calculate symmetric rigid rotor oscillating modes, which are treated as generalized rotations and are included in symmetry group - fractional $SO^\alpha(3)$, where α is Lévy index, $1 < \alpha \leq 2$. A comparison with the ground state band spectra of nuclei shows an agreement with experimental data better than 2% [75]. Another interesting nuclear physics application of fractional quantum mechanics is the concept of fractional q-deformed Lie algebras invented by Herrmann [76]. The corresponding q-number was found for the fractional harmonic oscillator and it has been shown that the energy spectrum is well suited to describe the ground state spectra of even-even nuclei.

Longhi [77] came up with a quantum optics set-up to test predictions of fractional quantum mechanics. The idea is to use transverse light dynamics in aspherical optical cavities to design a quantum optical resonator, in which the transverse modes and resonance frequencies correspond to the eigenfunctions and the eigenvalues of the fractional Schrödinger equation with oscillator potential. The laser system presented by Longhi provides a quantum optical realization of the fractional quantum harmonic oscillator, different modes of which can be selectively excited by suitable off-axis pumping. An option to implement fractional pseudo-differential operators in quantum optics could be exploited further to initiate other quantum physics models with involvement of fractional operators. The results of work [77] indicate that the field of quantum optics can be served as suitable polygon where fractional models developed in quantum physics can become experimentally accessible. Fractional optical models look promising from a stand point of developing new approaches to control light diffraction and to generate novel beam solutions [78].

Quantum field-theoretical application of fractional calculus was initiated by Lim [79]. He discovered a fractional generalization of the Parisi–Wu stochastic quantization approach [80] to quantize gauge fields. The idea Lim came up with [79], [81] is to use fractional Langevin stochastic differential equation[1] to model dynamics of fractional Euclidean Klein–Gordon field. The fractional Parisi–Wu approach developed by Lim with coauthors [79], [81] allows to quantize fractional gauge fields. Free energy of massless fractional Klein–Gordon field at finite temperature as well as the Casimir energy of electromagnetic field confined between parallel walls were calculated and analyzed in [79]. The fractional quantum field-theoretical approach developed by Lim with coauthors is interesting and prospective extension of fractional quantum mechanics to fractional quantum field theory.

1.6   Book outline

The book is organized as follows.

Chapter 1 presents an introduction to the subject with basic definitions. Some physical applications of fractional quantum mechanics are listed.

Chapter 2 introduces the concept of geometric and random fractals. The definition of the fractal dimension has been explained using the example of the geometric fractal - Koch curve. Two random fractals were studied: (i) the trajectory of Brownian motion, and (ii) the trajectory of Lévy flight. The fractal dimension of Brownian and Lévy paths has been calculated. The Holtsmark probability distribution has been studied as an example of the α-stable Lévy distribution.

Fundamentals of the fractional Schrödinger equation have been developed in Chapter 3, where the quantum Riesz fractional derivative has been introduced and analyzed. The velocity operator and fractional current density were obtained. The fractional Schrödinger equation in momentum representation has been developed. Hermiticity of the fractional Hamiltonian operator was proven. It has been shown that fractional quantum mechanics supports the parity conservation law.

The time independent fractional Schrödinger equation, which plays an important role for many physical applications, has been explored in Chapter 4. The fractional Schrödinger equation in a periodic potential field has been analyzed. It was shown that Bloch's theorem holds in the framework of fractional quantum mechanics.

In Chapter 5 a new mathematical formulation of the Heisenberg Uncertainty Principle has been developed in the framework of fractional quantum mechanics. The fundamental physical concept of the quantum Lévy wave

---

[1] The fractional Langevin stochastic differential equation was first developed by Laskin (2000) [82] in the field of financial mathematics as the first fractional stochastic dynamic model to explain long tail behavior in returns of financial market indices.

packet was introduced. Quantum mechanics vs fractional quantum mechanics fundamentals involved into the mathematical formulation of Heisenberg's Uncertainty Principle have been discussed. The path integral over Lévy flights has been introduced and elaborated in Chapter 6. The path integral over Lévy flights has been developed in phase space and momentum representations. The fractional Schrödinger equation, which is a manifestation of fractional quantum mechanics, has been derived from the path integral over Lévy flights. Deep insight has been provided into the fundamental relationship between the path integral over Lévy flights and the celebrated Feynman's path integral.

Chapters 7 and 8 present new equations for a free particle quantum kernel in the framework of fractional quantum mechanics. The scaling properties of a free particle kernel have been studied using the renormalization group technique. The Laplace, energy-time and Fourier transforms of a free particle quantum kernel were obtained.

A new physics model, the quantum fractional oscillator has been introduced and explored in Chapter 9. A non-equidistant energy spectrum of the 1D quantum fractional oscillator has been found in semiclassical approximation in coordinate and momentum representations. The symmetries of quantum fractional oscillator were studied.

Chapter 10 presents a few exactly solvable models of fractional quantum mechanics. They include a free particle solution to 1D and 3D fractional Schrödinger equations, quantum particle in the symmetric infinite potential well, bound state in δ-potential well, linear potential field and quantum kernel for a free particle in the box.

In Chapter 11 fractional nonlinear quantum dynamics is introduced based on Davydov's Hamiltonian with long-range exciton-exciton interaction. Fractional nonlinear Schrödinger equation, nonlinear Hilbert–Schrödinger equation, fractional generalization of Zakharov system and fractional Ginzburg–Landau equation have been discovered and explored.

Time fractional quantum mechanics and its applications were developed in Chapters 12 and 13. A new version of the space-time fractional Schrödinger equation was developed. Our space-time fractional Schrödinger equation involves two scale dimensional parameters, one of which can be considered as a time fractional generalization of the famous Planck's constant, while the other can be interpreted as a time fractional generalization of the scale parameter emerging in fractional quantum mechanics. The concept of energy in the framework of time fractional quantum mechanics was introduced and developed. It has been shown that in time fractional quantum mechanics a quantum system does not have stationary states, and eigenvalues of the pseudo-Hamilton operator are not energy levels. The solution to the space-time fractional Schrödinger equation was obtained in the case when the pseudo-Hamilton operator does not depend on time. It was found that time fractional quantum mechanics does not support a fundamental property of quantum mechanics - conservation of quantum mechanical probability. Depending on the particular choice of space and time fractality parameters the introduced space-time fractional Schrödinger equation covers the following three special cases: the Schrödinger equation, the fractional Schrödinger equation and the time fractional Schrödinger equation.

In Chapter 14 fractional statistical mechanics has been introduced based on the path integral over Lévy flights. The path integral representations were obtained for density matrix and partition function. Equation of state of many-particle quantum system was found in the framework of fractional statistical mechanics.

In Chapter 15 fractional classical mechanics has been introduced as a classical counterpart of fractional quantum mechanics. The Lagrange, the Hamilton, the Hamilton–Jacobi and the Poisson bracket approaches were developed and studied in the framework of fractional classical mechanics. A classical fractional oscillator model was introduced, and its exact analytical solution was found. Scaling analysis of fractional classical motion equations has been implemented based on the mechanical similarity. We discover and discuss fractional Kepler's third law which is a generalization of the well-known Kepler's third law. A map between the energy dependence of the period of classical oscillations and the non-equidistant distribution of the energy levels for the quantum fractional oscillator has been established. A classical fractional oscillator model was introduced. An exact analytical solution to the equation of motion of classical fractional oscillator was found in the framework of the Hamilton–Jacobi approach.

Fractional classical dynamics on two-dimensional sphere will be presented in Chapter 16. A class of fractional Hamiltonian systems, generalizing the classical mechanics problem of motion in a central field, has been introduced and analyzed. The analysis is based on transforming an integrable Hamiltonian system with two degrees of freedom on the plane into a dynamic system that is defined on the sphere and inherits the integrals of motion of the original fractional dynamic system. The passage to a dynamic system whose configurational manifold is a two-dimensional sphere is an alternative approach to study the classical fractional oscillator model.

It has been discovered that in the four-dimensional space of structural parameters, there exists a one-dimensional manifold, containing the case of the planar Kepler problem, along which the closedness of the orbits of all finite motions and the third Kepler law are saved. Similarly, it has been found that there exists a one-dimensional manifold, containing the case of the two-dimensional isotropic harmonic oscillator, along which the closedness of the orbits and the isochronism of oscillations are saved. Any deformation of orbits on these manifolds does not violate the hidden symmetry belonging to the two-dimensional isotropic oscillator and the planar Kepler problem. Two-dimensional manifolds, where all systems are characterized by the same rotation number for the finite motion orbits, have also been considered.

The Afterword summarizes fundamental fractional quantum physics concepts and results covering fractional quantum mechanics, time fractional quantum mechanics, fractional statistical mechanics and fractional classical mechanics.

Appendices include the properties of Fox's H-function and some definitions and formulas of fractional calculus for the readers' convenience.

Appendix A presents Fox's H-function definition and lists its properties, which are used to perform calculations in fractional quantum mechanics, time fractional quantum mechanics and fractional statistical mechanics.

In Appendix B a brief introduction to fractional calculus is presented including definitions of Riemann–Liouville fractional derivative, Riemann–Liouville fractional integral, Caputo fractional derivative and quantum Riesz fractional derivative.

Calculation of the integral arising in fractional quantum mechanical problem of finding bound energy and bound wave function for a particle in δ-potential well has been developed in Appendix C.

In Appendix D the polylogarithm function and its properties are presented. The polylogarithm function comes out in natural way while studying quantum lattice dynamics with long-range interaction.

In Appendix E two new functions related to the Mittag-Leffler function have been introduced to solve the space-time fractional Schrödinger equation. These two new functions can be considered as a fractional generalization of the well-known trigonometric functions sine and cosine. A fractional generalization of the celebrated Euler's formula is presented.

Chapter 2

Fractals

The term fractal is now used as a scientific concept, as well as a physical object. Fractal is a geometric shape that is self-similar on all scales. In other words, no matter how much you magnify a fractal, it always looks the same (or at least similar). Fractals are generally irregular (not smooth) in shape, and thus are not objects definable by traditional geometry. That means that fractals tend to have significant details, visible at any arbitrary scale. In other words, 'zooming in' to a fractal simply shows similar pictures, and it is so-called self-similarity. For example, a normal 'Euclidean' shape, such as a circle, looks flatter and flatter as it is magnified. At infinite magnification it is impossible to tell the difference between a circle and a straight line. Fractals are not like this. The conventional idea of curvature, which represents the reciprocal of the radius of an approximating circle, cannot be usefully applied because it scales away. Instead, in a fractal, increasing the magnification reveals details that you simply couldn't see before. The secondary characteristics of fractals, while intuitively appealing, are remarkably hard to condense into a mathematically precise definition. Strictly, a fractal should have a fractional, that is, non-integer dimension. Fractals can be deterministic or random. The examples of deterministic fractals include the Mandelbrot set [27], Lyapunov fractal, Cantor set, Sierpinski carpet and triangle, Peano curve, and the Koch snowflake [28]. Examples of random fractals are the Brownian motion and Lévy flights and their generalizations [27].

Thus, following Mandelbrot [27], a fractal is defined as an object with two properties: (a) self-similarity and (b) its fractal dimension strictly exceeds its topological dimension.

The relation between fractals and quantum (or statistical) mechanics is easily observed in the framework of the Feynman path integral formulation

[19], [83]. The background of the Feynman approach to quantum mechanics is a path integral over the Brownian paths. The Brownian paths are non-differentiable, self-similar curves whose fractal dimension is different from its topological dimension [27], [28]. Brownian motion was historically the first example of the fractal in physics.

First of all, let us describe how the fractal dimension can be defined. Suppose we want to measure the length of the Brownian path. We take a yardstick, representing a straight line of a given length Δx. To measure the length of the Brownian path we use the particular yardstick, starting a new step where the previous step leaves off. It is obvious that the number of steps N multiplied with the yardstick length gives a value $l = N\Delta x$ for the length of the Brownian path. Then we repeat the same procedure with a smaller yardstick say, for instance $\Delta x'$. Doing this for different resolutions Δx yields a function l versus Δx. Usually it is assumed a power law

$$l(\Delta x) \underset{\Delta x \to 0}{\to} l_0 (\Delta x)^{-\delta}, \qquad \delta \geq 0, \qquad (2.1)$$

where l_0 is the dimension factor.

The scaling index (or critical exponent) δ is related to fractal dimension d_{fract} as

$$\delta = d_{\text{fract}} - 1,$$

from which we obtain the definition of the fractal dimension

$$d_{\text{fract}} = 1 + \delta. \qquad (2.2)$$

Note that when $d_{\text{fract}} = 1$, then $\delta = 0$ and the definition (2.1) just reduces to the usual concept of length.

Let us calculate the fractal dimension with three examples of fractals: the Koch curve, the Brownian path and the Lévy flights trajectory.

2.1 The Koch curve

As an example of the above consideration let us find the fractal dimension of a deterministic fractal such as the Koch curve (see, for example, [28]). This curve is composed of 4 sub-segments each of which is scaled down by a factor of $1/3$ from its parent, see Fig. 1.

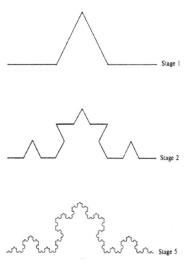

Fig. 1. The Koch curve

It is easy to see that each step in the construction of the Koch curve increases its length by 4/3. The Koch curve is a self-similar curve. Indeed, if we view the curve with a space resolution $\Delta x' = \frac{1}{3}\Delta x$ then we see just a scaled down version of the curve we saw before with a space resolution Δx.

Suppose we examine the Koch curve with the yardstick of the length Δx and measure its length to be l,

$$l = l_0(\Delta x)^{-\delta}. \tag{2.3}$$

Then, if we reduce the yardstick length so that $\Delta x' = \frac{1}{3}\Delta x$, the next level of wiggles in the curve will contribute to the total length,

$$l' = l_0(\Delta x')^{-\delta} = l_0(\frac{1}{3}\Delta x)^{-\delta}. \tag{2.4}$$

On the other hand, each next level of wiggles increases the length by a factor of 4/3,

$$l' = \frac{4}{3}l. \tag{2.5}$$

Thus, from Eqs. (2.4) and (2.5) we have

$$\frac{4}{3}l = l_0(\frac{1}{3}\Delta x)^{-\delta}$$

or

$$l = \frac{3}{4}l_0(\frac{1}{3}\Delta x)^{-\delta}. \tag{2.6}$$

By comparing Eq. (2.3) and Eq. (2.6) we conclude that

$$\frac{3}{4}(\frac{1}{3})^{-\delta} = 1.$$

This implies that

$$1 + \delta = \frac{\ln 4}{\ln 3},$$

or in accordance with the definition given by Eq. (2.2), the fractal dimension $d_{\text{fract}}^{\text{Koch}}$ of the Koch curve is

$$d_{\text{fract}}^{\text{Koch}} = \frac{\ln 4}{\ln 3} \simeq 1.2619.$$

Thus, the Koch curve has fractal dimension $d_{\text{fract}}^{\text{Koch}} \simeq 1.2619$.

2.2 Brownian motion

> *There are no things, only processes.*
> David Bohm

2.2.1 *Historical remarks*

There are two meanings of the term Brownian motion: the physical phenomenon that minute particles immersed in a fluid will experience a random movement, and the celebrated mathematical model used to describe it. The Brownian motion as a mathematical model can also be used to describe many phenomena not resembling the random movement of minute particles. The Brownian motion as a physical phenomenon was discovered by biologist Brown in 1827. Brown studied pollen particles floating in water under microscope as he observed minute particles within the vacuoles in the pollen grains executing the jittery motion that now bears his name. By doing the same with particles of dust, he was able to rule out that the motion was due to pollen being "alive", but it remained to explain the origin of the motion. The first to give a theory of Brownian motion was none other than Einstein in 1905 [84]. At that time the atomic nature of matter was still a controversial idea. Einstein observed that, if the kinetic theory of fluids was right, then the molecules of water would move at random and so a small particle would receive a random number of impacts of random strength and from random directions in any short period of time. This random bombardment by the molecules of the fluid would cause a sufficiently small particle to move in exactly the way described by Brown.

The mathematical theory of Brownian motion has been applied in contexts ranging far beyond the movement of particles in fluids. Mathematically, Brownian motion is a Wiener random process with the conditional probability distribution function of the Gaussian form [63], [68], [85]. Brownian motion is related to the random walk problem and it is generic in the sense that many different stochastic processes reduce to Brownian motion in suitable limits. These are all reasonable approximations to the physical properties of Brownian motion. More sophisticated formulations of the random walk problem have led to the mathematical theory of diffusion processes. The accompanying equation of motion is called the Langevin equation or the Fokker-Planck equation depending on whether it is formulated in terms of random trajectories or probability densities.

2.2.2 The Wiener process

Brownian motion as a mathematical model can be presented in terms of the well-known Wiener random process. The Wiener process $w(t)$ can be introduced by the following equation

$$w(t, w_0; \eta) = w_0 + \int\limits_{t_0}^{t} d\tau \eta(\tau), \qquad w(t = t_0) = w_0, \qquad (2.7)$$

where $\eta(t)$ is "white noise" with statistical properties defined by the set of equations

$$< \eta(t_1)...\eta(t_{2n+1}) >_\eta = 0, \qquad (2.8)$$

$$< \eta(t_1)...\eta(t_{2n}) >_\eta = \sigma^n \sum_{i_1,...,i_{2n} \in (1,...,2n)} \delta(t_{i_1} - t_{i_2})...\delta(t_{i_{2n-1}} - t_{i_{2n}}),$$

here the summation is going over all possible rearrangements $i_1, ..., i_{2n}$ by pairs (the total number of these rearragements is $(2n - 1)!! = (2n)!/2^n n!$), and $< ... >_\eta$ denotes the averaging over "white noise" realizations, and σ is the diffusion coefficient.

From random dynamics stand point, we may say that the Wiener process $w(t)$ is defined by the following stochastic dynamic equation

$$\dot{w}(t) = \eta(t), \quad w(t_0) = w_0.$$

Going from the stochastic dynamic description (in terms of realizations of the stochastic process $w(t)$) to statistical one (in terms of probability distributions), we define the probability density function (pdf) $p(wt|w_0t_0)$ that the stochastic process $w(t)$ will be found at w at time t under condition that it started at $t = t_0$ from $w(t_0) = w_0$,

$$p(wt|w_0t_0) = <\delta(w - w(t, w_0; \eta)) >_\eta. \tag{2.9}$$

Letting the δ-function be the Fourier integral

$$\delta(x) = \frac{1}{2\pi} \int\limits_{-\infty}^{\infty} dq e^{iqx},$$

and substituting it into Eq. (2.9) we have

$$p(wt|w_0t_0) = \frac{1}{2\pi} \int\limits_{-\infty}^{\infty} dq e^{iq(w-w_0)} < \exp\{iq \int\limits_{t_0}^{t} d\tau \eta(\tau)\} >_\eta. \tag{2.10}$$

Using Eqs. (2.8) we calculate

$$< \exp\{iq \int\limits_{t_0}^{t} d\tau \eta(\tau)\} >_\eta = \exp\{-\frac{q^2}{2} \int\limits_{t_0}^{t} d\tau \int\limits_{t_0}^{t} d\tau < \eta(\tau)\eta(\tau') >_\eta\} \tag{2.11}$$

$$= \exp\{-\frac{1}{2}\sigma q^2(t - t_0)\}.$$

Then for $p(wt|w_0t_0)$ defined by Eq. (2.10) we find

$$p(wt|w_0t_0) \equiv p(w - w_0, t - t_0)$$

$$= \frac{1}{2\pi} \int\limits_{-\infty}^{\infty} dq e^{iq(w-w_0)} \exp\{-\frac{1}{2}\sigma q^2(t - t_0)\} \tag{2.12}$$

$$= \frac{1}{\sqrt{2\pi\sigma(t-t_0)}} \exp\{-\frac{(w-w_0)^2}{2\sigma(t-t_0)}\}, \qquad t > t_0.$$

It is easy to see that the probability density $p(wt|w_0t_0)$ satisfies the differential equation

$$\frac{\partial p(wt|w_0t_0)}{\partial t} = \frac{\sigma}{2}\frac{\partial^2}{\partial w^2}p(wt|w_0t_0), \qquad (2.13)$$

$$p(wt_0|w_0t_0) = \delta(w-w_0),$$

and the Smoluchowski–Chapman–Kolmogorov equation

$$p(w_1t_1|w_2t_2) = \int dw' p(w_1t_1|w't')p(w't'|w_2t_2). \qquad (2.14)$$

A generalization to D-dimensional Wiener process $w_a(t)$ $(a = 1, ..., D)$ is straightforward. Let $\mathbf{w}$ be the D-vector. The averaging over the trajectories of D-dimensional "white noise" $\boldsymbol{\eta}$ is defined by the following equations

$$< \eta_{a_1}(t_1)...\eta_{a_{2n+1}}(t_{2n+1}) >_{\boldsymbol{\eta}} = 0 \qquad (2.15)$$

and

$$< \eta_{a_1}(t_1)...\eta_{a_{2n}}(t_{2n}) >_{\boldsymbol{\eta}} \qquad (2.16)$$

$$= \sum_{i_1,...,i_{2n}\in(1,...,2n)} \sigma_{a_{i_1}a_{i_2}}...\sigma_{a_{i_{2n-1}}a_{i_{2n}}} \delta(t_{i_1}-t_{i_2})...\delta(t_{i_{2n-1}}-t_{i_{2n}}),$$

where the summation is over all possible rearrangements $i_1, ..., i_{2n}$ by pairs (the total number of these rearrangements is $(2n-1)!! = (2n)!/2^n n!$).

Expressing the D-dimensional δ-function as the Fourier integral

$$\delta(\mathbf{x}) = \frac{1}{(2\pi)^D} \int d^D q e^{i\mathbf{q}\mathbf{x}}, \qquad d^D q = dq_1...dq_D,$$

with $\mathbf{x}$ and $\mathbf{q}$ being the D-dimensional vectors, and substituting $\delta(\mathbf{x})$ into the D-dimensional generalization of Eq. (2.9) we find

$$p(\mathbf{w}t|\mathbf{w}_0t_0) = \frac{1}{(2\pi)^D} \int d^D q e^{i\mathbf{q}(\mathbf{w}-\mathbf{w}_0)} < \exp\{iq_a \int_{t_0}^{t} d\tau \eta_a(\tau)\} >_{\boldsymbol{\eta}}, \quad (2.17)$$

here and further we adopt the following summation convention: a repeated index implies summation over the range of the index, that is the summation over index a goes from 1 to D.

Using Eqs. (2.15) and (2.16) we find

$$< \exp\{iq_a \int_{t_0}^{t} d\tau \eta_a(\tau)\} >_{\boldsymbol{\eta}} = \exp\{-\frac{q_a q_b}{2} \int_{t_0}^{t} d\tau \int_{t_0}^{t} d\tau < \eta_a(\tau)\eta_b(\tau') >_{\boldsymbol{\eta}}\}$$

$$= \exp\{-\frac{1}{2} q_a \sigma_{ab} q_b (t - t_0)\},$$

where σ_{ab} is the $D \times D$ diffusion matrix.

Then $p(\mathbf{w}t|\mathbf{w}_0 t_0)$ reads

$$p(\mathbf{w}t|\mathbf{w}_0 t_0) \equiv p(\mathbf{w} - \mathbf{w}_0, t - t_0) \qquad (2.18)$$

$$= \frac{1}{\sqrt{2\pi(\det \sigma)(t - t_0)}} \exp\{-\frac{(w_a - w_{0a})\sigma_{ab}^{-1}(w_b - w_{0b})}{2(t - t_0)}\}, \quad t > t_0,$$

where σ_{ab}^{-1} is the inverse matrix with respect to the matrix σ_{ab} and $\det \sigma$ is the determinant of the matrix σ_{ab}. It is easily seen that the probability density $p(\mathbf{w}t|\mathbf{w}_0 t_0)$ satisfies the D-dimensional diffusion equation

$$\frac{\partial p(\mathbf{w}t|\mathbf{w}_0 t_0)}{\partial t} = \frac{\sigma_{ab}}{2} \frac{\partial^2}{\partial w_a \partial w_b} p(\mathbf{w}t|\mathbf{w}_0 t_0), \qquad (2.19)$$

with the initial condition

$$p(\mathbf{w}t|\mathbf{w}_0 t_0)|_{t=t_0} = \delta(\mathbf{w} - \mathbf{w}_0),$$

and the D-dimensional generalization of the Smoluchowski–Chapman–Kolmogorov equation (2.13)

$$p(\mathbf{w}t|\mathbf{w}_0 t_0) = \int d^D w' p(\mathbf{w}t|\mathbf{w}'t') \cdot p(\mathbf{w}'t'|\mathbf{w}_0 t_0). \qquad (2.20)$$

2.2.3 Fractal dimension of Brownian path

Brownian motion is a random fractal. It follows from Eq. (2.12) that

$$(w - w_0) \propto (\sigma(t - t_0))^{1/2}. \tag{2.21}$$

This scaling relation between an increment of the Wiener process $\Delta w = w - w_0$ and a time increment $\Delta t = t - t_0$ allows one to find the fractal dimension of the Brownian path. Indeed, let us consider the diffusion path between two given space-time points. If we divide the given time interval T into N slices, such as $T = N\Delta t$, then for the space length of the diffusion path we have

$$L = N\Delta w = \frac{T}{\Delta t}\Delta w = \sigma T(\Delta w)^{-1}, \tag{2.22}$$

where the scaling relation, Eq. (2.21) was taken into account. The fractal dimension tells us about the length of the path when space resolution goes to zero. The fractional dimension d_{fractal} may be introduced in the following way [27], [28]

$$L \propto (\Delta w)^{1 - d_{\text{fractal}}},$$

where $\Delta w \to 0$. Letting in Eq. (2.22) $\Delta w \to 0$ and comparing with the definition of the fractal dimension d_{fractal} yields

$$d_{\text{fractal}}^{(Brownian)} = 2. \tag{2.23}$$

Thus, we conclude that the fractal dimension of the Brownian path is 2.

2.3 Lévy flight process

2.3.1 Lévy probability distribution

The Lévy flights are a natural generalization of Brownian motion. In the mid of '30s Lévy [61] and Khintchine and Lévy [62] posed the question: When will the sum of N independent identically distributed random quantities $X = X_1 + X_2 + ... + X_N$ has the same probability distribution function $p_N(X)$ (up to scale factor) as the individual steps $P_i(X_i)$, $i = 1, ..., N$? The answer is that each $p_i(X_i)$ should be a Gaussian because of a central limit theorem [63], [64]. In other words, a sum of N Gaussians is again a Gaussian, but with N times the variance of the original. Lévy and Khintchine proved that there exists a possibility to generalize the central limit theorem. They discovered the class of non-Gaussian α-stable probability distributions which possess the property of identity (up to scale factor) the probability distribution function of individual random quantity X_i, $P_i(X_i)$, $i = 1, ..., N$ to the probability distribution function $P(X)$ of sum of N quantities $X = X_1 + X_2 + ... + X_N$. So, the α-stable probability law (stable under summation) [63], [65] is the generalization of the well-known Gaussian law. In other words, the α-stable distribution with index α results from sums of independent identically distributed random variables with probability distribution function $p(x) \sim x^{-1-\alpha}$, $x \to \infty$ in the same way that the Gaussian distribution results from sums of independent identically distributed random variables with probability distribution function $p(x) \sim \exp\{-x^2/2\}$. Each α-stable distribution has a stability index α often called by the Lévy index $0 < \alpha \leq 2$.

The question "when does the sum of independent identically distributed random quantities have the same probability distribution function as each random quantity has" can be considered as an expression of self-similarity, "when does the whole object look like any of its part". The trajectories of Brownian motion and Lévy flights are self-similar. The Lévy index serves as a measure of self-similarity and is equal to the Hausdorff–Besicovitch dimension of these trajectories [29].

To understand the Lévy probability distribution function let's take an example with two independent random variables x_1 and x_2 and their linear combination

$$cx = c_1 x_1 + c_2 x_2,$$

where c, c_1 and c_2 are scale constants.

Suggesting that x_1, x_2 and x have the same probability distribution P yields

$$P(cx) = \int\limits_{-\infty}^{\infty} dx_1 \int\limits_{-\infty}^{\infty} dx_2 P(x_1) P(x_2) \delta(cx - c_1 x_1 - c_2 x_2), \qquad (2.24)$$

where $P(x)$ is normalized

$$\int\limits_{-\infty}^{\infty} dx P(x) = 1, \qquad (2.25)$$

and has the characteristic function $\Phi(k)$

$$\Phi(k) = \int\limits_{-\infty}^{\infty} dx e^{ikx} P(x). \tag{2.26}$$

The probability distribution function $P(x)$ is expressed by the Fourier transform

$$P(x) = \frac{1}{2\pi} \int\limits_{-\infty}^{\infty} dk e^{-ikx} \Phi(k). \tag{2.27}$$

In terms of the characteristic function Eq. (2.24) has the form

$$\Phi(k) = \Phi(c_1 k)\Phi(c_2 k)$$

or

$$\ln \Phi(k) = \ln \Phi(c_1 k) + \ln \Phi(c_2 k). \tag{2.28}$$

The solution of the functional equation (2.28) can be written as

$$\Phi(k) = \exp(-c|k|^{\alpha}), \tag{2.29}$$

with the condition

$$c^{\alpha} = c_1^{\alpha} + c_2^{\alpha},$$

where α is the Lévy index and condition $0 < \alpha \leq 2$ guarantees positivity of the probability distribution given by Eq. (2.27).

With the help of Eqs. (2.27) and (2.29) the Lévy probability distribution function $P_{\alpha}(x)$ is expressed as [61], [62]

$$P_{\alpha}(x) = \frac{1}{2\pi} \int\limits_{-\infty}^{\infty} dk e^{-ikx} \exp(-c|k|^{\alpha}), \qquad 0 < \alpha \leq 2. \tag{2.30}$$

It is easy to show that for $\alpha = 1$ Eq. (2.30) gives us the Cauchy probability distribution function

$$P_1(x) = \frac{c}{\pi} \frac{1}{x^2 + c^2}.$$

Recovering the Gaussian probability distribution from Eq. (2.30) when $\alpha = 2$ is straightforward

$$P_2(x) = \sqrt{\frac{1}{4\pi c}} \exp(-\frac{x^2}{4c}).$$

2.3.2   *Isotropic D-dimensional Lévy flights*

Isotropic D-dimensional Lévy probability distribution function can be defined identically to the one-dimensional case. The isotropic D-dimensional characteristic function has a form which is identical to the characteristic function (2.29) of the one-dimensional Lévy law

$$\Phi^D(\mathbf{k}) = \exp(-c|\mathbf{k}|^\alpha), \tag{2.31}$$

where $\mathbf{k}$ is the D-vector and c is a scale constant.

The probability distribution function is given by the D-dimensional generalization of Eq. (2.27)

$$P_\alpha^D(\mathbf{x}) = \frac{1}{(2\pi)^D} \int d^D k\, e^{-i\mathbf{k}\mathbf{x}} \Phi(\mathbf{k}) \tag{2.32}$$

$$= \frac{1}{(2\pi)^D} \int d^D k\, e^{-i\mathbf{k}\mathbf{x}} \exp(-c|\mathbf{k}|^\alpha), \qquad d^D k = dk_1...dk_D.$$

Due to Eq. (2.31) D-dimensional Fourier integral (2.32) is transformed into one-fold integral over the variable k

$$P_\alpha^D(\mathbf{x}) = P_\alpha^D(r) = \frac{r^{1-D/2}}{(2\pi)^{D/2}} \int\limits_0^\infty dk\, J_{D/2-1}(kr) k^{D/2} \exp(-ck^\alpha), \tag{2.33}$$

where $r = \sqrt{r_1^2 + r_2^2 + ... + r_D^2}$ and $J_{D/2-1}(z)$ is a Bessel function [86].

It is interesting to note that the recursion relation linking $P_\alpha^D(r)$ to $P_\alpha^{D+2}(r)$ immediately follows from Eq. (2.33)

$$-\frac{1}{2\pi r} \frac{\partial}{\partial r} P_\alpha^D(r) = P_\alpha^{D+2}(r). \tag{2.34}$$

The proof of this relation is straightforward, and relies only on the identity [86]

$$-\frac{d}{dz}[z^{-\nu} J_\nu(z)] = z^{-\nu} J_{\nu+1}(z). \tag{2.35}$$

The above relation is just a special case of the general relation that holds between the D-dimensional and $(D+2)$-dimensional Fourier transforms of any isotropic function $f(\mathbf{q}) = f(q)$.

The D-dimensional generalization of the Cauchy probability distribution function $P_1^D(r)$ is recovered[1] from Eq. (2.33) when $\alpha = 1$,

---

[1] We used the formula (see Eq. (6.623.2) in [86])

$$P_1^D(r) = P_\alpha^D(r)|_{\alpha=1} = \frac{\Gamma(\frac{D+1}{2})}{\pi^{(D+1)/2}} \frac{c}{(c^2+r^2)^{(D+1)/2}}.$$

From Eq. (2.33) we can obtain D-dimensional generalization of the Gaussian probability distribution function ($\alpha = 2$) by using the formula (see Eq. (10.22.51) in [87])

$$\int_0^\infty dt\, t^{\nu+1} J_\nu(bt) e^{-p^2 t^2} = \frac{b^\nu}{(2p^2)^{\nu+1}} \exp(-\frac{b^2}{4p^2}),$$

$$\mathrm{Re}\,\nu > -1 \quad \text{and} \quad \mathrm{Re}\,p^2 > 0.$$

It gives us D-dimensional generalization of the Gaussian probability distribution function $P_2^D(r)$,

$$P_2^D(r) = \frac{1}{(4\pi c)^{D/2}} \exp(-\frac{r^2}{4c}), \tag{2.37}$$

where r^2 is defined as

$$r^2 = r_1^2 + r_2^2 + ... + r_D^2.$$

2.3.3 Lévy random process

The Lévy random process $x(t)$ can be defined by the stochastic differential equation

$$\dot{x}(t) = \eta_L(t), \quad x(t_0) = x_0, \tag{2.38}$$

with the "Lévy white noise" $\eta_L(\tau)$ defined by the following characteristic functional

$$< \exp\{i \int_{t_0}^t d\tau q(\tau)\eta_L(\tau)\} >_{\eta_L} = \exp\{-\sigma_\alpha \int_{t_0}^t d\tau |q(\tau)|^\alpha\}, \tag{2.39}$$

$$\int_0^\infty dt\, J_\nu(\beta t) t^{\nu+1} \exp(-\alpha t) = \frac{2\alpha(2\beta)^\nu \Gamma(\nu + \frac{3}{2})}{\sqrt{\pi}(\alpha^2 + \beta^2)^{\nu+3/2}}, \tag{2.36}$$

where $\mathrm{Re}\,\nu > -1$ and $\mathrm{Re}\,\alpha > |\,\mathrm{Im}\,\beta\,|$.

where σ_α is the intensity of the Lévy noise, α is the Lévy index, $0 < \alpha \leq 2$ and $<...>_{\eta_L}$ denotes the averaging[2] over all possible realizations of the Lévy white noise $\eta_L(\tau)$.

Similarly to the definition given by Eq. (2.8) the probability distribution function $p_L(xt|x_0t_0)$ reads

$$p_L(xt|x_0t_0) = < \delta(x - x(t, x_0; \eta_L)) >_{\eta_L}, \qquad (2.42)$$

here $x(t, x_0; \eta_L)$ is the solution of Eq. (2.38) with the initial condition x_0,

$$x(t, x_0; \eta) = x_0 + \int_{t_0}^{t} d\tau \eta_L(\tau), \qquad x(t_0) = x_0. \qquad (2.43)$$

Expressing the δ-function as the Fourier integral $\delta(x) = \frac{1}{2\pi} \int_{-\infty}^{\infty} dq e^{iqx}$ and taking into account Eqs. (2.43) and (2.39) we find

$$p_L(xt|x_0t_0) = \frac{1}{2\pi} \int_{-\infty}^{\infty} dq e^{iq(x-x_0)} \exp\{-\sigma_\alpha|q|^\alpha(t - t_0)\}, \qquad (2.44)$$

$$t > t_0, \quad 0 < \alpha \leq 2,$$

where σ_α can be considered at this point as a generalized diffusion coefficient with units of $[\sigma_\alpha] = \text{cm}^\alpha \cdot \text{sec}^{-1}$.

Thus, we get the α-stable probability distribution function of the Lévy random process or the Lévy flights. The α-stable distribution with $0 < \alpha < 2$ possesses finite moments of order μ, $0 < \mu < \alpha$, but infinite moments of higher order. Note that the Gaussian probability distribution ($\alpha = 2$) is stable one and it possesses moments of all orders.

---

[2]The averaging is defined as the following functional integral

$$< ... >_{\eta_L} = \int D\eta_L(\tau) P(\eta_L(\tau))..., \qquad (2.40)$$

where $\int D\eta_L(\tau)$ stands for functional integral over the Lévy white noise $\eta_L(\tau)$ and $P(\eta_L(\tau))$ is the Lévy white noise probability distribution functional introduced by the functional integral,

$$P(\eta_L(\tau)) = \int Dk(\tau) \exp\{-i \int_{t_0}^{t} d\tau k(\tau)\eta_L(\tau)\} \exp\{-\sigma_\alpha \int_{t_0}^{t} d\tau|k(\tau)|^\alpha\}. \qquad (2.41)$$

The α-stable Lévy distribution defined by Eq. (2.44) satisfies the fractional diffusion equation

$$\frac{\partial p_L(xt|x_0t_0)}{\partial t} = \sigma_\alpha \nabla^\alpha p_L(xt|x_0t_0), \qquad \nabla^\alpha \equiv \frac{\partial^\alpha}{\partial x^\alpha}, \qquad (2.45)$$

$$p_L(xt|x_0t_0)|_{t=t_0} = \delta(x - x_0),$$

where ∇^α is the fractional Riesz derivative defined through its Fourier transform [30], [47], [63], [88] (see Appendix B for details),

$$\nabla^\alpha p(x,t) = -\frac{1}{2\pi} \int\limits_{-\infty}^{\infty} dk e^{ikx} |k|^\alpha \overline{p}(k,t). \qquad (2.46)$$

Here $p(x,t)$ and $\overline{p}(k,t)$ are related to each other by the Fourier transforms

$$p(x,t) = \frac{1}{2\pi} \int\limits_{-\infty}^{\infty} dk e^{ikx} \overline{p}(k,t), \qquad \overline{p}(k,t) = \int\limits_{-\infty}^{\infty} dx e^{-ikx} p(x,t). \qquad (2.47)$$

A generalization to the D-dimensional isotropic Lévy process is straightforward. Indeed, let $\mathbf{x}(t)$ be the D-vector $x_a(t)$ ($a = 1, ..., D$). Then the D-dimensional generalization of Eq. (2.38) becomes

$$\dot{\mathbf{x}}(t) = \boldsymbol{\eta}_L(t), \qquad (2.48)$$

where $\boldsymbol{\eta}_L(t)$ is the D-dimensional isotropic α-stable Lévy process with the characteristic functional $\Phi\{\mathbf{q}(\tau)\}$,

$$\Phi\{\mathbf{q}(\tau)\} = < \exp\left\{ i \int\limits_{t_0}^{t} d\tau \mathbf{q}(\tau) \boldsymbol{\eta}_L(\tau) \right\} >_{\eta_L} = \exp\left\{ -\sigma_\alpha \int\limits_{t_0}^{t} d\tau |\mathbf{q}(\tau)|^\alpha \right\}. \qquad (2.49)$$

Here σ_α is the generalized diffusion coefficient or, intensity of the Lévy white noise, α is the Lévy index, $0 < \alpha \leq 2$ and $<...>_{\eta_L}$ denotes the averaging over all possible realizations of D-dimensional isotropic Lévy white noise $\boldsymbol{\eta}_L(\tau)$.

Similarly to the definition given by Eq. (2.42) the probability distribution function $p_L(\mathbf{x}t|\mathbf{x}_0t_0)$ of the D-dimensional isotropic Lévy process $\mathbf{x}(t)$ reads

$$p_L(\mathbf{x}t|\mathbf{x}_0t_0) = < \delta(\mathbf{x} - \mathbf{x}(t, \mathbf{x}_0; \boldsymbol{\eta}_L)) >_{\boldsymbol{\eta}_L}, \qquad (2.50)$$

where $\mathbf{x}(t, \mathbf{x}_0; \boldsymbol{\eta}_L)$ is the solution of Eq. (2.48) with the initial conditions $\mathbf{x}_0$,

$$\mathbf{x}(t, \mathbf{x}_0; \boldsymbol{\eta}_L) = \mathbf{x}_0 + \int\limits_{t_0}^{t} d\tau \boldsymbol{\eta}_L(\tau), \qquad \mathbf{x}(t_0) = \mathbf{x}_0. \qquad (2.51)$$

Letting the D-dimensional δ-function be the Fourier integral

$$\delta(\mathbf{x}) = \frac{1}{(2\pi)^D} \int d^D q e^{i\mathbf{q}\mathbf{x}},$$

and taking into account Eqs. (2.50) and (2.49) we finally find

$$p_L(\mathbf{x}t|\mathbf{x}_0t_0) = \frac{1}{(2\pi)^D} \int\limits_{-\infty}^{\infty} d^D q e^{i\mathbf{q}(\mathbf{x}-\mathbf{x}_0)} \exp\{-\sigma_\alpha |\mathbf{q}|^\alpha (t - t_0)\}, \qquad (2.52)$$

where $t > t_0$, α is the Lévy index $0 < \alpha \leq 2$, and σ_α can be considered as a generalized isotropic diffusion coefficient with units of $[\sigma_\alpha] = \mathrm{cm}^\alpha \cdot \mathrm{sec}^{-1}$.

It is easy to see that the D-dimensional isotropic α-stable Lévy distribution defined by Eq. (2.52) satisfies the D-dimensional fractional diffusion equation

$$\frac{\partial p_L(\mathbf{x}t|\mathbf{x}_0t_0)}{\partial t} = \sigma_\alpha (-\Delta)^{\alpha/2} p_L(\mathbf{x}t|\mathbf{x}_0t_0), \qquad \Delta = \frac{\partial}{\partial \mathbf{x}} \frac{\partial}{\partial \mathbf{x}}, \qquad (2.53)$$

$$p_L(\mathbf{x}t|\mathbf{x}_0t_0)|_{t=t_0} = \delta(\mathbf{x} - \mathbf{x}_0),$$

where $(-\Delta)^{\alpha/2}$ is D-dimensional isotropic fractional Riesz derivative defined through its Fourier transform [30], [63], [67], [88] (see Appendix B for details),

$$(-\Delta)^{\alpha/2} p(\mathbf{x}, t) = \frac{1}{(2\pi)^D} \int d^D k e^{i\mathbf{k}\mathbf{x}} |\mathbf{k}|^\alpha \overline{p}(\mathbf{k}, t),$$

where $p(\mathbf{x}, t)$ and $\overline{p}(\mathbf{k}, t)$ are related to each other by the D-dimensional Fourier transforms

$$p(\mathbf{x}, t) = \frac{1}{(2\pi)^D} \int d^D k e^{i\mathbf{k}\mathbf{x}} \overline{p}(\mathbf{k}, t), \qquad \overline{p}(\mathbf{k}, t) = \int d^D x e^{-i\mathbf{k}\mathbf{x}} p(\mathbf{x}, t).$$

Thus, we found the D-dimensional generalization $p_L(\mathbf{x}t|\mathbf{x}_0t_0)$ of the α-stable probability distribution function of the Lévy random process or the Lévy flights.

2.3.4 *Fractal dimension of Lévy flight path*

Lévy flights bring us another example of a random fractal. It follows from Eq. (2.44) that

$$(x - x_0) \propto (\sigma_\alpha (t - t_0))^{1/\alpha}, \quad 1 < \alpha \leq 2. \tag{2.54}$$

This scaling relation between an increment of the Lévy process $\Delta x = x - x_0$ and a time increment $\Delta t = t - t_0$ allows one to find the fractal dimension of the trajectory of a Lévy path. Let us consider the length of the Lévy path between two given space-time points. By dividing the given time interval T into N slices, such as $T = N\Delta t$, and taking into account the scaling relation Eq. (2.54) we have

$$L = N\Delta x = \frac{T}{\Delta t}\Delta x = \sigma_\alpha T (\Delta x)^{1-\alpha}.$$

Letting $\Delta x \to 0$ and comparing it with the definition of the fractal dimension $d_{fractal}$ [27], [28] yields

$$d_{fractal}^{(Lévy)} = \alpha, \quad 1 < \alpha \leq 2. \tag{2.55}$$

Thus, the fractal dimension of the Lévy path is α. The Lévy flight pattern consists of a self-similar clustering of local sojourns, interrupted by long jumps, at whose end a new cluster starts, and so on. Zooming into a cluster, in turn reveals clusters interrupted by long jumps. Lévy flights intimately combine the local jumps properties stemming from the center part of the jumps distribution around zero jump length, with strongly non-local long-distance jumps, thereby creating slowly decaying spatial correlations, a signature of non-Gaussian processes with diverging variance.

By comparing the scaling laws Eq. (2.21) and Eq. (2.54) we conclude that the cluster of Lévy particles will spread faster than the cluster of Brownian particles. From Eq. (2.54) we see that the space scale $x(t)$ of the Lévy cluster follows $x(t) \sim t^{1/\alpha}$. It is easy to see that the lower bound of the space scale $x(t)$ is being reached for the Brownian cluster, $\alpha = 2$. This fact represents the difference between Brownian motion patterns and clustered patterns of Lévy flights displayed in Figs. 2 and 3.

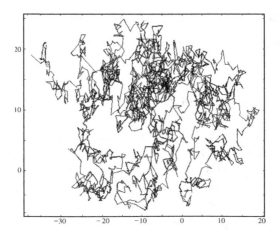

Fig. 2. An illustration of Brownian motion which corresponds to normal diffusion. It is obtained by iterating any random walk with identically independently distributed elementary steps having finite variance. The first 3000 steps of the walk are shown [89].

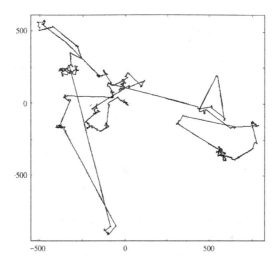

Fig. 3. This is the pattern of Lévy flight shown for the Lévy index $\alpha = 1.5$. Contrary to Brownian motion, the variance and any moment of order μ, $\mu \geq \alpha$ are infinite. As a consequence, any Lévy flight path is almost surely not continuous, and the figure indeed displays jumps. The first 3000 steps are presented [89].

2.3.5 *Holtsmark distribution*

One of the examples of appearance of the α-stable Lévy distribution with $\alpha = 3/2$ comes from astrophysics. It is the Holtsmark distribution [90], [91], which is the probability distribution function of the gravitational force created at a randomly chosen point by a given system of stars.

Let us consider a stellar system as a sphere of radius R where N stars are randomly located at $\mathbf{r}_i$ with masses M_i $(i = 1, 2, ..., N)$ that exert the gravitational force $\mathbf{F}_g$

$$\mathbf{F}_g = G \sum_{i=1}^{N} \frac{M_i}{r_i^3} \mathbf{r}_i \tag{2.56}$$

on a star, per unit mass, located at the origin $\mathbf{r} = 0$ of the system. Here G is the gravitational constant and $\mathbf{r}_i$ are 3D vectors.

We assume that stars are uniformly distributed in space with density $\rho = \text{const}$. Therefore, the random number N of stars in the sphere of radius R follows the Poisson probability distribution

$$P(N) = \frac{\left((4/3)\pi R^3 \rho\right)^N}{N!} \exp\{-(4/3)\pi R^3 \rho\}, \tag{2.57}$$

here $(4/3)\pi R^3$ is the volume of sphere of the radius R. Thus, the probability to find N stars in the sphere of radius R is given by $P(N)$. The normalization condition is

$$\sum_{N=0}^{\infty} P(N) = 1. \tag{2.58}$$

Further, it is assumed that the masses M_i of stars are identically distributed random variables with probability distribution function $\phi(M)$ with the normalization condition

$$\int_0^{\infty} dM \phi(M) = 1.$$

Then by definition (see Eq. (2.27)) the probability distribution function $P(\mathbf{F})$ of the gravitational force is

$$P(\mathbf{F}) = \frac{1}{(2\pi)^3} \int d^3 k \, e^{i\mathbf{k}\mathbf{F}} \Phi(\mathbf{k}), \tag{2.59}$$

where $\Phi(\mathbf{k})$ is the characteristic function defined by,

$$\Phi(\mathbf{k}) = < \exp(-i\mathbf{k}\mathbf{F}_g) >, \qquad (2.60)$$

here $\mathbf{F}_g$ is given by Eq. (2.56) and $< ... >$ means the averaging over three random impacts:

1. random masses M_i;
2. random locations $\mathbf{r}_1,...,\mathbf{r}_N$;
3. random number N of stars.

Let's define these three averaging procedures.

1. Averaging over random masses M_i, $< ... >_M$ is defined as,

$$< ... >_M = \int\limits_0^\infty dM_1... \int\limits_0^\infty dM_N \prod_{i=1}^N \phi(M_i)... . \qquad (2.61)$$

2. Averaging over random locations $\mathbf{r}_1, ..., \mathbf{r}_N$, $< ... >_R$ is given by,

$$< ... >_R = \frac{1}{(4/3)\pi R^3} \int\limits_{(4/3)\pi R^3} d^3r_1 ... \frac{1}{(4/3)\pi R^3} \int\limits_{(4/3)\pi R^3} d^3r_N ... . \qquad (2.62)$$

3. Averaging over random number N of stars, $< ... >_N$ in the volume $(4/3)\pi R^3 \rho$ is performed as,

$$< ... >_N = \sum_{N=0}^\infty \frac{\left((4/3)\pi R^3\rho\right)^N}{N!} \exp\{-(4/3)\pi R^3\rho\}.$$

With the help of Eqs. (2.56), (2.61) - (2.63) we transform Eqs. (2.59) and (2.60) for $P(\mathbf{F})$ as follows

$$P(\mathbf{F}) = \frac{1}{(2\pi)^3} \int d^3k \, e^{i\mathbf{k}\mathbf{F}} \sum_{N=0}^\infty \frac{\left((4/3)\pi R^3\rho\right)^N}{N!} \exp\{-(4/3)\pi R^3\rho\} \qquad (2.63)$$

$$\times \prod_{i=1}^N \left(\rho \int\limits_{(4/3)\pi R^3} d^3r_i \int\limits_0^\infty dM_i\phi(M_i)\exp(-iG\frac{M_i}{r_i^3}\mathbf{k}\mathbf{r}_i) \right).$$

Performing summation over N gives us

$$P(\mathbf{F}) = \frac{1}{(2\pi)^3} \int d^3k \, e^{i\mathbf{k}\mathbf{F}} \, \exp\{-\Lambda(\mathbf{k})\}, \qquad (2.64)$$

where $\Lambda(\mathbf{k})$ has been introduced as

$$\Lambda(\mathbf{k}) = \rho \int_{(4/3)\pi R^3} d^3r \left\{ 1 - \int_0^\infty dM\phi(M)\exp\left(-iG\frac{M}{r^3}\mathbf{kr}\right) \right\}. \qquad (2.65)$$

Because the integral over d^3r is convergent we can go to $R \to \infty$, then in the spherical coordinate system $\Lambda(\mathbf{k})$ becomes

$$\Lambda(\mathbf{k}) = \rho \int_0^\infty dr r^2 \int_0^{2\pi} d\varphi \int_{-\pi}^{\pi} d\vartheta \sin\vartheta \qquad (2.66)$$

$$\times \left\{ 1 - \int_0^\infty dM\phi(M)\exp(-iG\frac{M}{r^3}kr\cos\vartheta) \right\}.$$

Integrating over $d\varphi$ and introducing new integration variables y and x

$$y = \frac{GM}{r^2}, \qquad x = \cos\vartheta,$$

yield

$$\Lambda(\mathbf{k}) = \rho\pi G^{3/2} < M^{3/2} >_M \int_0^\infty dy y^{-5/2} \int_{-1}^1 dx\{1 - \exp(-ikyx)\} \qquad (2.67)$$

$$= 2\rho\pi G^{3/2} < M^{3/2} >_M \int_0^\infty dy y^{-5/2} \left(1 - \frac{\sin ky}{ky} \right),$$

with

$$< M^{3/2} >_M = \int_0^\infty dM M^{3/2}\phi(M).$$

To evaluate the integral Eq. (2.67) let's introduce a new integration variable $\eta = ky$. Then we have

$$\Lambda(\mathbf{k}) = 2\rho\pi G^{3/2} < M^{3/2} >_M k^{3/2} \int_0^\infty d\eta \eta^{-7/2} (\eta - \sin\eta). \qquad (2.68)$$

After a few integrations by parts, $\Lambda(\mathbf{k})$ is reduced to

$$\Lambda(\mathbf{k}) = \rho \frac{16}{15} \pi G^{3/2} < M^{3/2} >_M k^{3/2} \int\limits_0^\infty d\eta \eta^{-1/2} \cos\eta \qquad (2.69)$$

$$= \frac{4}{15} \rho (2\pi G)^{3/2} < M^{3/2} >_M k^{3/2},$$

where the formula for the Fresnel integral

$$\int\limits_0^\infty d\eta \eta^{-1/2} \cos\eta = \int\limits_0^\infty d\varsigma \cos(\varsigma^2) = \sqrt{\pi}/(2\sqrt{2}),$$

has been used.

By combining Eqs. (2.64), (2.65) and (2.69) we finally obtain the probability distribution function $P(\mathbf{F})$ of the gravitational force

$$P(\mathbf{F}) = \frac{1}{(2\pi)^3} \int d^3k \; e^{i\mathbf{k}\mathbf{F}} \; \exp\{-c|\mathbf{k}|^{3/2}\}, \qquad (2.70)$$

where the constant c is given by

$$c = (\frac{4}{15}) \rho (2\pi G)^{3/2} < M^{3/2} >_M, \qquad (2.71)$$

and the averaging over random star masses $< \ldots >_M$ is defined by Eq. (2.61).

From Eq. (2.70), one can see that probability distribution $P(\mathbf{F})$ of the gravitational force $\mathbf{F}_g$ has the 3-dimensional α-stable law with the Lévy index $\alpha = 3/2$.

To simplify Eq. (2.70) we go to the spherical reference frame $d^3k = k^2 dk \sin\vartheta d\vartheta d\varphi$ with polar axis along $\mathbf{F}$, and perform integration over angle variables ϑ and φ. Hence, we have

$$P(\mathbf{F}) = \frac{1}{(2\pi)^3} \int\limits_0^\infty k^2 dk \int\limits_{-\pi}^\pi \sin\vartheta d\vartheta \int\limits_0^{2\pi} d\varphi \cdot e^{ikF\cos\vartheta} \cdot \exp\{-ck^{3/2}\}$$

$$= \frac{1}{2\pi^2 F} \int\limits_0^\infty dk \cdot k \sin(kF) \exp\{-ck^{3/2}\}.$$

Using new variable $x = kF$ yields

$$P(\mathbf{F}) = \frac{1}{2\pi^2 F^3} \int_0^\infty dx \cdot x \sin(x) \exp\{-cx^{3/2}/F^{3/2}\}. \qquad (2.72)$$

With the help of the equation

$$P(\mathbf{F})d^3 F = 4\pi P(F)F^2 dF, \qquad (2.73)$$

we introduce the probability distribution function $Q(F)$, which satisfies

$$P(\mathbf{F})d^3 F = Q(F)dF.$$

Further, from Eqs. (2.72) and (2.73) it follows

$$Q(F) = \frac{2}{\pi F} \int_0^\infty dx \cdot x \sin(x) \exp\{-cx^{3/2}/F^{3/2}\}. \qquad (2.74)$$

Now, if we put $f = F/c^{2/3}$, then we obtain

$$Q(F)dF = c^{2/3}Q(f)df = H(f)df, \qquad (2.75)$$

where the probability distribution function $H(f)$ is given by [91]

$$H(f) = \frac{2}{\pi f} \int_0^\infty dx\, x \sin(x) \exp\{-(x/f)^{3/2}\}. \qquad (2.76)$$

The probability distribution function $H(f)$ is the Holtsmark distribution.

Thus, the 3-dimensional α-stable probability distribution function $P(\mathbf{F})$ with the Lévy index $\alpha = 3/2$ given by Eq. (2.70) has been expressed in terms of the Holtsmark distribution given by Eq. (2.76)

$$P(\mathbf{F})d^3 F = H(f)df.$$

The Holtsmark distribution was discovered in 1919 [90], before Lévy and Khintchine [61], [62] developments on α-stable probability distributions.

Chapter 3

Fractional Schrödinger Equation

3.1 Fundamentals

Classical mechanics and quantum mechanics are based on the assumption that the Hamiltonian function has the form

$$H(\mathbf{p}, \mathbf{r}) = \frac{\mathbf{p}^2}{2m} + V(\mathbf{r}), \qquad (3.1)$$

where $\mathbf{p}$ and $\mathbf{r}$ are the momentum and space coordinate of a particle with mass m, and $V(\mathbf{r})$ is the potential energy. In quantum mechanics, $\mathbf{p}$ and $\mathbf{r}$ should be considered as quantum mechanical operators $\widehat{\mathbf{p}}$ and $\widehat{\mathbf{r}}$. Then the Hamiltonian function $H(\mathbf{p}, \mathbf{r})$ becomes the Hamiltonian operator $\widehat{H}(\widehat{\mathbf{p}}, \widehat{\mathbf{r}})$,

$$\widehat{H}(\widehat{\mathbf{p}}, \widehat{\mathbf{r}}) = \frac{\widehat{\mathbf{p}}^2}{2m} + V(\widehat{\mathbf{r}}). \qquad (3.2)$$

The square dependence on the momentum in Eqs. (3.1) and (3.2) is an empirical physical fact. However, an attempt to get insight on the fundamentals behind this fact posts the question: are there other forms of the kinematic term in Eqs. (3.1) and (3.2) which do not contradict the fundamental principles of classical mechanics and quantum mechanics? A convenient theoretical physics approach to answer this question is the Feynman path integral approach to quantum mechanics [12], as it was first observed by Laskin [67]. Indeed, the Feynman path integral is the integral over Brownian-like paths. Brownian motion is a special case of the so-called α-stable probability distributions developed by Lévy [61] and Khintchine [62]. In mid '30s they posed the question: Does the sum of N independent identically distributed random quantities $X = X_1 + X_2 + ... + X_N$ have the same probability distribution $p_N(X)$ (up to scale factor) as the individual steps $P_i(X_i)$, $i = 1, ..., N$? The traditional answer is that each $P_i(X_i)$

should be a Gaussian, because of the central limit theorem. In other words, a sum of N Gaussians is again a Gaussian, but with N times the variance of the original. Lévy and Khintchine proved that there exists the possibility to generalize the central limit theorem. They discovered a class of non-Gaussian α-stable (stable under summation) probability distributions. Each α-stable distribution has a stability index α, often called the Lévy index $0 < \alpha \leq 2$. When $\alpha = 2$ Lévy motion is transformed into Brownian motion.

An option to develop the path integral over Lévy paths was discussed by Kac [23], who pointed out that the Lévy path integral generates a functional measure in the space of left (or right) continuous functions (paths) having only first kind discontinuities. The path integral over Lévy paths has first been introduced and elaborated with applications to fractional quantum mechanics and fractional statistical mechanics by Laskin (see [67], [92], [96]). He followed the framework of the Feynman space-time vision of quantum mechanics, but instead of the Brownian-like quantum mechanical trajectories, Laskin used the Lévy-like ones. If the fractal dimension (for definition of fractal dimension, see [27], [28]) of the Brownian path is $d_{\text{fractal}}^{(Brownian)} = 2$, then the Lévy path has fractal dimension $d_{\text{fractal}}^{(Lévy)} = \alpha$, where α is so-called the Lévy index, $1 < \alpha \leq 2$. The Lévy index α becomes a new fundamental parameter in fractional quantum and fractional classical mechanics similar to $d_{\text{fractal}}^{(Brownian)} = 2$ being a fundamental parameter in standard quantum and classical mechanics. The difference between the fractal dimensions of Brownian and Lévy paths leads to different physics. In fact, fractional quantum mechanics is generated by the Hamiltonian operator of the form [67], [92], [96]

$$\widehat{H}_\alpha(\widehat{\mathbf{p}}, \widehat{\mathbf{r}}) = D_\alpha |\widehat{\mathbf{p}}|^\alpha + V(\widehat{\mathbf{r}}), \qquad 1 < \alpha \leq 2, \qquad (3.3)$$

which is originated from the classical mechanics Hamiltonian function $H_\alpha(\mathbf{p}, \mathbf{r})$

$$H_\alpha(\mathbf{p}, \mathbf{r}) = D_\alpha |\mathbf{p}|^\alpha + V(\mathbf{r}), \qquad 1 < \alpha \leq 2, \qquad (3.4)$$

with substitutions $\mathbf{p} \to \widehat{\mathbf{p}}$, $\mathbf{r} \to \widehat{\mathbf{r}}$, and D_α being the scale coefficient with units of $[D_\alpha] = \text{erg}^{1-\alpha} \cdot \text{cm}^\alpha \cdot \text{sec}^{-\alpha}$. One can say that Eq. (3.4) is a natural generalization of the well-known Eq. (3.1). When $\alpha = 2$, $D_\alpha = 1/2m$ and Eq. (3.4) is transformed into Eq. (3.1) [67]. As a result, the fractional quantum mechanics based on the Lévy path integral generalizes the standard quantum mechanics based on the well-known Feynman path integral.

Indeed, if the path integral over Brownian trajectories leads to the well-known Schrödinger equation, then the path integral over Lévy trajectories leads to the fractional Schrödinger equation. The fractional Schrödinger equation is a new fundamental equation of quantum physics, and it includes the space derivative of order α instead of the second ($\alpha = 2$) order space derivative in the standard Schrödinger equation. Thus, the fractional Schrödinger equation is the fractional differential equation in accordance with modern terminology (see, for example, [30], [31], [37]-[43], [48], [69], [70]). This is the main point of the term, *fractional Schrödinger equation*, and for the more general term, *fractional quantum mechanics* [67], [92]. When Lévy index $\alpha = 2$, Lévy motion becomes Brownian motion. Thus, fractional quantum mechanics includes standard quantum mechanics as a particular Gaussian case at $\alpha = 2$. The quantum mechanical path integral over the Lévy paths [67] at $\alpha = 2$ becomes the Feynman path integral [12], [20].

In the limit case $\alpha = 2$ the fundamental equations of fractional quantum mechanics are transformed into the well-known equations of standard quantum mechanics [12], [20], [94].

3.2 Fractional Schrödinger equation in coordinate representation

3.2.1 Quantum Riesz fractional derivative

It follows from Eq. (3.1) that the energy E of a particle of mass m under the influence of the potential $V(\mathbf{r})$ is given by

$$E = \frac{\mathbf{p}^2}{2m} + V(\mathbf{r}). \tag{3.5}$$

To obtain the Schrödinger equation we introduce the operators following the well-known procedure,

$$E \to i\hbar \frac{\partial}{\partial t}, \qquad \mathbf{p} \to -i\hbar \boldsymbol{\nabla}, \tag{3.6}$$

where $\boldsymbol{\nabla} = \partial/\partial \mathbf{r}$ and $\hbar$ is Planck's constant. Further, substituting transformation (3.6) into Eq. (3.1) and applying it to the wave function $\psi(\mathbf{r}, t)$ yields

$$i\hbar \frac{\partial \psi(\mathbf{r}, t)}{\partial t} = -\frac{\hbar^2}{2m} \Delta \psi(\mathbf{r}, t) + V(\mathbf{r})\psi(\mathbf{r}, t), \tag{3.7}$$

here $\Delta = \nabla \cdot \nabla$ is the Laplacian. Thus, we obtained the Schrödinger equation [94].

By repeating the same consideration to Eq. (3.4) we find the fractional Schrödinger equation [67], [95], [96]

$$i\hbar \frac{\partial \psi(\mathbf{r}, t)}{\partial t} = D_\alpha (-\hbar^2 \Delta)^{\alpha/2} \psi(\mathbf{r}, t) + V(\mathbf{r}) \psi(\mathbf{r}, t), \qquad (3.8)$$

$$1 < \alpha \leq 2,$$

where D_α is scale coefficient D_α with units of $[D_\alpha] = \mathrm{erg}^{1-\alpha} \cdot \mathrm{cm}^\alpha \cdot \mathrm{sec}^{-\alpha}$, and the 3D generalization of the quantum Riesz fractional derivative $(-\hbar^2 \Delta)^{\alpha/2}$ has been introduced first by Laskin [67], [92]

$$(-\hbar^2 \Delta)^{\alpha/2} \psi(\mathbf{r}, t) = \frac{1}{(2\pi\hbar)^3} \int d^3 p \, e^{i\frac{\mathbf{p}\mathbf{r}}{\hbar}} |\mathbf{p}|^\alpha \varphi(\mathbf{p}, t), \qquad (3.9)$$

where the wave functions in space $\psi(\mathbf{r}, t)$ and momentum $\varphi(\mathbf{p}, t)$ representations are related to each other by the 3D Fourier transforms

$$\psi(\mathbf{r}, t) = \frac{1}{(2\pi\hbar)^3} \int d^3 p \, e^{i\frac{\mathbf{p}\mathbf{r}}{\hbar}} \varphi(\mathbf{p}, t), \qquad \varphi(\mathbf{p}, t) = \int d^3 r \, e^{-i\frac{\mathbf{p}\mathbf{r}}{\hbar}} \psi(\mathbf{r}, t).$$
$$(3.10)$$

The 3D fractional Schrödinger equation Eq. (3.8) can be rewritten in the operator form

$$i\hbar \frac{\partial \psi(\mathbf{r}, t)}{\partial t} = \widehat{H}_\alpha \psi(\mathbf{r}, t), \qquad 1 < \alpha \leq 2, \qquad (3.11)$$

if we introduce the fractional Hamiltonian $\widehat{H}_\alpha$ operator defined by

$$\widehat{H}_\alpha = D_\alpha (-\hbar^2 \Delta)^{\alpha/2} + V(\mathbf{r}). \qquad (3.12)$$

This operator can be expressed in the form given by Eq. (3.3) by introducing quantum-mechanical operator of momentum $\widehat{\mathbf{p}} = -i\hbar\nabla = -i\hbar\partial/\partial\mathbf{r}$ and operator of coordinate $\widehat{\mathbf{r}} = \mathbf{r}$.

Therefore, the 3D fractional Schrödinger equation Eq. (3.8) has the following operator form

$$i\hbar \frac{\partial \psi(\mathbf{r}, t)}{\partial t} = \widehat{H}_\alpha(\widehat{\mathbf{p}}, \widehat{\mathbf{r}}) \psi(\mathbf{r}, t),$$

with $\widehat{H}_\alpha(\widehat{\mathbf{p}}, \widehat{\mathbf{r}})$ introduced by Eq. (3.3).

Commutation relationships for operators $\widehat{\mathbf{p}}$ and $\widehat{\mathbf{r}}$ are [94]

$$[\widehat{r}_k, \widehat{p}_j] = i\hbar\delta_{kj},$$

$$[\widehat{r}_k, \widehat{r}_j] = 0,$$

$$[\widehat{p}_k, \widehat{p}_j] = 0,$$

where $k, j = 1, 2, 3$, $\hbar$ is Planck's constant and δ_{kj} is the Kronecker symbol.

The 1D fractional Schrödinger equation first introduced by Laskin has the form [67], [92], [96]

$$i\hbar\frac{\partial\psi(x,t)}{\partial t} = -D_\alpha(\hbar\nabla)^\alpha\psi(x,t) + V(x)\psi(x,t), \qquad 1 < \alpha \le 2, \qquad (3.13)$$

where $\psi(x,t)$ is wave function and the 1D quantum Riesz fractional derivative[1] $(\hbar\nabla)^\alpha$ has been defined in the following way [67], [92]

$$(\hbar\nabla)^\alpha\psi(x,t) = -\frac{1}{2\pi\hbar}\int\limits_{-\infty}^{\infty} dp\, e^{i\frac{px}{\hbar}}|p|^\alpha\varphi(p,t), \qquad (3.14)$$

where $\varphi(p,t)$ is the Fourier transform of the wave function $\psi(x,t)$ given by

$$\varphi(p,t) = \int\limits_{-\infty}^{\infty} dx\, e^{-i\frac{px}{\hbar}}\psi(x,t), \qquad (3.15)$$

and respectively

$$\psi(x,t) = \frac{1}{2\pi\hbar}\int\limits_{-\infty}^{\infty} dp\, e^{i\frac{px}{\hbar}}\varphi(p,t). \qquad (3.16)$$

When $\alpha = 2$, Eq. (3.13) is transformed into the well-known 1D Schrödinger equation

$$i\hbar\frac{\partial\psi(x,t)}{\partial t} = -\frac{\hbar^2}{2m}\frac{\partial^2}{\partial x^2}\psi(x,t) + V(x)\psi(x,t), \qquad (3.17)$$

---

[1]The Riesz fractional derivative was originally introduced in [88].

where m is a particle mass.

It is easy to see that Eq. (3.13) can be rewritten in the operator form, namely

$$i\hbar\frac{\partial\psi}{\partial t} = \widehat{\mathcal{H}}_\alpha\psi, \qquad (3.18)$$

here $\widehat{\mathcal{H}}_\alpha$ is the 1D fractional Hamiltonian operator

$$\widehat{\mathcal{H}}_\alpha = -D_\alpha(\hbar\nabla)^\alpha + V(x). \qquad (3.19)$$

Alternatively, 1D fractional Schrödinger equation (3.13) can be presented as

$$i\hbar\frac{\partial\psi(x,t)}{\partial t} = D_\alpha(-\hbar^2\Delta)^{\alpha/2}\psi(x,t) + V(x)\psi(x,t), \qquad 1 < \alpha \leq 2, \quad (3.20)$$

where 1D the quantum Riesz fractional derivative $(-\hbar^2\Delta)^{\alpha/2}$ has been introduced by

$$(-\hbar^2\Delta)^{\alpha/2}\psi(x,t) = \frac{1}{(2\pi\hbar)^3}\int d^3p\, e^{i\frac{pr}{\hbar}}|p|^\alpha\varphi(p,t), \qquad (3.21)$$

where the wave functions in space $\psi(x,t)$ and momentum $\varphi(p,t)$ representations are related to each other by the Fourier transforms given by Eqs. (3.15) and (3.16).

The operator form of Eq. (3.20) is

$$i\hbar\frac{\partial\psi(x,t)}{\partial t} = \widehat{H}_\alpha\psi(x,t), \qquad 1 < \alpha \leq 2, \quad (3.22)$$

where the Hamilton operator $\widehat{H}_\alpha$ is given by

$$\widehat{H}_\alpha = D_\alpha(-\hbar^2\Delta)^{\alpha/2} + V(x), \qquad 1 < \alpha \leq 2. \quad (3.23)$$

The classical mechanics Hamilton function H_α corresponding to the Hamilton operator $\widehat{H}_\alpha$ is

$$H_\alpha(p,x) = D_\alpha|p|^\alpha + V(x), \qquad 1 < \alpha \leq 2. \quad (3.24)$$

For the special case when $\alpha = 2$ and $D_2 = 1/2m$ (see, for details [67], [92]), where m is the particle mass, Eqs. (3.8), (3.13) and (3.20) are transformed into the well-known 3D and 1D Schrödinger equations [94].

3.3 Fractional Schrödinger equation in momentum representation

Substituting the wave function $\psi(x,t)$ from Eq. (3.16) into Eq. (3.13), multiplying by $\exp(-ipx/\hbar)$ and integrating over dx bring us the 1D fractional Schrödinger equation in momentum representation

$$i\hbar\frac{\partial\varphi(p,t)}{\partial t} = D_\alpha|p|^\alpha\varphi(p,t) + \int d^3p'\,U_{p,p'}\varphi(p',t), \qquad 1 < \alpha \le 2, \quad (3.25)$$

where $U_{p,p'}$ is given by

$$U_{p,p'} = \int dx\exp(-i(p-p')x/\hbar)V(x), \qquad (3.26)$$

and we used the following representation for the delta function $\delta(x)$

$$\delta(x) = \frac{1}{2\pi\hbar}\int dp\exp(ipx/\hbar). \qquad (3.27)$$

It is assumed that the integral in Eq. (3.26) exists.

Equation (3.25) is the 1D fractional Schrödinger equation in momentum representation for the wave function $\varphi(p,t)$.

To obtain the 3D fractional Schrödinger equation in momentum representation, let us substitute wave function $\psi(\mathbf{r},t)$ from Eq. (3.10) into Eq. (3.8),

$$\frac{i\hbar}{(2\pi\hbar)^3}\frac{\partial}{\partial t}\int d^3p'\,e^{i\frac{\mathbf{p}'\mathbf{r}}{\hbar}}\varphi(\mathbf{p}',t) = \frac{D_\alpha}{(2\pi\hbar)^3}\int d^3p'\,e^{i\frac{\mathbf{p}'\mathbf{r}}{\hbar}}|\mathbf{p}'|^\alpha\varphi(\mathbf{p}',t) \qquad (3.28)$$

$$+\frac{V(\mathbf{r})}{(2\pi\hbar)^3}\int d^3p'\,e^{i\frac{\mathbf{p}'\mathbf{r}}{\hbar}}\varphi(\mathbf{p}',t).$$

Further, by multiplying Eq. (3.28) by $\exp(-i\mathbf{p}\mathbf{r}/\hbar)$ and integrating over d^3r we obtain the equation for the wave function $\varphi(\mathbf{p},t)$ in momentum representation

$$i\hbar\frac{\partial}{\partial t}\varphi(\mathbf{p},t) = D_\alpha|\mathbf{p}|^\alpha\varphi(\mathbf{p},t) + \int d^3p'\,U_{\mathbf{p},\mathbf{p}'}\varphi(\mathbf{p}',t), \qquad (3.29)$$

where $U_{\mathbf{p},\mathbf{p}'}$ has been defined by

$$U_{\mathbf{p},\mathbf{p}'} = \int d^3r\exp(-i(\mathbf{p} - \mathbf{p}')\mathbf{r}/\hbar)V(\mathbf{r}), \qquad (3.30)$$

and we used the following representation for the delta function $\delta(\mathbf{r})$

$$\delta(\mathbf{r}) = \frac{1}{(2\pi\hbar)^3} \int d^3p \exp(i\mathbf{pr}/\hbar). \tag{3.31}$$

It is assumed that $U_{\mathbf{p},\mathbf{p}'}$ introduced by Eq. (3.30) exists.

Equation (3.29) is the 3D fractional Schrödinger equation in momentum representation for the wave function $\varphi(\mathbf{p}, t)$.

If the normalization condition for wave function in coordinate space $\psi(\mathbf{r}, t)$ has the form

$$\int d^3r |\psi(\mathbf{r}, t)|^2 = 1, \tag{3.32}$$

then the wave function in momentum space $\varphi(\mathbf{p}, t)$ is normalized as

$$\frac{1}{(2\pi\hbar)^3} \int d^3p |\varphi(\mathbf{p}, t)|^2 = 1. \tag{3.33}$$

Indeed, substituting $\psi(\mathbf{r}, t)$ from Eq. (3.10) into Eq. (3.32), performing integration over d^3r, and using Eq. (3.31) yields Eq. (3.33).

3.4   Hermiticity of the fractional Hamiltonian operator

The fractional Hamiltonian $\widehat{\mathcal{H}}_\alpha$ given by Eq. (3.19) is the Hermitian operator in the space with scalar product

$$(\phi, \chi) = \int\limits_{-\infty}^{\infty} dx \phi^*(x, t) \chi(x, t), \tag{3.34}$$

where $\phi^*(x, t)$ is the complex conjugate wave function of $\phi(x, t)$.

To prove the hermiticity of $\widehat{\mathcal{H}}_\alpha$ let us note that in accordance with the definition of the quantum Riesz fractional derivative given by Eq. (3.14) there exists the integration by parts formula

$$(\phi, (\hbar\nabla)^\alpha \chi) = ((\hbar\nabla)^\alpha \phi, \chi). \tag{3.35}$$

The average energy E_α of a fractional quantum system with Hamiltonian H_α is

$$E_\alpha = \int\limits_{-\infty}^{\infty} dx \psi^*(x, t) \widehat{\mathcal{H}}_\alpha \psi(x, t). \tag{3.36}$$

Taking into account Eqs. (3.19) and (3.35) we have

$$E_\alpha = \int\limits_{-\infty}^{\infty} dx \psi^*(x,t) \widehat{\mathcal{H}}_\alpha \psi(x,t) = \int\limits_{-\infty}^{\infty} dx (\widehat{\mathcal{H}}_\alpha^+ \psi(x,t))^* \psi(x,t) = E_\alpha^*.$$

As a physical consequence, the energy of the system is real. Thus, the fractional Hamiltonian $\widehat{\mathcal{H}}_\alpha$ defined by Eq. (3.19) is the Hermitian or self-adjoint operator in the space with the scalar product defined by Eq. (3.34) [95], [96]

$$(\widehat{\mathcal{H}}_\alpha^+ \phi, \chi) = (\phi, \widehat{\mathcal{H}}_\alpha \chi). \tag{3.37}$$

Note that the 1D fractional Schrödinger equation (3.13) leads to the fundamental equation

$$\frac{\partial}{\partial t} \int dx \psi^*(x,t) \psi(x,t) = 0, \tag{3.38}$$

which shows that the wave function remains normalized, if it is normalized once. Multiplying Eq. (3.13) from the left by $\psi^*(x,t)$, and multiplying the conjugate complex of Eq. (3.13) by $\psi(x,t)$, and then subtracting the two resultant equations finally yield

$$i\hbar \frac{\partial}{\partial t} (\psi^*(x,t)\psi(x,t)) = \psi^*(x,t)\widehat{H}_\alpha \psi(x,t) - \psi(x,t)\widehat{H}_\alpha^+ \psi^*(x,t).$$

Integrating this relation over the space variable x and using the fact that the operator $\widehat{H}_\alpha$ is self-adjoint, we come to Eq. (3.38).

The 3D generalization of the proof of hermiticity is straightforward. The fractional Hamiltonian $\widehat{H}_\alpha(\widehat{\mathbf{p}}, \widehat{\mathbf{r}})$ given by Eq. (3.3) is the Hermitian operator in the space with scalar product

$$(\phi, \chi) = \int d^3r \phi^*(\mathbf{r},t)\chi(\mathbf{r},t), \tag{3.39}$$

where $\phi^*(\mathbf{r},t)$ is the complex conjugate wave function of $\phi(\mathbf{r},t)$.

To prove the hermiticity of $\widehat{H}_\alpha(\widehat{\mathbf{p}}, \widehat{\mathbf{r}})$ let us note that in accordance with the definition of the 3D quantum Riesz fractional derivative given by Eq. (3.9) there exists the integration by parts formula

$$(\phi, (-\hbar^2 \Delta)^{\alpha/2} \chi) = ((-\hbar^2 \Delta)^{\alpha/2} \phi, \chi). \tag{3.40}$$

With the help of Eqs. (3.3) and (3.40) the average energy of quantum system E_α becomes

$$E_\alpha = \int\limits_{-\infty}^{\infty} d^3r \psi^*(\mathbf{r},t)\widehat{H}_\alpha(\widehat{\mathbf{p}},\widehat{\mathbf{r}})\psi(\mathbf{r},t) = \int\limits_{-\infty}^{\infty} dr(\widehat{H}_\alpha^+(\widehat{\mathbf{p}},\widehat{\mathbf{r}})\psi(\mathbf{r},t))^*\psi(\mathbf{r},t) = E_\alpha^*.$$

Hence, the fractional Hamiltonian $\widehat{H}_\alpha(\widehat{\mathbf{p}},\widehat{\mathbf{r}})$ given by Eq. (3.3) is the Hermitian or self-adjoint operator in the space with the scalar product defined by Eq. (3.39).

Defined by Eq. (3.8) the 3D fractional Schrödinger equation leads to the important equation

$$\frac{\partial}{\partial t} \int d^3r \psi^*(\mathbf{r},t)\psi(\mathbf{r},t) = 0, \tag{3.41}$$

which is the conservation law for normalization condition of the wave function in 3D case. The proof of Eq. (3.41) can be done by the straightforward generalization of the proof of Eq. (3.38) in 1D case.

3.5   The parity conservation law in fractional quantum mechanics

From definition (3.14) of the quantum Riesz fractional derivative it follows that

$$(\hbar\nabla)^\alpha \exp\left\{i\frac{px}{\hbar}\right\} = -|p|^\alpha \exp\left\{i\frac{px}{\hbar}\right\}. \tag{3.42}$$

In other words, the function $\exp\{ipx/\hbar\}$ is the eigenfunction of the quantum Riesz fractional operator $(\hbar\nabla)^\alpha$ with eigenvalue $-|p|^\alpha$.

The 3D generalization is straightforward as it follows from Eq. (3.9),

$$(-\hbar^2\Delta)^{\alpha/2} \exp\left\{i\frac{px}{\hbar}\right\} = |\mathbf{p}|^\alpha \exp\left\{i\frac{px}{\hbar}\right\}, \tag{3.43}$$

which means that the function $\exp\{i\mathbf{px}/\hbar\}$ is the eigenfunction of the 3D quantum Riesz fractional operator $(-\hbar^2\Delta)^{\alpha/2}$ with eigenvalue $|\mathbf{p}|^\alpha$.

Thus, the operators $(\hbar\nabla)^\alpha$ and $(-\hbar^2\Delta)^{\alpha/2}$ are the symmetrized fractional derivative, that is

$$(\hbar\nabla_x)^\alpha... = (\hbar\nabla_{-x})^\alpha..., \tag{3.44}$$

and

$$(-\hbar^2 \Delta_{\mathbf{r}})^{\alpha/2}\dots = (-\hbar^2 \Delta_{-\mathbf{r}})^{\alpha/2}\dots \qquad (3.45)$$

Due to properties (3.35) and (3.36), the kinetic part of fractional Hamiltonian H_α (see, for example Eqs. (3.3) or (3.19)) remains invariant under *inversion* transformation. Inversion, or to be precise, spatial inversion inverts the sign of all three spatial coordinates

$$\mathbf{r} \to -\mathbf{r}, \qquad x \to -x, \quad y \to -y, \quad z \to -z. \qquad (3.46)$$

Let us denote the inversion operator by $\widehat{P}$, which acts as

$$\widehat{P}\psi(\mathbf{r}, t) = \psi(-\mathbf{r}, t). \qquad (3.47)$$

Obviously, $P^2 = \mathbb{I}$, here $\mathbb{I}$ is the unit operator. So, eigenvalues of parity operator P are ± 1.

Taking into account Eqs. (3.45) and (3.47) we have

$$\widehat{P}(D_\alpha(-\hbar^2\Delta)^{\alpha/2})\psi(\mathbf{r}, t) = (D_\alpha(-\hbar^2\Delta)^{\alpha/2})\widehat{P}\psi(\mathbf{r}, t) \qquad (3.48)$$

or

$$[\widehat{P}, (-\hbar^2\Delta)^{\alpha/2}] = 0, \qquad (3.49)$$

where the notation $[\dots, \dots]$ stands for the commutator of two operators,

$$[\widehat{A}, \widehat{B}] = \widehat{A}\widehat{B} - \widehat{B}\widehat{A}.$$

Hence, the kinetic part of fractional Hamiltonian H_α (see, for example Eqs. (3.3) or (3.19)) is invariant under *inversion* transformation.

Further, if the potential energy in the fractional Hamiltonian (see, for example, Eq. (3.3)) is even, that is

$$V(\mathbf{r}) = V(-\mathbf{r}), \qquad (3.50)$$

then the chain of transformations holds

$$PV(\widehat{\mathbf{r}})\psi(\mathbf{r}, t) = V(-\widehat{\mathbf{r}})\psi(-\mathbf{r}, t) = V(\widehat{\mathbf{r}})P\psi(\mathbf{r}, t), \qquad (3.51)$$

which can be expressed in operator form

$$PV(\widehat{\mathbf{r}}) - V(\widehat{\mathbf{r}})P = [P, V(\widehat{\mathbf{r}})] = 0. \tag{3.52}$$

It follows from Eqs. (3.48) and (3.52) that the inversion operator $\widehat{P}$ and the fractional Hamiltonian $\widehat{H}_\alpha$ commute,

$$[\widehat{P}, \widehat{H}_\alpha] = 0. \tag{3.53}$$

We can divide the wave functions of quantum mechanical states with a well-defined eigenvalue of the operator $\widehat{P}$ into two classes; (i) functions which are not changed when acted upon by the inversion operator, $\widehat{P}\psi_+(\mathbf{r}) = \psi_+(\mathbf{r})$, the corresponding states are called even states; (ii) functions which change sign under action of the inversion operator, $\widehat{P}\psi_-(\mathbf{r}) = -\psi_-(\mathbf{r})$, the corresponding states are called odd states. Equation (3.53) represents the "parity conservation law" for fractional quantum mechanics [96], that is, if the state of a closed fractional quantum mechanical system has a given parity (i.e. if it is even or odd), then the parity is conserved.

3.6   Velocity operator

The quantum mechanical velocity operator is defined as follows: $\widehat{\mathbf{v}} = d\widehat{\mathbf{r}}/dt$, where $\widehat{\mathbf{r}}$ is the operator of coordinate. Using the general quantum mechanical rule for differentiation of operators

$$\frac{d}{dt}\widehat{\mathbf{r}} = \frac{i}{\hbar}[\widehat{H}_\alpha, \widehat{\mathbf{r}}],$$

we find

$$\widehat{\mathbf{v}} = \frac{i}{\hbar}(\widehat{H}_\alpha\widehat{\mathbf{r}} - \widehat{\mathbf{r}}\widehat{H}_\alpha).$$

Further, with the help of equation $f(\widehat{\mathbf{p}})\mathbf{r} - \mathbf{r}f(\widehat{\mathbf{p}}) = -i\hbar\partial f/\partial\mathbf{p}$, which holds for any function $f(\widehat{\mathbf{p}})$ of the momentum operator $\widehat{\mathbf{p}}$, and taking into account Eq. (3.3) for the Hamiltonian operator $\widehat{H}_\alpha(\widehat{\mathbf{p}}, \widehat{\mathbf{r}})$ we obtain the equation for the velocity operator in fractional quantum mechanics [96]

$$\widehat{\mathbf{v}} = \alpha D_\alpha(\widehat{\mathbf{p}}^2)^{\alpha/2-1}\widehat{\mathbf{p}}, \tag{3.54}$$

here $\widehat{\mathbf{p}}$ is the momentum operator, $\widehat{\mathbf{p}} = -i\hbar\partial/\partial\mathbf{r}$.

The velocity operator $\widehat{\mathbf{v}}$ plays a fundamental role in fractional quantum mechanics as well as in standard quantum mechanics.

When $\alpha = 2$, the fractional quantum mechanics velocity operator Eq. (3.54) becomes the well-known quantum velocity operator $\widehat{\mathbf{v}} = \widehat{\mathbf{p}}/m$ of standard quantum mechanics [94].

In 1D space the fractional quantum mechanics velocity operator is

$$\widehat{v} = \alpha D_\alpha (\widehat{p}^2)^{\alpha/2-1}\widehat{p}, \qquad (3.55)$$

where $\widehat{p} = -i\hbar\partial/\partial x$ is the 1D momentum operator.

3.7 Fractional current density

Multiplying Eq. (3.8) from the left by $\psi^*(\mathbf{r}, t)$, and multiplying the complex conjugate of Eq. (3.8) by $\psi(\mathbf{r}, t)$, then subtracting the two resultant equations yield

$$\frac{\partial}{\partial t}\int d^3r\,(\psi^*(\mathbf{r},t)\psi(\mathbf{r},t)) \qquad (3.56)$$

$$= \frac{D_\alpha}{i\hbar}\int d^3r\,\left(\psi^*(\mathbf{r},t)(-\hbar^2\Delta)^{\alpha/2}\psi(\mathbf{r},t) - \psi(\mathbf{r},t)(-\hbar^2\Delta)^{\alpha/2}\psi^*(\mathbf{r},t)\right).$$

From this equation we are led to the following differential equation [97]

$$\frac{\partial\rho(\mathbf{r},t)}{\partial t} + \mathrm{div}\mathbf{j}(\mathbf{r},t) + \mathrm{K}(\mathbf{r},t) = 0, \qquad (3.57)$$

where $\rho(\mathbf{r},t) = \psi^*(\mathbf{r},t)\psi(\mathbf{r},t)$ is the quantum mechanical probability density, the vector $\mathbf{j}(\mathbf{r},t)$ introduced in [96] is called the fractional probability current density vector

$$\mathbf{j}(\mathbf{r},t) = \frac{D_\alpha\hbar}{i}(\psi^*(\mathbf{r},t)(-\hbar^2\Delta)^{\alpha/2-1}\boldsymbol{\nabla}\psi(\mathbf{r},t) \qquad (3.58)$$

$$-\psi(\mathbf{r},t)(-\hbar^2\Delta)^{\alpha/2-1}\boldsymbol{\nabla}\psi^*(\mathbf{r},t)),$$

and $\mathrm{K}(\mathbf{r},t)$ was originally found in [97] while developing the system of fractional two-fluid hydrodynamic equations to model superfluid hydrodynamic phenomena in nanoporous environments,

$$K(\mathbf{r}, t) = -\frac{D_\alpha \hbar}{i} (\boldsymbol{\nabla}\psi^*(\mathbf{r}, t)(-\hbar^2 \Delta)^{\alpha/2-1}\boldsymbol{\nabla}\psi(\mathbf{r}, t) \qquad (3.59)$$

$$-\boldsymbol{\nabla}\psi(\mathbf{r}, t)(-\hbar^2 \Delta)^{\alpha/2-1}\boldsymbol{\nabla}\psi^*(\mathbf{r}, t)),$$

here the notation $\boldsymbol{\nabla} = \partial/\partial\mathbf{r}$ has been used.

In terms of the momentum operator $\widehat{\mathbf{p}} = -i\hbar\partial/\partial\mathbf{r}$, vector $\mathbf{j}(\mathbf{r}, t)$ given by Eq. (3.58) can be expressed in the form [96]

$$\mathbf{j} = D_\alpha(\psi^*(\mathbf{r}, t)(\widehat{\mathbf{p}}^2)^{\alpha/2-1}\widehat{\mathbf{p}}\psi(\mathbf{r}, t) \qquad (3.60)$$

$$+\psi(\mathbf{r}, t)(\widehat{\mathbf{p}}^2)^{\alpha/2-1}\widehat{\mathbf{p}}^*\psi^*(\mathbf{r}, t)),$$

which is a fractional generalization of the well-known equation for the probability current density vector of standard quantum mechanics [94].

Introduced by Eq. (3.59), $K(\mathbf{r}, t)$ can be expressed in the form

$$K(\mathbf{r}, t) = \frac{iD_\alpha}{\hbar} (\widehat{\mathbf{p}}^*(\mathbf{r}, t)\psi^*(\mathbf{r}, t)(\widehat{\mathbf{p}}^2)^{\alpha/2-1}\widehat{\mathbf{p}}\psi(\mathbf{r}, t) \qquad (3.61)$$

$$-\widehat{\mathbf{p}}\psi(\mathbf{r}, t)(\widehat{\mathbf{p}}^2)^{\alpha/2-1}\widehat{\mathbf{p}}^*\psi^*(\mathbf{r}, t)).$$

In terms of the velocity operator defined by Eq. (3.54), the fractional probability current density vector $\mathbf{j}$ reads

$$\mathbf{j} = \frac{1}{\alpha} \left(\psi^*(\mathbf{r}, t)\widehat{\mathbf{v}}\psi(\mathbf{r}, t) + \psi(\mathbf{r}, t)\widehat{\mathbf{v}}^*\psi^*(\mathbf{r}, t)\right), \qquad 1 < \alpha \le 2, \qquad (3.62)$$

the extra term $K(\mathbf{r}, t)$ can be expressed as [97], [98]

$$K(\mathbf{r}, t) = \frac{i}{\alpha\hbar} \left(\widehat{\mathbf{p}}^*\psi^*(\mathbf{r}, t)\widehat{\mathbf{v}}\psi(\mathbf{r}, t) - \widehat{\mathbf{p}}\psi(\mathbf{r}, t)\widehat{\mathbf{v}}^*\psi^*(\mathbf{r}, t)\right). \qquad (3.63)$$

In standard quantum mechanics, when $\alpha = 2$, and $D_\alpha = 1/2m$, with m being a particle mass, the extra term $K(\mathbf{r}, t)$ is vanished, velocity operator given by Eq. (3.54) becomes $\widehat{\mathbf{v}} = \widehat{\mathbf{p}}/m$. Hence, from Eq. (3.62) and we come to the well-known expression for the probability current density vector [94]

$$\mathbf{j}|_{\alpha=2} = \frac{1}{2m} \left(\psi^*(\mathbf{r}, t)\widehat{\mathbf{p}}\psi(\mathbf{r}, t) + \psi(\mathbf{r}, t)\widehat{\mathbf{p}}^*\psi^*(\mathbf{r}, t)\right), \qquad (3.64)$$

here $\psi(\mathbf{r}, t)$ stands for a solution to standard Schrödinger equation (3.7). We conclude, that in the case $\alpha = 2$ Eq. (3.57) turns into the well-known quantum mechanical probability conservation law

$$\frac{\partial \rho(\mathbf{r}, t)}{\partial t} + \mathrm{div} \mathbf{j}(\mathbf{r}, t)|_{\alpha=2} = 0. \qquad (3.65)$$

The wave function of a free particle in the framework of fractional quantum mechanics can be normalized to get a probability current density equal to 1 (the current when one particle passes through a unit area per unit time). The normalized wave function of a free particle is

$$\psi(\mathbf{r}, t) = \sqrt{\frac{\alpha}{2\mathrm{v}}} \exp\{\frac{i}{\hbar}\mathbf{p}\mathbf{r} - \frac{i}{\hbar}Et\}, \qquad E = D_\alpha |\mathbf{p}|^\alpha, \qquad (3.66)$$

$$1 < \alpha \le 2,$$

where v is the particle velocity, $\mathrm{v} = \alpha D_\alpha p^{\alpha-1}$. Then we have

$$\mathbf{j} = \frac{\mathbf{v}}{\mathrm{v}}, \qquad \mathbf{v} = \alpha D_\alpha (\mathbf{p}^2)^{\frac{\alpha}{2}-1}\mathbf{p}, \qquad (3.67)$$

that is, the vector $\mathbf{j}$ is indeed the unit vector.

Chapter 4

Time-Independent Fractional
Schrödinger Equation

The special case when Hamiltonian H_α does not depend explicitly on time is of great importance for physical applications. It is easy to see that in this case there exists a special solution to the 3D fractional Schrödinger equation (3.8) in the form

$$\psi(\mathbf{r}, t) = e^{-(i/\hbar)Et}\phi(\mathbf{r}), \qquad (4.1)$$

where E is the energy of a quantum particle,

$$E = D_\alpha|\mathbf{p}|^\alpha, \qquad 1 < \alpha \leq 2. \qquad (4.2)$$

The time-independent wave function $\phi(\mathbf{r})$ is the eigenvalue of fractional Hamilton operator (3.3),

$$\widehat{H}_\alpha(\widehat{\mathbf{p}}, \widehat{\mathbf{r}})\phi(\mathbf{r}) = E\phi(\mathbf{r}), \qquad (4.3)$$

with the eigenvalue given by (4.2).

With the help of Eq. (3.3) we find

$$D_\alpha|\widehat{\mathbf{p}}|^\alpha\phi(\mathbf{r}) + V(\widehat{\mathbf{r}})\phi(\mathbf{r}) = E\phi(\mathbf{r}), \qquad (4.4)$$

or, in terms of the quantum Riesz fractional derivative $(-\hbar^2\Delta)^{\alpha/2}$ introduced by Eq. (3.9),

$$D_\alpha(-\hbar^2\Delta)^{\alpha/2}\phi(\mathbf{r}) + V(\widehat{\mathbf{r}})\phi(\mathbf{r}) = E\phi(\mathbf{r}), \qquad 1 < \alpha \leq 2, \qquad (4.5)$$

which is the time-independent 3D fractional Schrödinger equation.

In 1D case we have

$$\psi(x,t) = e^{-(i/\hbar)Et}\phi(x), \qquad (4.6)$$

where E is the energy of a quantum particle,

$$E = D_\alpha |p|^\alpha. \qquad (4.7)$$

The wave function $\phi(x)$ satisfies

$$\widehat{\mathcal{H}}_\alpha \phi(x) = E\phi(x), \qquad (4.8)$$

where $\widehat{\mathcal{H}}_\alpha$ has been introduced by Eq. (3.19) and E in this case is defined by Eq. (4.7).

Hence, the time-independent 1D fractional Schrödinger equation reads

$$-D_\alpha(\hbar\nabla)^\alpha \phi(x) + V(x)\phi(x) = E\phi(x), \qquad 1 < \alpha \le 2 \qquad (4.9)$$

or

$$D_\alpha(-\hbar^2\Delta)^{\alpha/2}\phi(x) + V(x)\phi(x) = E\phi(x), \qquad 1 < \alpha \le 2, \qquad (4.10)$$

where $(\hbar\nabla)^\alpha$ is 1D the quantum Riesz fractional derivative introduced by Eq. (3.14), and $(-\hbar^2\Delta)^{\alpha/2}$ is an alternative expression for 1D the quantum Riesz fractional derivative following from Eq. (3.9) with $\Delta = \partial^2/\partial x^2$.

We call equations (4.4), (4.5) and (4.9) the time-independent (or stationary) fractional Schrödinger equation [95], [96]. From Eqs. (4.1) and (4.6) we see that the fractional quantum mechanical wave function oscillates with a definite frequency. The frequency of oscillations of a wave function corresponds to the energy. Therefore, we say that when the fractional wave function is of the special form given by Eqs. (4.1) or (4.6), the state has a definite energy E. The probability density to find a particle at $\mathbf{r}$ is the absolute square of the wave function $|\psi(\mathbf{r},t)|^2$. In view of Eq. (4.1) the probability density is equal to $|\phi(\mathbf{r})|^2$ and does not depend upon the time. That is, the probability density of finding the particle in any location is independent of the time. In other words, the system is in a stationary state - stationary in the sense that there is no variation in the quantum mechanical probability as a function of time.

4.1 Continuity of derivative of wave function

Following Dong and Xu [99], we integrate both sides of Eq. (4.10) over x ranging from $a - \varepsilon$ to $a + \varepsilon$, with ε being a small positive quantity. The result of the integration is

$$-\hbar^2 D_\alpha(-\hbar^2\Delta)^{\alpha/2-1}\nabla\phi(x)|_{a-\varepsilon}^{a+\varepsilon} = \int\limits_{a-\varepsilon}^{a+\varepsilon} dx(E - V(x))\phi(x), \qquad (4.11)$$

$$1 < \alpha \le 2,$$

where $\nabla = \partial/\partial x$.

If potential energy $V(x)$ is finite, we have

$$\lim_{\varepsilon\to 0}(-\hbar^2\Delta)^{\alpha/2-1}\nabla\phi(x)|_{a-\varepsilon}^{a+\varepsilon} = 0, \qquad (4.12)$$

which means that $(-\hbar^2\Delta)^{\alpha/2-1}\nabla\phi(x)$ is continuous at $x = a$.

By introducing the operator [99]

$$\mathcal{V}_\alpha = (-\hbar^2\Delta)^{\alpha/2-1}\nabla, \qquad (4.13)$$

we rewrite Eq. (4.12)

$$\lim_{\varepsilon\to 0}\mathcal{V}_\alpha\phi(x)|_{a-\varepsilon}^{a+\varepsilon} = 0. \qquad (4.14)$$

This equation is a fractional generalization of the continuity condition of space derivative of wave function. This continuity condition is a restriction to the wave function of a fractional quantum system when the potential function is bounded. If $V(x)$ is infinite at some boundaries, then a jump (discontinuity) condition for $\mathcal{V}_\alpha\phi(x)$ has to be taken into account. Obviously, when $\alpha = 2$ we obtain from Eq. (4.13)

$$\mathcal{V}_2 = \mathcal{V}_\alpha|_{\alpha=2} = \nabla,$$

and Eq. (4.12) becomes the well-known continuity condition for the first order space derivative of wave function of standard quantum mechanics.

The operator $\mathcal{V}_\alpha$ first introduced by Dong and Xu [99] can be expressed in terms of 1D velocity operator $\widehat{v}$ defined by Eq. (3.55)

$$\mathcal{V}_\alpha = i\frac{\widehat{v}}{\alpha\hbar D_\alpha}. \qquad (4.15)$$

The operator $\mathcal{V}_\alpha$ has been applied by Dong and Xu [99] to find exact solutions to the fractional Schrödinger equation for a finite square well, periodic potential, the δ-potential well, for the problem of penetration through a δ-potential barrier, and for the Dirac comb.

4.2   The time-independent fractional Schrödinger equation in momentum representation

To obtain the fractional Schrödinger equation in momentum representation we define the Fourier transforms for the steady state wave functions in space $\phi(\mathbf{r})$ and momentum $\phi(\mathbf{p})$ representations,

$$\phi(\mathbf{r}) = \frac{1}{(2\pi\hbar)^3} \int d^3p\, e^{i\frac{\mathbf{p}\mathbf{r}}{\hbar}} \phi(\mathbf{p}), \qquad \phi(\mathbf{p}) = \int d^3r\, e^{-i\frac{\mathbf{p}\mathbf{r}}{\hbar}} \phi(\mathbf{r}). \qquad (4.16)$$

Substituting $\phi(\mathbf{r})$ given by Eq. (4.16) into Eq. (4.5) yields

$$D_\alpha |\mathbf{p}^2|^{\alpha/2} \phi(\mathbf{p}) + \int d^3p' U_{\mathbf{p},\mathbf{p}'} \varphi(\mathbf{p}'), = E\phi(\mathbf{p}), \qquad 1 < \alpha \leq 2, \qquad (4.17)$$

where $U_{\mathbf{p},\mathbf{p}'}$ has been introduced by Eq. (3.30).

Equation (4.17) is the 3D time-independent fractional Schrödinger equation in momentum representation.

In the 1D case the time-independent fractional Schrödinger equation in momentum representation can be obtained in a similar way. We define the Fourier transforms for the steady state wave functions,

$$\phi(x) = \frac{1}{2\pi\hbar} \int\limits_{-\infty}^{\infty} dp\, e^{ipx/\hbar} \varphi(p), \qquad \varphi(p) = \int\limits_{-\infty}^{\infty} dx\, e^{-ipx/\hbar} \phi(x). \qquad (4.18)$$

Then substituting $\phi(x)$ given by Eq. (4.18) into Eq. (4.17) yields

$$D_\alpha |p|^\alpha \varphi(p) + \int\limits_{-\infty}^{\infty} dp' U_{p,p'} \varphi(p') = E\varphi(p), \qquad 1 < \alpha \leq 2, \qquad (4.19)$$

where we introduced

$$U_{p,p'} = \frac{1}{2\pi\hbar} \int\limits_{-\infty}^{\infty} dx\, e^{-i(p-p')/\hbar} V(x).$$

Equation (4.19) is the 1D time-independent fractional Schrödinger equation in momentum representation.

4.3 Orthogonality of the wave functions

Suppose that E_1 is a possible energy for which Eq. (4.10) has a solution ϕ_1, and that E_2 is another value for energy for which this equation has some other solution ϕ_2. Then we know two special solutions of the fractional Schrödinger equation, namely

$$\psi_1(x,t) = e^{-(i/\hbar)E_1 t}\phi_1(x) \quad \text{and} \quad \psi_2(x,t) = e^{-(i/\hbar)E_2 t}\phi_2(x).$$

Since the fractional Schrödinger equation is linear, it is clear that if ψ is a solution, then ψ multiplied by any constant c is also a solution, $c\psi$. Furthermore, if ψ_1 is a solution and ψ_2 is a solution, then the sum of ψ_1 and ψ_2 is also a solution. Obviously, then, the wave function

$$\psi = c_1 e^{-(i/\hbar)E_1 t}\phi_1 + c_2 e^{-(i/\hbar)E_2 t}\phi_2 \qquad (4.20)$$

is a solution to the fractional Schrödinger equation.

It can be shown that if all of the possible values of E and the corresponding functions ϕ are worked out, any solution to Eq. (4.10) can be written as a linear combination of these special solutions of definite energy.

It is clear that the total probability to be anywhere is constant. With the help of Eq. (4.20) for ψ, we have

$$\int\limits_{-\infty}^{\infty} dx\,\psi^*\psi = c_1^* c_1 \int\limits_{-\infty}^{\infty} dx|\phi_1|^2 + c_1^* c_2 e^{(i/\hbar)(E_1 - E_2)} \int\limits_{-\infty}^{\infty} dx\,\phi_1^*\phi_2$$

$$+ c_1 c_2^* e^{-(i/\hbar)(E_1 - E_2)} \int\limits_{-\infty}^{\infty} dx\,\phi_1\phi_2^* + c_2^* c_2 \int\limits_{-\infty}^{\infty} dx|\phi_2|^2.$$

Since it has to be a constant, the time-variable terms (i.e. terms including $e^{\pm(i/\hbar)(E_1 - E_2)}$) must vanish for all possible choices of c_1 and c_2. This means

$$\int\limits_{-\infty}^{\infty} dx\,\phi_1^*\phi_2 = \int\limits_{-\infty}^{\infty} dx\,\phi_1\phi_2^* = 0. \qquad (4.21)$$

Thus, Eq. (4.21) says that two states with different energies are orthogonal. In other words, if a particle is known to have an energy E_1 and the wave function $\psi_1(x, t) = e^{-(i/\hbar)E_1 t}\phi_1(x)$, then the amplitude that it is found to have a different energy E_2 and the wave function $\psi_2(x, t) = e^{-(i/\hbar)E_2 t}\phi_2(x)$ must be 0.

On the other hand, it is useful to show directly from the fractional Schrödinger equation (4.4) the orthogonality of the wave functions for stationary quantum mechanical states with different energies, $E_m \neq E_n$ ($m \neq n$). Let $\phi_m(\mathbf{r})$ and $\phi_n(\mathbf{r})$ be two wave functions which satisfy the 3D fractional Schrödinger equations

$$D_\alpha(-\hbar^2\Delta)^{\alpha/2}\phi_m(\mathbf{r}) + V(\mathbf{r})\phi_m(\mathbf{r}) = E_m\phi_m(\mathbf{r}),$$

and

$$D_\alpha(-\hbar^2\Delta)^{\alpha/2}\phi_n^*(\mathbf{r}) + V(\mathbf{r})\phi_n^*(\mathbf{r}) = E_n^*\phi_n^*(\mathbf{r}),$$

respectively.

Multiplying the first one by $\phi_n^*(\mathbf{r})$ and the second one by $\phi_m(\mathbf{r})$, and then subtracting the second one from the first one, yield

$$(E_m - E_n)\phi_m(\mathbf{r})\phi_n^*(\mathbf{r}) \tag{4.22}$$

$$= D_\alpha(\phi_n^*(\mathbf{r})(-\hbar^2\Delta)^{\alpha/2}\phi_m(\mathbf{r}) - \phi_m(\mathbf{r})(-\hbar^2\Delta)^{\alpha/2}\phi_n^*(\mathbf{r})).$$

We integrate both sides of Eq. (4.22) over d^3r. Integral over d^3r from the right of Eq. (4.22) is 0 because of the hermiticity of fractional operator $(-\hbar^2\Delta)^{\alpha/2}$, and we obtain

$$(E_m - E_n)\int d^3r\psi_m(\mathbf{r})\psi_n^*(\mathbf{r}) = 0, \tag{4.23}$$

which results in the orthogonality condition

$$\int d^3r\psi_m(\mathbf{r})\psi_n^*(\mathbf{r}) = 0, \tag{4.24}$$

because of the condition that $E_m \neq E_n$.

4.4   Linear combination of steady-state functions

Let the functions corresponding to the set of energy levels E_n be not only orthogonal but also normalized, i.e. that the integral of the absolute square over all x is 1. Then we will have

$$\int\limits_{-\infty}^{\infty} dx \phi_n^*(x)\phi_m(x) = \delta_{nm}, \tag{4.25}$$

where $\phi_n^*(x)$ is the complex conjugate wave function of $\phi_n(x)$, δ_{nm}, the Kronecker symbol, is defined by $\delta_{nm} = 0$ if $n \neq m$ and $\delta_{nn} = 1$. Any function which is likely to arise as a wave function can be expressed as a linear combination of such ϕ_n's. That is,

$$\chi(x) = \sum_{n=1}^{\infty} a_n \phi_n(x). \tag{4.26}$$

The coefficients a_n are easily obtained with the help of Eq. (4.25),

$$a_n = \int\limits_{-\infty}^{\infty} dx \phi_n^*(x)\chi(x).$$

Substituting a_n into Eq. (4.26) yields

$$\chi(x) = \sum_{n=1}^{\infty} \left(\int\limits_{-\infty}^{\infty} dx' \phi_n^*(x')\chi(x') \right) \phi_n(x)$$

$$= \int\limits_{-\infty}^{\infty} dx' \left(\sum_{n=1}^{\infty} \phi_n(x)\phi_n^*(x') \right) \chi(x'),$$

and we conclude that

$$\sum_{n=1}^{\infty} \phi_n(x)\phi_n^*(x') = \delta(x - x'),$$

where $\delta(x)$ is the delta function.

It is clear that we can express the fractional kernel $K(x_b t_b | x_a t_a)$ in terms of the fractional wave functions $\phi_n(x)$ and the energy values E_n. Suppose that we know the wave function at time t_1. Then, what is the wave function at time t_2? At time t_1 the wave function $\psi(x, t_1)$ reads

$$\psi(x, t_1) = \sum_{n=1}^{\infty} c_n e^{-(i/\hbar) E_n t_1} \phi_n(x), \qquad (4.27)$$

while at time t_2 the wave function $\psi(x, t_2)$ reads

$$\psi(x, t_2) = \sum_{n=1}^{\infty} c_n e^{-(i/\hbar) E_n t_2} \phi_n(x). \qquad (4.28)$$

Now, using Eq. (4.25) we find that

$$c_m = e^{(i/\hbar) E_m t_1} \int_{-\infty}^{\infty} dx \psi(x, t_1) \phi_m(x),$$

and as a consequence $\psi(x, t_2)$ is

$$\psi(x, t_2) = \int_{-\infty}^{\infty} dx' \sum_{n=1}^{\infty} \phi_n(x) \phi_n^*(x') e^{-(i/\hbar) E_n (t_2 - t_1)} \psi(x', t_1). \qquad (4.29)$$

This equation determines the wave function $\psi(x, t_2)$ at time t_2 completely in terms of the wave function $\psi(x, t_1)$ at previous time t_1, $t_1 < t_2$.

4.5 Variational principle

The time-independent fractional Schrödinger equation (4.9) can be derived from the quantum variational principle. That is, the problem of finding the minimum of the functional

$$\int dx \phi^*(x) \widehat{H}_\alpha \phi(x), \qquad (4.30)$$

with $\widehat{H}_\alpha$ defined by Eq. (3.23) at the additional condition

$$\int dx \phi^*(x) \phi(x) = 1, \qquad (4.31)$$

leads to the stationary fractional Schrödinger equation (4.9).

The variational principle is

$$\delta \int dx \phi^*(x) \widehat{H}_\alpha \phi(x) = 0 \qquad (4.32)$$

with the condition given by Eq. (4.31). It follows from Eq. (4.32) that

$$\int dx \delta \phi^*(x) \widehat{H}_\alpha \phi(x) + \int dx \phi^*(x) \widehat{H}^*_\alpha \phi(x) = 0. \qquad (4.33)$$

Using the hermiticity of the fractional Hamiltonian $\widehat{H}_\alpha$ given by Eq. (3.23), we rewrite Eq. (4.33) in the form

$$\int dx \delta \phi^*(x) \widehat{H}_\alpha \phi(x) + \int dx \delta \phi(x) \widehat{H}^*_\alpha \phi^*(x) = 0. \qquad (4.34)$$

The variations $\delta \phi^*(x)$ and $\delta \phi(x)$ in Eq. (4.33) should satisfy the addition condition

$$\int dx \delta \phi^*(x) \phi(x) + \int dx \phi^*(x) \delta \phi(x) = 0, \qquad (4.35)$$

which follows from Eq. (4.31).

By applying the method of Lagrange multipliers we can rewrite two equations (4.34) and (4.35) as one equation

$$\int dx \delta \phi^*(x) (\widehat{H}_\alpha - E) \phi(x) + \int dx \delta \phi(x) (\widehat{H}^*_\alpha - E) \phi^*(x) = 0, \qquad (4.36)$$

where E is the Lagrange multiplier and the variations $\delta \phi^*(x)$ and $\delta \phi(x)$ are now independent.

The variational equation (4.36) is satisfied for all possible variations $\delta \phi^*(x)$ and $\delta \phi(x)$ if wave functions $\phi(x)$ and $\phi^*(x)$ satisfy the fractional Schrödinger equations

$$(\widehat{H}_\alpha - E) \phi(x) = 0, \qquad (4.37)$$

and

$$(\widehat{H}^*_\alpha - E) \phi^*(x) = 0. \qquad (4.38)$$

Equations (4.37) and (4.38) are two fractional eigenvalue problems.

Thus, we conclude that the fractional time-independent Schrödinger equations $\widehat{H}_\alpha \phi(x) = E \phi(x)$ with $\widehat{H}_\alpha$ defined by Eq. (3.23) is equivalent to the quantum variational principle

$$\delta \int dx \phi^*(x) (\widehat{H}_\alpha - E) \phi(x) = 0. \qquad (4.39)$$

with $\phi(x)$ being normalized by the condition (4.31).

4.6   Fractional Schrödinger equation in periodic potential

One of the fundamental problems of solid state physics is the problem of a particle moving in a periodic structure - crystal. The quantum problem of a particle moving in a one-dimensional periodic structure brings us the 1D fractional Schrödinger equation in a periodic potential

$$V(x + a) = V(x),$$

where a is the lattice constant.

4.6.1   *Periodic potential*

We consider the 1D time-independent fractional Schrödinger equation (4.9) in the case when potential $V(x)$ is periodic function

$$V(x) = V(x + L), \tag{4.40}$$

where due to translational periodicity we have

$$L = la, \tag{4.41}$$

here a is the period of the function $V(x)$ and l is an integer.

Given the potential (4.40) one can think of a 1D periodic structure with period a as a 1D crystal with lattice constant a.

It is well known that any periodical function can be expressed by the Fourier series. Thus, we write

$$V(x) = \sum_n V_n e^{2\pi i n x/a}, \tag{4.42}$$

with n being an integer. By rewriting this equation as

$$V(x) = \sum_g V_g e^{igx}, \tag{4.43}$$

we introduce *reciprocal lattice* with period g defined as

$$g_n = n\frac{2\pi}{a}. \tag{4.44}$$

The coefficients V_g in Eq. (4.43) are given by

$$V_g = \frac{1}{a} \int\limits_0^a dx V(x) e^{-igx}. \tag{4.45}$$

It is easy to see that Eq. (4.43) is in agreement with Eq. (4.40). Indeed, with the help of Eq. (4.43) we have

$$V(x+L) = \sum_g V_g e^{ig(x+L)} = \sum_g V_g e^{igx} = V(x), \tag{4.46}$$

if we note that $\exp(igL) = 1$, because of

$$gL = n\frac{2\pi}{a}la = 2\pi nl, \tag{4.47}$$

here nl is an integer.

4.6.2 Bloch theorem

To search for the solution to the 1D time-independent fractional Schrödinger equation

$$-D_\alpha(\hbar\nabla)^\alpha\phi(x) + V(x)\phi(x) = E\phi(x), \qquad 1 < \alpha \le 2, \tag{4.48}$$

with periodic potential $V(x)$ we exploit the ansatz

$$\phi(x) = \sum_k C_k e^{ikx}, \tag{4.49}$$

where C_k and k have to be defined.

Substituting wave function $\phi(x)$ introduced by Eq. (4.49) and $V(x)$ given by Eq. (4.43) into Eq. (4.9) gives

$$\sum_k D_\alpha |k|^\alpha C_k e^{ikx} + \sum_{g,k} V_g e^{igx} C_k e^{ikx} = E\sum_k C_k e^{ikx}. \tag{4.50}$$

The sum over all possible values of g and k can be rewritten as

$$\sum_{g,k} V_g e^{igx} C_k e^{ikx} = \sum_{g,k} V_g C_{k-g} e^{ikx}. \tag{4.51}$$

Then Eq. (4.50) reads

$$\sum_k e^{ikx}\{(D_\alpha|k|^\alpha C_k - E) + \sum_g V_g C_{k-g}\} = 0, \qquad 1 < \alpha \le 2. \qquad (4.52)$$

The plane waves belong to an orthogonal set of functions. Therefore, the coefficient of each term in the sum over k must vanish,

$$(D_\alpha|k|^\alpha - E)C_k + \sum_g V_g C_{k-g} = 0, \qquad 1 < \alpha \le 2. \qquad (4.53)$$

Substituting $k = q - g'$ yields

$$(D_\alpha|q - g'|^\alpha - E)C_{q-g'} + \sum_g V_g C_{q-g'-g} = 0. \qquad (4.54)$$

Introducing new summation variable $\tilde{g} = g' + g$ results in

$$(D_\alpha|q - g'|^\alpha - E)C_{q-g'} + \sum_{\tilde{g}} V_{\tilde{g}-g'} C_{q-\tilde{g}} = 0. \qquad (4.55)$$

This equation shows that at a particular value of q the C_{q-g} depends on $q - g$, where g is reciprocal lattice period. It follows from Eq. (4.49) that the wave function $\phi(x)$ takes the form

$$\phi_q(x) = \sum_g C_{q-g} e^{i(q-g)x} = e^{iqx} u_q(x)$$

or

$$\phi_q(x) = e^{iqx} u_q(x), \qquad (4.56)$$

where function $u_q(x)$

$$u_q(x) = \sum_g C_{q-g} e^{-igx}, \qquad (4.57)$$

has periodicity of the lattice

$$u_q(x) = u_q(x + L),$$

with L given by Eq. (4.41).

The function $\phi_q(x)$ is Bloch's wave function. Equation (4.56) is a mathematical formulation of Bloch's theorem [100]. In the framework of fractional quantum mechanics Bloch's theorem has the same form as in standard quantum mechanics.

Bloch's theorem: *Every solution to the fractional Schrödinger equation in a periodic potential is a plane wave modulated by some function with the periodicity of the lattice.*

We conclude that the fractional Schrödinger equation supports Bloch's theorem.

Chapter 5

Fractional Uncertainty Relation

The more precise the measurement of position, the more imprecise the measurement of momentum, and vice versa.

Werner Heisenberg

Heisenberg's Uncertainty Principle [101], which is a fundamental aspect of quantum mechanics, has many different mathematical formulations. The uncertainty relation is a mathematical formulation of Heisenberg's Uncertainty Principle. The first uncertainty relation in the framework of standard quantum mechanics was developed by Kennard [102] in 1927. Since then many other mathematical approaches have been developed to formulate Heisenberg's Uncertainty Principle.

Here we study the fractional uncertainty relation, which is a new mathematical formulation of Heisenberg's Uncertainty Principle in the framework of fractional quantum mechanics. The fractional uncertainty relation was formulated for the first time in [92].

5.1 Quantum Lévy wave packet

The 1D fractional Schrödinger equation for a free particle has the following plane wave solution

$$\psi(x,t) = C \exp\left\{ i\frac{px}{\hbar} - i\frac{D_\alpha |p|^\alpha t}{\hbar} \right\}, \tag{5.1}$$

where C is a normalization constant. In the special Gaussian case ($\alpha = 2$ and $D_2 = 1/2m$ with m being a particle mass) Eq. (5.1) gives a plane wave of standard quantum mechanics. Localized states are obtained by a superposition of plane waves

$$\psi_L(x,t) = \frac{1}{2\pi\hbar} \int\limits_{-\infty}^{\infty} dp\varphi(p) \exp\left\{ i\frac{px}{\hbar} - i\frac{D_\alpha |p|^\alpha t}{\hbar} \right\}, \qquad (5.2)$$

here $\varphi(p)$ is a "weight" function. We will study Eq. (5.2) for a one-dimensional Lévy wave packet $\psi_L(x,t)$ which has been introduced in [92]

$$\psi_L(x,t) = \frac{A_\nu}{2\pi\hbar} \int\limits_{-\infty}^{\infty} dp \exp\left\{ -\frac{|p-p_0|^\nu l^\nu}{2\hbar^\nu} \right\} \exp\left\{ i\frac{px}{\hbar} - i\frac{D_\alpha |p|^\alpha t}{\hbar} \right\}, \qquad (5.3)$$

with the "weight" function $\varphi_L(p)$ of the form

$$\varphi_L(p) = A_\nu \exp\left\{ -\frac{|p-p_0|^\nu l^\nu}{2\hbar^\nu} \right\},$$

$$p_0 > 0, \qquad \nu \le \alpha, \qquad 1 < \alpha \le 2,$$

where A_ν is a normalization constant, l is a characteristic space scale, p_0 is the center of the packet in momentum space and α is the Lévy index, $1 < \alpha \le 2$.

The Lévy wave packet $\psi_L(x,t)$ is a generalization of the well-known Gaussian wave packet in the framework of standard quantum mechanics. Indeed, in the special case when $\alpha = 2$ and $\nu = 2$ the quantum mechanical Lévy wave packet $\psi_L(x,t)$ turns into the Gaussian wave packet $\psi_G(x,t)$ introduced by

$$\psi_G(x,t) = \frac{1}{2\pi\hbar} \int\limits_{-\infty}^{\infty} dp\varphi_G(p) \exp\left\{ i\frac{px}{\hbar} - i\frac{p^2 t}{2m\hbar} \right\}, \qquad (5.4)$$

with function $\varphi(p)$ of the form

$$\varphi_G(p) = A_2 \exp\left\{ -\frac{|p-p_0|^2 l^2}{2\hbar^2} \right\}, \qquad p_0 > 0,$$

where $A_2 = A_\nu|_{\nu=2}$ is the Gaussian wave packet normalization constant, and $l/\hbar$ can be thought of as the width of the Gaussian wave packet in the momentum space.

We are interested in the probability density $\rho(x,t)$ that a particle occupies the position x, and the probability density $w(p,t)$ that a particle has particular momentum value p. The wave packet $\psi_L(x,t)$ defined by Eq. (5.3) gives the probability density $\rho(x,t)$

$$\rho(x,t) = |\psi_L(x,t)|^2 = \frac{A_\nu^2}{(2\pi\hbar)^2} \int\limits_{-\infty}^{\infty} dp_1 dp_2 \exp\left\{-\frac{|p_1 - p_0|^\nu l^\nu}{2\hbar^\nu}\right\} \quad (5.5)$$

$$\times \exp\left\{-\frac{|p_2 - p_0|^\nu l^\nu}{2\hbar^\nu}\right\} \exp\left\{i\frac{(p_1 - p_2)x}{\hbar} - i\frac{D_\alpha(|p_1|^\alpha - |p_2|^\alpha)t}{\hbar}\right\}.$$

Now, we can fix the factor A_ν such that $\int dx\rho(x,t) = \int dx|\psi_L(x,t)|^2 = 1$ with the result

$$A_\nu = \sqrt{\frac{\pi\nu l}{\Gamma(1/\nu)}}, \quad (5.6)$$

where $\Gamma(1/\nu)$ is the Gamma function[1]. It follows immediately from (5.6) that the normalization constant A_2 for Gaussian wave packet is

$$A_2 = A_\nu|_{\nu=2} = \sqrt{2}l\pi^{1/4}.$$

Having A_ν we present the Lévy wave packet (5.3) in the form

$$\psi_L(x,t) = \frac{1}{\hbar}\sqrt{\frac{\nu l}{4\pi\Gamma(1/\nu)}} \int\limits_{-\infty}^{\infty} dp \exp\left\{-\frac{|p - p_0|^\nu l^\nu}{2\hbar^\nu}\right\}$$

$$\times \exp\left\{i\frac{px}{\hbar} - i\frac{D_\alpha|p|^\alpha t}{\hbar}\right\}, \quad (5.8)$$

$$\nu \leq \alpha, \qquad 1 < \alpha \leq 2,$$

and the Gaussian wave packet in the form

$$\psi_G(x,t) = \frac{1}{\hbar}\sqrt{\frac{l}{2\pi^{3/2}}} \int\limits_{-\infty}^{\infty} dp \exp\left\{-\frac{|p - p_0|^2 l^2}{2\hbar^2}\right\} \exp\left\{i\frac{px}{\hbar} - i\frac{p^2 t}{2m\hbar}\right\},$$

$$(5.9)$$

---

[1]The gamma function $\Gamma(z)$ has the familiar integral representation

$$\Gamma(z) = \int\limits_0^\infty dt\, t^{z-1} e^{-t}, \qquad \mathrm{Re}\, z > 0. \quad (5.7)$$

where $p_0 > 0$.

The probability density $w(p,t)$ that a particle has momentum p is defined by

$$w(p,t) = |\phi_L(p,t)|^2,$$

where $\phi_L(p,t)$ is the Lévy wave packet in momentum $\phi_L(p,t)$ representation. The Lévy wave packets in space $\psi_L(x,t)$ and momentum $\phi_L(p,t)$ representations are related to each other by the Fourier transforms

$$\psi_L(x,t) = \frac{1}{2\pi\hbar} \int\limits_{-\infty}^{\infty} dp \exp\left\{ i\frac{px}{\hbar} \right\} \phi_L(p,t) \tag{5.10}$$

and

$$\phi_L(p,t) = \int\limits_{-\infty}^{\infty} dp \exp\{ -i\frac{px}{\hbar} \}\psi_L(x,t).$$

By comparing Eqs. (5.10) and (5.8) we conclude that

$$\phi_L(p,t) = \sqrt{\frac{\pi\nu l}{\Gamma(1/\nu)}} \exp\left\{ -\frac{|p-p_0|^\nu l^\nu}{2\hbar^\nu} \right\} \exp\left\{ -i\frac{D_\alpha |p|^\alpha t}{\hbar} \right\}, \tag{5.11}$$

$$\nu \leq \alpha, \qquad 1 < \alpha \leq 2,$$

which can be considered as the Lévy wave packet $\phi_L(p,t)$ in the momentum representation.

Note that $\phi_L(p,t)$ satisfies the fractional free particle Schrödinger equation in the momentum representation

$$i\hbar\frac{\partial\phi_L(p,t)}{\partial t} = D_\alpha |p|^\alpha \phi_L(p,t),$$

with the initial condition

$$\phi_L(p,0) = \sqrt{\frac{\pi\nu l}{\Gamma(1/\nu)}} \exp\left\{ -\frac{|p-p_0|^\nu l^\nu}{2\hbar^\nu} \right\}.$$

We have

$$\int\limits_{-\infty}^{\infty} dx |\psi_L(x,t)|^2$$

$$= \frac{A_\nu^2}{(2\pi\hbar)^2} \int\limits_{-\infty}^{\infty} dx \int\limits_{-\infty}^{\infty} dp dp' \exp\left\{ i \frac{(p-p')x}{\hbar} \right\} \phi_L(p,t)\phi_L^*(p',t) \qquad (5.12)$$

$$= \frac{A_\nu^2}{(2\pi\hbar)} \int\limits_{-\infty}^{\infty} dp |\phi_L(p,t)|^2 = 1,$$

because of

$$\frac{1}{(2\pi\hbar)} \int\limits_{-\infty}^{\infty} dx \exp\left\{ i \frac{(p-p')x}{\hbar} \right\} = \delta(p-p').$$

From Eq. (5.12) we come to the following definition for the probability density $w(p,t)$ in momentum space

$$w(p,t) = \frac{A_\nu^2}{2\pi\hbar} |\phi_L(p,t)|^2 \qquad (5.13)$$

with A_ν defined by Eq. (5.6). The probability density $w(p,t)$ in momentum space reads

$$w(p,t) \equiv w(p) = \frac{\nu l}{2\hbar\Gamma(1/\nu)} \exp\left\{ -\frac{|p-p_0|^\nu l^\nu}{\hbar^\nu} \right\}. \qquad (5.14)$$

It is time independent, since we are considering a free particle.

In coordinate space the probability to find a particle at position x in the "box" dx is given by $\rho(x,t)dx$ with the probability density $\rho(x,t)$ defined by Eq. (5.5). Correspondingly, the probability to find the particle with momentum p in the "box" dp is represented by $w(p,t)dp$.

5.2   Expectation values and μ-deviations of position and momentum

Expectation values and the μ-deviations for a free particle position and momentum can be estimated with the help of the probability densities defined by Eqs. (5.5) and (5.14). The expectation value of the position is defined by

$$< x >= \int\limits_{-\infty}^{\infty} dx\, x \rho(x,t) \tag{5.15}$$

$$= \frac{A_\nu^2}{(2\pi\hbar)^2} \int\limits_{-\infty}^{\infty} dx\, x \int\limits_{-\infty}^{\infty} dp\, dp'\, \exp\left\{ i\frac{(p-p')x}{\hbar} \right\} \phi_L(p,t)\phi_L^*(p',t).$$

By substituting

$$x \rightarrow \frac{\hbar}{i}\frac{\partial}{\partial p},$$

we have

$$< x >= \frac{A_\nu^2}{(2\pi\hbar)^2} \int\limits_{-\infty}^{\infty} dx \int\limits_{-\infty}^{\infty} dp\, dp' \left(\frac{\hbar}{i}\frac{\partial}{\partial p} \exp\left\{ i\frac{(p-p')x}{\hbar} \right\} \right) \phi_L(p,t)\phi_L^*(p',t).$$

Integrating by parts yields

$$< x >= -\frac{A_\nu^2}{(2\pi\hbar)}\frac{\hbar}{i} \int\limits_{-\infty}^{\infty} dp \left(\frac{l^\nu}{\hbar^\nu}\frac{\partial}{\partial p}|p-p_0|^\nu - i\frac{D_\alpha t}{\hbar}\frac{\partial}{\partial p}|p|^\alpha \right)$$

$$\times \exp\left\{ -\frac{|p-p_0|^\nu l^\nu}{\hbar^\nu} \right\}.$$

It is easy to check that the first term in the brackets vanishes, and we find that the position expectation value is

$$< x >= \alpha D_\alpha |p_0|^{\alpha-1} t. \tag{5.16}$$

Using the dispersion relation given by Eq. (12.42) we may rewrite $< x >$ as follows

$$< x >= \frac{\partial E_p}{\partial p}|_{p=p_0} \cdot t = v_0 t, \qquad (5.17)$$

here

$$v_0 = \frac{\partial E_p}{\partial p}|_{p=p_0} = \alpha D_\alpha |p_0|^{\alpha-1} \mathrm{sign} p \qquad (5.18)$$

is the group velocity of the wave packet. We see that the maximum of the Lévy wave packet (5.2) moves with the group velocity v_0 like a classical particle.

The μ-deviation ($\mu < \nu$) of position $< |\Delta x|^\mu >$ is defined by

$$< |\Delta x|^\mu >=< |x- < x > |^\mu >= \int\limits_{-\infty}^{\infty} dx|x- < x > |^\mu \rho(x,t)$$

$$= \frac{A_\nu^2}{(2\pi\hbar)^2} \int\limits_{-\infty}^{\infty} dx|x- < x > |^\mu \int\limits_{-\infty}^{\infty} dpdp' \exp\left\{ i\frac{(p-p')x}{\hbar} \right\} \phi_L(p,t)\phi_L^*(p',t).$$

This equation can be rewritten as

$$< |\Delta x|^\mu >= \frac{l^\mu}{2}\mathcal{N}(\alpha,\mu,\nu;\tau,\eta_0), \qquad (5.19)$$

where we introduce the following notations

$$\mathcal{N}(\alpha,\mu,\nu;\tau,\eta_0) = \frac{2^{1/\nu}\nu}{4\pi\Gamma(1/\nu)} \int\limits_{-\infty}^{\infty} d\varsigma|\varsigma|^\mu$$

$$\times \int\limits_{-\infty}^{\infty} d\eta \int\limits_{-\infty}^{\infty} d\eta' \exp\{i(\eta-\eta')(\varsigma+\alpha\tau\eta_0^{\alpha-1})\} \qquad (5.20)$$

$$\times \exp\{-i\tau(|\eta|^\alpha - |\eta'|^\alpha) - |\eta-\eta_0|^\nu - |\eta'-\eta_0|^\nu\}$$

and

$$\eta_0 = \frac{p_0 l}{2^{1/\nu}\hbar}, \qquad \tau = \frac{D_\alpha t}{\hbar}\left(\frac{2^{1/\nu}\hbar}{l} \right)^\alpha.$$

For the μ-root of the μ-deviation of position we obtain

$$< |\Delta x|^\mu >^{1/\mu} = \frac{l}{2^{1/\mu}} \mathcal{N}^{1/\mu}(\alpha, \mu, \nu; \tau, \eta_0). \quad (5.21)$$

Thus, we found the uncertainty $< |\Delta x|^\mu >^{1/\mu}$ in the position for the Lévy wave packet.

Further, with the help of Eq. (5.14) the expectation value of the momentum is calculated as

$$< p > = \int\limits_{-\infty}^{\infty} dp\, p\, w(p) = \int\limits_{-\infty}^{\infty} dp (p - p_0) w(p) + \int\limits_{-\infty}^{\infty} dp\, p_0 w(p). \quad (5.22)$$

The first integral vanishes, since $w(p)$ is an even function of $(p - p_0)$ and the momentum expectation value is

$$< p > = p_0. \quad (5.23)$$

The μ-deviation of the momentum is

$$< |\Delta p|^\mu > = \int\limits_{-\infty}^{\infty} dp\, |p - < p >|^\mu w(p) = \left(\frac{\hbar}{l}\right)^\mu \frac{\Gamma((\mu + 1)/\nu)}{\Gamma(1/\nu)}. \quad (5.24)$$

Then the uncertainty in the momentum (the μ-root of the μ-deviation of momentum) is

$$< |\Delta p|^\mu >^{1/\mu} = \frac{\hbar}{l} \left(\frac{\Gamma((\mu + 1)/\nu)}{\Gamma(1/\nu)}\right)^{1/\mu}. \quad (5.25)$$

5.3   Fractional uncertainty relation

> *What is not surrounded by uncertainty cannot be the truth.*
> Richard P. Feynman

Equations (5.21) and (5.25) yield

$$< |\Delta x|^\mu >^{1/\mu} < |\Delta p|^\mu >^{1/\mu}$$

$$= \frac{\hbar}{2^{1/\mu}} \left(\frac{\Gamma((\mu + 1)/\nu)}{\Gamma(1/\nu)}\right)^{1/\mu} \mathcal{N}^{1/\mu}(\alpha, \mu, \nu; \tau, \eta_0), \quad (5.26)$$

where $\mathcal{N}(\alpha, \mu, \nu; \tau, \eta_0)$ is given by Eq. (5.20) and $\mu < \nu \le \alpha$. We obtain the fractional quantum mechanics uncertainty relation. This relation implies that a spatially extended Lévy wave packet corresponds to a narrow momentum distribution whereas a sharp Lévy wave packet corresponds to a broad momentum distribution.

Since $\mathcal{N}(\alpha, \mu, \nu; \tau, \eta_0) > 1$ and $\Gamma((\mu + 1)/\nu)/\Gamma(1/\nu) \approx 1/\nu$, Eq. (5.26) with $\nu = \alpha$ becomes

$$< |\Delta x|^\mu >^{1/\mu} < |\Delta p|^\mu >^{1/\mu} > \frac{\hbar}{(2\alpha)^{1/\mu}}, \tag{5.27}$$

$$\mu < \alpha, \qquad 1 < \alpha \le 2.$$

Note that for the special case when $\alpha = 2$ we have $\mu = \alpha = 2$. That is, for standard quantum mechanics ($\alpha = 2$) with the definitions of position and momentum uncertainties as the square-root of the square deviation, Eq. (5.27) is transformed into the well-known uncertainty relation of standard quantum mechanics (see, for instance, [94]). The uncertainty relation given by Eq. (5.27) can be considered as a generalization of the well-known mathematical formulation of the fundamental Heisenberg Uncertainty Principle for quantum systems described by the Hamilton operator Eq. (3.3). In other words, when the fractal dimension of a quantum mechanical path is less than 2, $d_{\text{fractal}}^{(Lévy)} = \alpha < 2$, then the uncertainty relation takes the form given by Eq. (5.27).

In standard quantum mechanics the energy-momentum relationship has the form

$$E_p = \frac{p^2}{2m}, \tag{5.28}$$

where E_p is the energy, p is the momentum, and m is the mass of a quantum particle. In fractional quantum mechanics the energy-momentum relationship has the form

$$E_p = D_\alpha |p|^\alpha, \qquad 1 < \alpha \le 2, \tag{5.29}$$

where D_α is the scale coefficient which first appeared in Eq. (3.3),

$$D_\alpha |_{\alpha=2} = D_2 = \frac{1}{2m}.$$

The difference between Eq. (5.28) and Eq. (5.29) impacts the quantum mechanics fundamentals, including the Schrödinger equation and mathematical formulation of Heisenberg's Uncertainty Principle - the uncertainty relation.

5.4   Uncertainty relation: Heisenberg vs fractional

To emphasize the new features which fractional quantum mechanics brings into the mathematical formulation of Heisenberg's Uncertainty Principle, we present a map for the fundamentals involved in the well-known Heisenberg's uncertainty relation (see, for example, [94]) and fractional uncertainty relation (5.27). Table 1 from [98] presents two sets of equations (one for standard quantum mechanics and another for fractional quantum mechanics) involved in the mathematical formulation of Heisenberg's Uncertainty Principle: the energy-momentum relationship E_p, Schrödinger equation, initial state wave function $\psi(x, 0)$, the Gaussian wave packet defined by Eq. (5.9) and the Lévy wave packet defined by Eq. (5.8), position mean $< x >$, the definition of uncertainty, and, finally, the uncertainty relation.

Quantum Mechanics	Fractional Quantum Mechanics
$E_p = p^2/2m$	$E_p = D_\alpha \lvert p \rvert^\alpha, 1 < \alpha \leq 2$
$i\hbar \frac{\partial \psi(x,t)}{\partial t} = -\frac{\hbar^2}{2m} \nabla^2 \psi + V(x)\psi$	$i\hbar \frac{\partial \psi(x,t)}{\partial t} = -D_\alpha (\hbar \nabla)^\alpha \psi + V(x)\psi$
$\psi_G(x,0) = (1/\hbar)(l/2\pi^{3/2})^{1/2}$ $\times \int\limits_{-\infty}^{\infty} dp\, e^{-\frac{\lvert p - p_0 \rvert^2 l^2}{2\hbar^2}} e^{i\frac{px}{\hbar}}$	$\psi_L(x,0) = (1/\hbar)(\nu l/4\pi\Gamma(1/\nu))^{1/2}$ $\times \int\limits_{-\infty}^{\infty} dp\, e^{-\frac{\lvert p - p_0 \rvert^\nu l^\nu}{2\hbar^\nu}} e^{i\frac{px}{\hbar}}$
Gaussian wave packet Eq. (5.9)	Lévy wave packet Eq. (5.8)
$< x >_G = p_0 t/m$	$< x >_L = \alpha D_\alpha \lvert p_0 \rvert^{\alpha-1} t$
$< \Delta x >_G = (< (x - < x >)^2 >)^{1/2}$	$< \Delta x >_L = (< \lvert x - < x > \rvert^\mu >)^{1/\mu}$
$< \Delta x >_G \cdot < \Delta p >_G > \frac{\hbar}{2}$	$< \Delta x >_L \cdot < \Delta p >_L > \frac{\hbar}{(2\alpha)^{1/\mu}}$

Table 1. *Quantum Mechanics and Fractional Quantum Mechanics fundamentals*[2] *involved into the mathematical formulation of Heisenberg's Uncertainty Principle.*

The right column in Table 1 shows the new features, which fractional quantum mechanics brings into the mathematical formulation of Heisenberg's Uncertainty Principle. Note that the restrictions on possible values of the physical parameters α, ν, and μ come from quantum physics [67], [92] and fundamentals of the Lévy α-stable probability distribution [65], [66].

---

[2] All definitions and equations related to the fractional quantum mechanics are taken from [92].

Chapter 6

Path Integral over Lévy Flights

Thirty-one years ago [1949!], Dick Feynman told me about his "sum over histories" version of quantum mechanics. "The electron does anything it likes", he said. "It just goes in any direction at any speed, forward or backward in time, however it likes, and then you add up the amplitudes and it gives you the wave-function."

I said to him, "You're crazy."

But he wasn't.

Freeman Dyson, 1980

6.1 Quantum kernel

If a particle at an initial time t_a starts from the point x_a and goes to a final point x_b at time t_b, we will simply say that the particle goes from a to b and its trajectory (path)[1] $x(t)$ will have the property that $x(t_a) = x_a$ and $x(t_b) = x_b$. In quantum mechanics, then, we have a quantum-mechanical amplitude or a kernel to get from the point a to the point b, which we write as $K(x_b t_b | x_a t_a)$. The kernel $K(x_b t_b | x_a t_a)$ is the sum of contributions from all trajectories that go between the end points [12].

In the 1D case when the fractional Hamilton operator has a form given by Eq. (3.19), we come to the definition of the kernel $K(x_b t_b | x_a t_a)$ in terms of the path integral over Lévy-like quantum paths in the phase space

---

[1]We first consider 1D case, and then will generalize our developments to 3D case.

representation first introduced by Laskin in [92]

$$K(x_b t_b | x_a t_a) = \lim_{N \to \infty} \int_{-\infty}^{\infty} dx_1 ... dx_{N-1} \frac{1}{(2\pi\hbar)^N} \int_{-\infty}^{\infty} dp_1 ... dp_N$$

$$\times \exp\left\{ \frac{i}{\hbar} \sum_{j=1}^{N} p_j (x_j - x_{j-1}) \right\} \qquad (6.1)$$

$$\times \exp\left\{ -\frac{i}{\hbar} D_\alpha \varepsilon \sum_{j=1}^{N} |p_j|^\alpha - \frac{i}{\hbar} \varepsilon \sum_{j=1}^{N} V(x_j, j\varepsilon) \right\},$$

here $\varepsilon = (t_b - t_a)/N$, $x_j = x(t_a + j\varepsilon)$, $p_j = p(t_a + j\varepsilon)$ and $x(t_a + j\varepsilon)|_{j=0} = x_a$, $x(t_a + j\varepsilon)|_{j=N} = x(t_b) = x_b$.

Equation (6.1) is the definition of the Laskin path integral in the phase space representation.

Then in the continuum limit $N \to \infty$, $\varepsilon \to 0$ we obtain

$$K(x_b t_b | x_a t_a) \qquad (6.2)$$

$$= \int_{x(t_a)=x_a}^{x(t_b)=x_b} Dx(\tau) \int Dp(\tau) \exp\left\{ \frac{i}{\hbar} \int_{t_a}^{t_b} d\tau [p(\tau)\dot{x}(\tau) - H_\alpha(p(\tau), x(\tau), \tau)] \right\},$$

where $\dot{x}(\tau)$ denotes the time derivative $\dot{x}(\tau) = dx/d\tau$, $H_\alpha(p(\tau), x(\tau), \tau)$ is the fractional Hamiltonian given by Eq. (3.24) with the substitutions $p \to p(\tau)$, $x \to x(\tau)$, and $\{p(\tau), x(\tau)\}$ is the particle trajectory in phase space, and $\int_{x(t_a)=x_a}^{x(t_b)=x_b} Dx(\tau) \int Dp(\tau)$...stands for the path integral "measure" formally introduced as

$$\int_{x(t_a)=x_a}^{x(t_b)=x_b} Dx(\tau) \int Dp(\tau)... \qquad (6.3)$$

$$= \lim_{N \to \infty} \int_{-\infty}^{\infty} dx_1 ... dx_{N-1} \frac{1}{(2\pi\hbar)^N} \int_{-\infty}^{\infty} dp_1 ... dp_N ....$$

The exponential in Eq. (6.2) can be written as $\exp\{iS_\alpha(p,x)/\hbar\}$ if we introduce the fractional classical mechanics action $S_\alpha(p,x)$ as a functional of trajectory $\{p(\tau), x(\tau)\}$ in phase space [92]

$$S_\alpha(p,x) = \int_{t_a}^{t_b} d\tau (p(\tau)\dot{x}(\tau) - H_\alpha(p(\tau), x(\tau), \tau)). \qquad (6.4)$$

Therefore, we have

$$K(x_b t_b | x_a t_a) = \int_{x(t_a)=x_a}^{x(t_b)=x_b} Dx(\tau) \int Dp(\tau) \exp\{iS_\alpha(p,x)/\hbar\} \qquad (6.5)$$

$$= \int_{x(t_a)=x_a}^{x(t_b)=x_b} Dx(\tau) \int Dp(\tau) \exp\left\{\frac{i}{\hbar} \int_{t_a}^{t_b} d\tau [p(\tau)\dot{x}(\tau) - H_\alpha(p(\tau), x(\tau), \tau)]\right\}.$$

This equation introduces the quantum mechanical kernel $K(x_b t_b | x_a t_a)$ as a phase space path integral of $\exp\{iS_\alpha(p,x)/\hbar\}$ over Lévy-like quantum paths.

Since the coordinates x_a and x_b in definition (6.3) are fixed, all possible trajectories in Eq. (6.2) satisfy the boundary conditions $x(t_b) = x_b$, $x(t_a) = x_a$. We see that the definition given by Eq. (6.3) includes one more p_j-integrals than x_j-integrals. Indeed, while x_a and x_b are held fixed and the x_j-integrals are done for $j = 1, ..., N-1$, each increment $x_j - x_{j-1}$ is accompanied by one p_j-integral for $j = 1, ..., N$. The above observed asymmetry is a consequence of the particular boundary conditions. Namely, the end points x_a and x_b are fixed in the coordinate space. There exists the possibility of proceeding in a conjugate way keeping the initial p_a and final p_b momenta fixed. The associated kernel can be derived by following the same steps as before but working in the momentum representation (see, for example, [20]).

The kernel $K(x_b t_b | x_a t_a)$ introduced by Eq. (6.2) describes the evolution of the quantum mechanical particle

$$\psi(x_b, t_b) = \int_{-\infty}^{\infty} dx_a K(x_b t_b | x_a t_a) \psi(x_a, t_a), \qquad (6.6)$$

where $\psi(x_a, t_a)$ is the wave function of the initial state (at $t = t_a$ a particle is in position x_a) and $\psi(x_b, t_b)$ is the wave function of the final state (at

$t = t_b$ a particle is in position x_b). The wave function $\psi(x_a, t_a)$ is a quantum mechanical amplitude of probability to find a particle at (x_a, t_a). Hence, the evolution equation (6.6) says that the amplitude $\psi(x_b, t_b)$ to get (x_b, t_b) is integral over all possible x_a of amplitude $\psi(x_a, t_a)$ to find the particle at (x_a, t_a) multiplied by transition amplitude $K(x_b t_b | x_a t_a)$ to go from (x_a, t_a) to $(x_b, t_b, t_b > t_a)$. In other words, knowledge of kernel $K(x_b t_b | x_a t_a)$ allows us to study the evolution of a quantum particle by means of Eq. (6.6).

6.2 Phase space representation

When the fractional Hamilton operator has the form given by Eq. (3.3) the phase space path integral over Lévy-like quantum paths is introduced as

$$K(\mathbf{r}_b t_b | \mathbf{r}_a t_a) = \lim_{N \to \infty} \int d\mathbf{r}_1 ... d\mathbf{r}_{N-1} \frac{1}{(2\pi\hbar)^{3N}} \int d\mathbf{p}_1 ... d\mathbf{p}_N$$

$$\times \exp\left\{ \frac{i}{\hbar} \sum_{j=1}^{N} \mathbf{p}_j (\mathbf{r}_j - \mathbf{r}_{j-1}) \right\} \tag{6.7}$$

$$\times \exp\left\{ -\frac{i}{\hbar} D_\alpha \varepsilon \sum_{j=1}^{N} |\mathbf{p}_j|^\alpha - \frac{i}{\hbar} \varepsilon \sum_{j=1}^{N} V(\mathbf{r}_j, j\varepsilon) \right\},$$

which is the generalization of Eq. (6.1) for 3D coordinate and 3D momentum spaces. Here $\varepsilon = (t_b - t_a)/N$, $\mathbf{r}_j = \mathbf{r}(t_a + j\varepsilon)$, $\mathbf{p}_j = \mathbf{p}(t_a + j\varepsilon)$ and $\mathbf{r}(t_a + j\varepsilon)|_{j=0} = \mathbf{r}_a$, $\mathbf{r}(t_a + j\varepsilon)|_{j=N} = \mathbf{r}_b$, with $\mathbf{r}_a$ and $\mathbf{r}_b$ being the initial and final points of particle paths, $\mathbf{r}_j$ and $\mathbf{p}_j$ are 3D vectors. We adopt the notations $d\mathbf{r}_i = d^3 r_i$, $(i = 1, 2, ..., N - 1)$ and $d\mathbf{p}_j = d^3 p_j$, $(j = 1, 2, ..., N)$ while working with the path integral over Lévy-like quantum paths in phase-space representation. Then in the continuum limit $N \to \infty$, $\varepsilon \to 0$ we obtain

$$K(\mathbf{r}_b t_b | \mathbf{r}_a t_a) \tag{6.8}$$

$$= \int_{\mathbf{r}(t_a) = \mathbf{r}_a}^{\mathbf{r}(t_b) = \mathbf{r}_b} \mathrm{D}\mathbf{r}(\tau) \int \mathrm{D}\mathbf{p}(\tau) \exp\left\{ \frac{i}{\hbar} \int_{t_a}^{t_b} d\tau [\mathbf{p}(\tau)\dot{\mathbf{r}}(\tau) - H_\alpha(\mathbf{p}(\tau), \mathbf{r}(\tau), \tau)] \right\},$$

where $\dot{\mathbf{r}}$ denotes the time derivative $d/d\tau$, $H_\alpha(\mathbf{p}(\tau), \mathbf{r}(\tau), \tau)$ is the fractional Hamiltonian[2] given by Eq. (3.3) with the substitutions $\mathbf{p} \to \mathbf{p}(\tau)$, $\mathbf{r} \to$

---

[2]In general, the Hamiltonian $H_\alpha(\mathbf{p}, \mathbf{r}, \tau)$ can depend on τ through τ-dependency of potential energy term $V(\mathbf{r}, \tau)$.

$\mathbf{r}(\tau)$, and $\{\mathbf{p}(\tau), \mathbf{r}(\tau)\}$ is the particle trajectory in 6D phase space and $\int_{\mathbf{r}(t_a)=\mathbf{r}_a}^{\mathbf{r}(t_b)=\mathbf{r}_b} \mathrm{D}\mathbf{r}(\tau) \int \mathrm{D}\mathbf{p}(\tau)...$ stands for the path integral "measure" formally introduced as

$$\int_{\mathbf{r}(t_a)=\mathbf{r}_a}^{\mathbf{r}(t_b)=\mathbf{r}_b} \mathrm{D}\mathbf{r}(\tau) \int \mathrm{D}\mathbf{p}(\tau)... \qquad (6.9)$$

$$= \lim_{N\to\infty} \int d\mathbf{r}_1...d\mathbf{r}_{N-1} \frac{1}{(2\pi\hbar)^{3N}} \int d\mathbf{p}_1...d\mathbf{p}_N....$$

For a free particle $V(\mathbf{r}) = 0$, and equation (6.8) gives us a free particle quantum mechanical kernel $K^{(0)}(\mathbf{r}_b t_b | \mathbf{r}_a t_a)$

$$K^{(0)}(\mathbf{r}_b t_b | \mathbf{r}_a t_a) \qquad (6.10)$$

$$= \int_{\mathbf{r}(t_a)=\mathbf{r}_a}^{\mathbf{r}(t_b)=\mathbf{r}_b} \mathrm{D}\mathbf{r}(\tau) \int \mathrm{D}\mathbf{p}(\tau) \exp\left\{ \frac{i}{\hbar} \int_{t_a}^{t_b} d\tau [\mathbf{p}(\tau)\dot{\mathbf{r}}(\tau) - D_\alpha |\mathbf{p}(\tau)|^\alpha] \right\}.$$

The exponential in Eq. (6.8) can be written as $\exp\{iS_\alpha(\mathbf{p},\mathbf{r})/\hbar\}$ if we introduce the fractional classical mechanics action $S_\alpha(\mathbf{p}, \mathbf{r})$ as a functional of trajectory $\{\mathbf{p}(\tau), \mathbf{r}(\tau)$ in phase space

$$S_\alpha(\mathbf{p}, \mathbf{r}) = \int_{t_a}^{t_b} d\tau (\mathbf{p}(\tau)\dot{\mathbf{r}}(\tau) - H_\alpha(\mathbf{p}(\tau), \mathbf{r}(\tau), \tau)). \qquad (6.11)$$

Then we have

$$K(\mathbf{r}_b t_b | \mathbf{r}_a t_a) = \int_{\mathbf{r}(t_a)=\mathbf{r}_a}^{\mathbf{r}(t_b)=\mathbf{r}_b} \mathrm{D}\mathbf{r}(\tau) \int \mathrm{D}\mathbf{p}(\tau) \exp\{iS_\alpha(\mathbf{p}, \mathbf{r})/\hbar\}. \qquad (6.12)$$

This is quantum mechanical kernel $K(\mathbf{r}_b t_b | \mathbf{r}_a t_a)$ defined as phase space path integral of $\exp\{iS_\alpha(\mathbf{p}, \mathbf{r})/\hbar\}$ over Lévy-like quantum paths.

Given that the 3D vectors $\mathbf{r}_a$ and $\mathbf{r}_b$ in Eqs. (6.8) and (6.9) are fixed, all possible trajectories in Eqs. (6.8) and (6.9) satisfy the boundary condition $\mathbf{r}(t_b) = \mathbf{r}_b$, $\mathbf{r}(t_a) = \mathbf{r}_a$. We see that the definition given by Eq. (6.9) includes one more $\mathbf{p}_j$-integrals than $\mathbf{r}_j$-integrals. Indeed, while $\mathbf{r}_a$ and $\mathbf{r}_b$ are held fixed and the $\mathbf{r}_j$-integrals are done for $j = 1, ..., N - 1$, each increment $\mathbf{r}_j - \mathbf{r}_{j-1}$ is accompanied by one $\mathbf{p}_j$-integral for $j = 1, ..., N$. The above observed asymmetry is a consequence of the particular boundary condition, namely, the end points $\mathbf{r}_a$ and $\mathbf{r}_b$ are fixed in the 3D coordinate space.

The kernel $K(\mathbf{r}_b t_b | \mathbf{r}_a t_a)$ introduced by Eq. (6.12) describes the evolution of the quantum mechanical system

$$\psi(\mathbf{r}_b, t_b) = \int d^3 r_a K(\mathbf{r}_b t_b | \mathbf{r}_a t_a) \psi(\mathbf{r}_a, t_a), \qquad (6.13)$$

where $\psi(\mathbf{r}_a, t_a)$ is the wave function of the initial state (at $t = t_a$ a particle is in position $\mathbf{r}_a$) and $\psi(\mathbf{r}_b, t_b)$ is the wave function of the final state (at $t = t_b$ a particle is in position $\mathbf{r}_b$). The kernel $K(\mathbf{r}_b t_b | \mathbf{r}_a t_a)$ satisfies

$$K(\mathbf{r}_b t_b | \mathbf{r}_a t_a) = \int d^3 r' K(\mathbf{r}_b t_b | \mathbf{r}' t') K(\mathbf{r}' t' | \mathbf{r}_a t_a). \qquad (6.14)$$

This is a fundamental quantum mechanics equation, which establishes the transformation law for the kernels, when two quantum transitions $(\mathbf{r}_a t_a \to \mathbf{r}' t')$ and $(\mathbf{r}' t' \to \mathbf{r}_b t_b)$ occur in succession.

6.3   Coordinate representation

The path integral in coordinate space representation can be introduced by performing the integration in Eq. (6.1) over momentums involved. To calculate the integrals over $dp_1...dp_N$ in Eq. (6.1) we introduce the Lévy probability distribution function $L_\alpha(z)$ by means of the following equation

$$L_\alpha(z) = \frac{1}{2\pi} \int\limits_{-\infty}^{\infty} d\varsigma \exp\{iz\varsigma - |\varsigma|^\alpha\}. \qquad (6.15)$$

In terms of $L_\alpha(z)$ we have

$$\frac{1}{2\pi\hbar} \int\limits_{-\infty}^{\infty} dp \exp\left\{ \frac{i}{\hbar} px - \frac{i}{\hbar} D_\alpha \tau |p|^\alpha \right\} \qquad (6.16)$$

$$= \frac{1}{\hbar} \left(\frac{i D_\alpha \tau}{\hbar} \right)^{-1/\alpha} L_\alpha \left\{ \frac{1}{\hbar} \left(\frac{\hbar}{i D_\alpha \tau} \right)^{1/\alpha} |x| \right\}.$$

Then the kernel $K(x_b t_b | x_a t_a)$ introduced by Eq. (6.1) can be expressed as

$$K(x_b t_b | x_a t_a) = \lim_{N \to \infty} \int_{-\infty}^{\infty} dx_1...dx_{N-1} \hbar^{-N} \left(\frac{i D_\alpha \varepsilon}{\hbar} \right)^{-N/\alpha} \quad (6.17)$$

$$\times \prod_{j=1}^{N} L_\alpha \left\{ \frac{1}{\hbar} \left(\frac{\hbar}{i D_\alpha \varepsilon} \right)^{1/\alpha} |x_j - x_{j-1}| \right\} \exp \left\{ -\frac{i}{\hbar} \varepsilon \sum_{j=1}^{N} V(x_j, j\varepsilon) \right\},$$

where $\varepsilon = (t_b - t_a)/N$, $x_j = x(t_a + j\varepsilon)$, and $x(t_a + j\varepsilon)|_{j=0} = x_a$, $x(t_a + j\varepsilon)|_{j=N} = x_b$.

Equation (6.17) is definition of the Laskin path integral in coordinate representation.

In the continuum limit $N \to \infty$, $\varepsilon \to 0$ we obtain

$$K(x_b t_b | x_a t_a) = \int_{x(t_a)=x_a}^{x(t_b)=x_b} \mathcal{D}x(\tau) \exp\{ -\frac{i}{\hbar} \int_{t_a}^{t_b} d\tau V(x(\tau), \tau) \}, \quad (6.18)$$

where $V(x(\tau), \tau)$ is the potential energy as a functional of the Lévy flight path $x(\tau)$ and time τ, and $\displaystyle\int_{x(t_a)=x_a}^{x(t_b)=x_b} \mathcal{D}x(\tau)...$ is the path integral measure in coordinate space, first introduced by Laskin [67]

$$\int_{x(t_a)=x_a}^{x(t_b)=x_b} \mathcal{D}x(\tau)... \quad (6.19)$$

$$= \lim_{N \to \infty} \int_{-\infty}^{\infty} dx_1...dx_{N-1} \hbar^{-N} \left(\frac{i D_\alpha \varepsilon}{\hbar} \right)^{-N/\alpha}$$

$$\times \prod_{j=1}^{N} L_\alpha \left\{ \frac{1}{\hbar} \left(\frac{\hbar}{i D_\alpha \varepsilon} \right)^{1/\alpha} |x_j - x_{j-1}| \right\} ...,$$

here $\hbar$ denotes Planck's constant, $x(t_a + j\varepsilon)|_{j=0} = x_a$, $x(t_a + j\varepsilon)|_{j=N} = x_b$, $\varepsilon = (t_b - t_a)/N$, and the Lévy probability distribution function L_α is defined by Eq. (6.15).

The Lévy probability distribution function L_α can be expressed in terms of Fox's H-function [92], [103], [104]

$$\frac{1}{\hbar}(\frac{iD_\alpha t}{\hbar})^{-1/\alpha} L_\alpha \left\{ \frac{1}{\hbar} \left(\frac{\hbar}{iD_\alpha t} \right)^{1/\alpha} |x| \right\} \tag{6.20}$$

$$= \frac{1}{\alpha|x|} H_{2,2}^{1,1} \left[\frac{1}{\hbar} \left(\frac{\hbar}{iD_\alpha t} \right)^{1/\alpha} |x| \left| \begin{array}{c} (1,1/\alpha),(1,1/2) \\ (1,1),(1,1/2) \end{array} \right. \right],$$

with D_α being the scale coefficient, and α being the Lévy index. Hence, the path integral measure introduced by Eq. (6.19) can be alternatively presented as

$$\int_{x(t_a)=x_a}^{x(t_b)=x_b} \mathcal{D}x(\tau)...$$

$$= \lim_{N \to \infty} \int_{-\infty}^{\infty} dx_1...dx_{N-1} \prod_{j=1}^{N} \frac{1}{\alpha|x_j - x_{j-1}|} \tag{6.21}$$

$$\times H_{2,2}^{1,1} \left[\frac{1}{\hbar} \left(\frac{\hbar}{iD_\alpha(t_b - t_a)} \right)^{1/\alpha} |x_j - x_{j-1}| \left| \begin{array}{c} (1,1/\alpha),(1,1/2) \\ (1,1),(1,1/2) \end{array} \right. \right]....$$

The fractional path integral measure defined by Eq. (6.19) is generated by the Lévy flights stochastic process. Indeed, from Eq. (6.19) we can find that the scaling relation between a length increment $(x_j - x_{j-1})$ and a time increment Δt has the Lévy scaling

$$|x_j - x_{j-1}| \propto \left(\hbar^{\alpha-1} D_\alpha \right)^{1/\alpha} (\Delta t)^{1/\alpha}.$$

The scaling $1/\alpha$ implies that the fractal dimension of the Lévy-like quantum-mechanical path is $d_{fractal}^{(Lévy)} = \alpha$. We conclude that the Lévy flights quantum background leads to fractional quantum mechanics. Equations (6.18)-(6.21) introduce fractional quantum mechanics via a newly invented path integral over Lévy-like flights.

The kernel $K(x_b t_b | x_a t_a)$ introduced by Eq. (6.18) describes the evolution of the quantum mechanical system

$$\psi(x_b, t_b) = \int dx_a K(x_b t_b | x_a t_a) \psi(x_a, t_a), \tag{6.22}$$

where $\psi(x_a, t_a)$ is the wave function of the initial state (at $t = t_a$ the particle is in position x_a) and $\psi(x_b, t_b)$ is the wave function of the final state (at $t = t_b$ the particle is in position x_b). By comparing Eq. (4.29) with Eq. (6.22), we come to the following expression for the kernel $K(x_b t_b | x_a t_a)$

$$K(x_b t_b | x_a t_a) = \sum_{n=1}^{\infty} \phi_n(x_b) \phi_n^*(x_a) e^{-(i/\hbar) E_n (t_b - t_a)}, \quad \text{for } t_b > t_a, \quad (6.23)$$

and $K(x_b t_b | x_a t_a) = 0$ for $t_b < t_a$.

To interpret Eq. (6.23) we will follow Feynman [12]. The kernel $K(x_b t_b | x_a t_a)$ is given by the path integral (6.18) over all possible paths between two points (x_a, t_a) and (x_b, t_b). Equation (6.23) defines the kernel in terms of all possible energy states for the quantum mechanical transition between two points (x_a, t_a) and (x_a, t_a). Hence, we have to sum the product of the following terms over all possible energy states E_n labeled by n:

1. $\phi_n(x_b)$, which is a quantum mechanical amplitude that the quantum system is in the energy state n with energy E_n and at space point x_b;

2. $\phi_n^*(x_a)$, which is a quantum mechanical amplitude that the quantum system is in the energy state n with energy E_n and at space point x_a;

3. $e^{-(i/\hbar) E_n (t_b - t_a)}$ which is a quantum mechanical amplitude to occupy the energy state n, with energy E_n at time t_b, if at time t_a the quantum system was in the energy state[3] n with the same energy E_n.

On the other hand, we introduced the kernel $K(x_b t_b | x_a t_a)$ as the path integral defined by Eq. (6.18). By equating expressions (6.18) and (6.23) we obtain an important identity, which presents the path integral (6.18) in terms of solutions to the time-independent 1D fractional Schrödinger equation (4.10)

$$\int_{x(t_a)=x_a}^{x(t_b)=x_b} \mathcal{D}x(\tau) \exp\{-\frac{i}{\hbar} \int_{t_a}^{t_b} d\tau V(x(\tau), \tau)\} \quad (6.24)$$

$$= \sum_{n=1}^{\infty} \phi_n(x_b) \phi_n^*(x_a) e^{-(i/\hbar) E_n (t_b - t_a)}.$$

If we note that the left-hand side of this equation can be written as the path integral (6.5) in phase space representation, then we obtain another

---

[3] The energy of the state is not changed while the quantum system goes from (x_1, t_1) to (x_2, t_2). This is a fundamental property of quantum stationary states.

important identity, which presents the phase space path integral (6.5) in terms of solutions to the time-independent 1D fractional Schrödinger equation (4.10)

$$\int_{x(t_a)=x_a}^{x(t_b)=x_b} \mathcal{D}x(\tau) \int \mathcal{D}p(\tau) \exp\{iS_\alpha(p,x)/\hbar\} \qquad (6.25)$$

$$= \sum_{n=1}^{\infty} \phi_n(x_b)\phi_n^*(x_a)e^{-(i/\hbar)E_n(t_b-t_a)}.$$

As an example, let us calculate a free particle kernel $K^{(0)}(x_bt_b|x_at_a)$. For a free particle we have $V(x) = 0$, and Eqs. (6.18) and (6.19) yield

$$K^{(0)}(x_bt_b|x_at_a) = \int_{x(t_a)=x_a}^{x(t_b)=x_b} \mathcal{D}x(\tau)\cdot 1 \qquad (6.26)$$

$$= \hbar^{-1}\left(\frac{iD_\alpha(t_b-t_a)}{\hbar}\right)^{-1/\alpha} L_\alpha\left\{\frac{1}{\hbar}\left(\frac{\hbar}{iD_\alpha(t_b-t_a)}\right)^{1/\alpha}|x_b-x_a|\right\},$$

or in terms of Fox's H-function [67],

$$K^{(0)}(x_bt_b|x_at_a) = \int_{x(t_a)=x_a}^{x(t_b)=x_b} \mathcal{D}x(\tau)\cdot 1 \qquad (6.27)$$

$$= \frac{1}{\alpha|x_b-x_a|}H_{2,2}^{1,1}\left[\frac{1}{\hbar}\left(\frac{\hbar}{iD_\alpha(t_b-t_a)}\right)^{1/\alpha}|x_b-x_a|\left|\begin{array}{c}(1,1/\alpha),(1,1/2)\\(1,1),(1,1/2)\end{array}\right.\right].$$

Equations (6.26) and (6.27) present a new family of free particle quantum mechanical kernels $K^{(0)}(x_bt_b|x_at_a)$ parametrized by the parameter α.

6.3.1 *3D coordinate representation*

The path integral in coordinate space representation can be introduced by performing the integration in Eq. (6.1) over momentums involved. To calculate the integrals over $d\mathbf{p}_1...d\mathbf{p}_N$ in Eq. (6.7), we introduce the Lévy probability distribution function $L_\alpha(z)$ by means of the following equation

$$L_\alpha(|\mathbf{r}|) = \frac{1}{(2\pi)^3}\int d^3k \exp\{i\mathbf{k}\mathbf{r}-|\mathbf{k}|^\alpha\}. \qquad (6.28)$$

Hence, 3D generalization of Eq. (6.16) in terms of $L_\alpha(|\mathbf{r}|)$ has the form

$$K(\mathbf{r}_b t_b | \mathbf{r}_a t_a) = \lim_{N \to \infty} \int d\mathbf{r}_1 ... d\mathbf{r}_{N-1} \hbar^{-3N} \left(\frac{iD_\alpha \varepsilon}{\hbar} \right)^{-3N/\alpha} \qquad (6.29)$$

$$\times \prod_{i=1}^{N} L_\alpha \left\{ \frac{1}{\hbar} \left(\frac{\hbar}{iD_\alpha \varepsilon} \right)^{1/\alpha} |\mathbf{r}_i - \mathbf{r}_{i-1}| \right\} \exp \left\{ -\frac{i}{\hbar} \varepsilon \sum_{i=1}^{N} V(\mathbf{r}_i, i\varepsilon) \right\},$$

which is the generalization of Eq. (6.17) for 3D coordinate and 3D momentum spaces. Here $\varepsilon = (t_b - t_a)/N$, $\mathbf{r}_i = \mathbf{r}(t_a + i\varepsilon)$, and $\mathbf{r}(t_a + i\varepsilon)|_{i=0} = \mathbf{r}_a$, $\mathbf{r}(t_a + i\varepsilon)|_{i=N} = \mathbf{r}_b$, with $\mathbf{r}_a$ and $\mathbf{r}_b$ being initial and final points of particle paths and $L_\alpha(|\mathbf{r}|)$ is the Lévy probability distribution function given by Eq. (6.28). We adopt the notations $d\mathbf{r}_i = d^3 r_i$, $(i = 1, 2, ..., N-1)$ while working with the path integral over Lévy-like quantum paths in 3D space. In the continuum limit $N \to \infty$, $\varepsilon \to 0$ we obtain

$$K(\mathbf{r}_b t_b | \mathbf{r}_a t_a) = \int_{\mathbf{r}(t_a)=\mathbf{r}_a}^{\mathbf{r}(t_b)=\mathbf{r}_b} \mathcal{D}\mathbf{r}(\tau) \exp \left\{ -\frac{i}{\hbar} \int_{t_a}^{t_b} d\tau V(\mathbf{r}(\tau), \tau) \right\}, \qquad (6.30)$$

where $V(\mathbf{r}(\tau), \tau)$ is the potential energy as a functional of the 3D Lévy flights path $\mathbf{r}(\tau)$ and time τ, and $\displaystyle\int_{\mathbf{r}(t_a)=\mathbf{r}_a}^{\mathbf{r}(t_b)=\mathbf{r}_b} \mathcal{D}\mathbf{r}(\tau)..$ is the path integral measure in coordinate space, first introduced by Laskin [92]

$$\int_{\mathbf{r}(t_a)=\mathbf{r}_a}^{\mathbf{r}(t_b)=\mathbf{r}_b} \mathcal{D}\mathbf{r}(\tau)... \qquad (6.31)$$

$$= \lim_{N \to \infty} \int d\mathbf{r}_1 ... d\mathbf{r}_{N-1} \hbar^{-3N} \left(\frac{iD_\alpha \varepsilon}{\hbar} \right)^{-3N/\alpha}$$

$$\times \prod_{i=1}^{N} L_\alpha \left\{ \frac{1}{\hbar} \left(\frac{\hbar}{iD_\alpha \varepsilon} \right)^{1/\alpha} |\mathbf{r}_i - \mathbf{r}_{i-1}| \right\} ...,$$

here $\hbar$ denotes Planck's constant, $\mathbf{r}(t_a + i\varepsilon)|_{i=0} = \mathbf{r}_a$, $\mathbf{r}(t_a + i\varepsilon)|_{i=N} = \mathbf{r}_b$, $\varepsilon = (t_b - t_a)/N$ and L_α is the Lévy probability distribution function defined by Eq. (6.28).

The Lévy probability distribution function L_α involved in Eq. (6.31) can be expressed in terms of Fox's H-function [92], [103], [104]

$$\hbar^{-3}(\frac{iD_\alpha t}{\hbar})^{-3/\alpha} L_\alpha \left\{ \frac{1}{\hbar} \left(\frac{\hbar}{iD_\alpha(t_b - t_a)} \right)^{1/\alpha} |\mathbf{r}_i - \mathbf{r}_{i-1}| \right\}$$

$$= -\frac{1}{2\pi\alpha} \frac{1}{|\mathbf{r}_i - \mathbf{r}_{i-1}|^3} \tag{6.32}$$

$$\times H^{1,2}_{3,3} \left[\frac{1}{\hbar} \left(\frac{\hbar}{iD_\alpha(t_b - t_a)} \right)^{1/\alpha} |\mathbf{r}_i - \mathbf{r}_{i-1}| \, \Bigg| \, \begin{matrix} (1,1),(1,1/\alpha),(1,1/2) \\ (1,1),(1,1/2),(2,1) \end{matrix} \right].$$

with D_α being the scale coefficient with units of $[D_\alpha] = \mathrm{erg}^{1-\alpha} \cdot \mathrm{cm}^\alpha \cdot \mathrm{sec}^{-\alpha}$ and α being the Lévy index.

Therefore, the path integral measure introduced by Eq. (6.31) can be alternatively presented in terms of Fox's $H^{1,2}_{3,3}$-function

$$\int_{\mathbf{r}(t_a)=\mathbf{r}_a}^{\mathbf{r}(t_b)=\mathbf{r}_b} \mathcal{D}\mathbf{r}(\tau)\dots$$

$$= \lim_{N\to\infty} \int d\mathbf{r}_1 \dots d\mathbf{r}_{N-1} \prod_{i=1}^{N} \left(-\frac{1}{2\pi\alpha|\mathbf{r}_i - \mathbf{r}_{i-1}|^3} \right) \tag{6.33}$$

$$\times H^{1,2}_{3,3} \left[\frac{1}{\hbar} \left(\frac{\hbar}{iD_\alpha\varepsilon} \right)^{1/\alpha} |\mathbf{r}_i - \mathbf{r}_{i-1}| \, \Bigg| \, \begin{matrix} (1,1),(1,1/\alpha),(1,1/2) \\ (1,1),(1,1/2),(2,1) \end{matrix} \right] \dots,$$

where $\varepsilon = (t_b - t_a)/N$.

The kernel $K(\mathbf{r}_b t_b | \mathbf{r}_a t_a)$ defined by Eq. (6.30) can be expressed in terms of solutions $\phi_n(\mathbf{r})$ to the time-independent 3D fractional Schrödinger equation (4.5) by means of 3D generalization of Eq. (6.23),

$$K(\mathbf{r}_b t_b | \mathbf{r}_a t_a) = \begin{cases} \sum\limits_{n=1}^{\infty} \phi_n(\mathbf{r}_b)\phi_n^*(\mathbf{r}_a)e^{-(i/\hbar)E_n(t_b - t_a)}, & \text{for } t_b > t_a, \\ 0, & \text{for } t_b < t_a. \end{cases} \tag{6.34}$$

By equating expressions (6.30) and (6.34) we obtain an important identity, which presents the path integral (6.30) in terms of solutions to 3D time-independent fractional Schrödinger equation (4.5)

$$\int_{\mathbf{r}(t_a)=\mathbf{r}_a}^{\mathbf{r}(t_b)=\mathbf{r}_b} \mathcal{D}\mathbf{r}(\tau) \exp\left\{-\frac{i}{\hbar}\int_{t_a}^{t_b} d\tau V(\mathbf{r}(\tau),\tau)\right\} \qquad (6.35)$$

$$= \sum_{n=1}^{\infty} \phi_n(\mathbf{r}_b)\phi_n^*(\mathbf{r}_a)e^{-(i/\hbar)E_n(t_b-t_a)}.$$

A free particle 3D time-independent fractional Schrödinger equation has the plane wave solution

$$\phi_{\mathbf{p}}(\mathbf{r}) = \exp\{\frac{i}{\hbar}\mathbf{pr}\},$$

with the energy $E_{\mathbf{p}} = D_\alpha|\mathbf{p}|^\alpha$. By considering the vector $\mathbf{p}$ as a label for energy state we write the orthogonality condition for the wave functions $\phi_{\mathbf{p}}(\mathbf{r})$ in the form

$$\int d^3r\phi_{\mathbf{p}}^*(\mathbf{r})\phi_{\mathbf{p}'}(\mathbf{r}) = 0 \qquad E_{\mathbf{p}} \neq E_{\mathbf{p}'},$$

where $\phi_{\mathbf{p}}^*(\mathbf{r})$ is wave function complex conjugate of wave function $\phi_{\mathbf{p}}(\mathbf{r})$. Hence, a free particle 3D fractional quantum-mechanical kernel $K^{(0)}(\mathbf{r}_bt_b|\mathbf{r}_at_a)$ can be written as

$$K^{(0)}(\mathbf{r}_bt_b|\mathbf{r}_at_a) = \sum_{\mathbf{p}} \phi_{\mathbf{p}}(\mathbf{r}_b)\phi_{\mathbf{p}}^*(\mathbf{r}_a)\exp\{-\frac{i}{\hbar}D_\alpha|\mathbf{p}|^\alpha(t_b-t_a)\}. \qquad (6.36)$$

Since the momentums are distributed over a continuum, the sum over the $\mathbf{p}$ is really equivalent to an integral over the values of $\mathbf{p}$, namely,

$$\sum_{\mathbf{p}} ... \to \frac{1}{(2\pi\hbar)^3}\int d^3p....$$

Using this substitution we see that Eq. (6.36) becomes the Fourier representation of a free particle fractional 3D kernel $K(\mathbf{r}_bt_b|\mathbf{r}_at_a)$

$$K^{(0)}(\mathbf{r}_bt_b|\mathbf{r}_at_a) = \frac{1}{(2\pi\hbar)^3}\int d^3p\exp\{\frac{i}{\hbar}\mathbf{p}(\mathbf{r}_b-\mathbf{r}_a)\}\exp\{-\frac{i}{\hbar}D_\alpha|\mathbf{p}|^\alpha(t_b-t_a)\}.$$

6.4   Feynman's path integral

It is simple, therefore it is beautiful.

Richard P. Feynman

In the special case, when $\alpha = 2$ we rediscover the Feynman path integral. When $\alpha = 2$, the Lévy probability distribution function $L_\alpha(z)$ defined by Eq. (6.15) becomes the normal probability distribution function $N(z)$

$$N(z) = L_\alpha(z)|_{\alpha=2} = \frac{1}{2\pi} \int\limits_{-\infty}^{\infty} d\varsigma\, e^{iz\varsigma - |\varsigma|^2} = \frac{1}{\sqrt{4\pi}} e^{-z^2/4}. \qquad (6.37)$$

When $\alpha = 2$, then $D_2 = 1/2m$, with m being a particle mass. In this case Eq. (6.16) reads

$$\frac{1}{2\pi\hbar} \int\limits_{-\infty}^{\infty} dp \exp\left\{ \frac{i}{\hbar} px - i\frac{p^2}{2m\hbar}\tau \right\} \qquad (6.38)$$

$$= \frac{1}{\hbar} \left(\frac{iD_2\tau}{\hbar} \right)^{-1/2} L_2 \left\{ \frac{1}{\hbar} \left(\frac{\hbar}{iD_2\tau} \right)^{1/2} |x| \right\} = \sqrt{\frac{m}{2\pi i \hbar \tau}} \exp\left\{ i\frac{mx^2}{2\hbar\tau} \right\}.$$

The kernel $K(x_b t_b | x_a t_a)$ introduced by Eq. (6.18) is transformed into Feynman's kernel

$$K_F(x_b t_b | x_a t_a) = \int\limits_{x(t_a)=x_a}^{x(t_b)=x_b} \mathcal{D}_F x(\tau) \exp\left\{ -\frac{i}{\hbar} \int\limits_{t_a}^{t_b} d\tau V(x(\tau), \tau) \right\}, \qquad (6.39)$$

where $V(x(\tau), \tau)$ is the potential energy as a functional of quantum Brownian motion path $x(\tau)$ and time τ, and $\int\limits_{x(t_a)=x_a}^{x(t_b)=x_b} \mathcal{D}_F x(\tau)$ is Feynman's path integral measure introduced as (see, Eq. (3-2) in [12])

$$\int\limits_{x(t_a)=x_a}^{x(t_b)=x_b} \mathcal{D}_F x(\tau) ... \qquad (6.40)$$

$$= \lim_{N \to \infty} \int\limits_{-\infty}^{\infty} dx_1 ... dx_{N-1} \left(\frac{2\pi i \hbar \varepsilon}{m} \right)^{-N/2} \prod_{j=1}^{N} \exp\{ i\frac{m(x_j - x_{j-1})^2}{2\hbar\varepsilon} \}...,$$

here $\hbar$ denotes Planck's constant, $x(j\varepsilon)|_{j=0} = x_0 = x_a$, $x(j\varepsilon)|_{j=N} = x_N = x_b$, $\varepsilon = (t_b - t_a)/N$.

The Feynman's path integral measure defined by Eq. (6.40) is generated by the Brownian-like motion process. Indeed, from Eq. (2.21) we can find that the scaling relation between a length increment $(x_j - x_{j-1})$ and a time increment Δt has the well-known $1/2$ law

$$|x_j - x_{j-1}| \propto (\hbar/2)^{1/2} (\Delta t)^{1/2}.$$

The diffusion scaling $1/2$ implies that the fractal dimension of the Brownian-like quantum-mechanical path is $d_{\text{fractal}}^{(B\,row\,nian)} = 2$. We conclude that the Brownian motion quantum background leads to quantum mechanics.

Equations (6.39) and (6.40) introduce quantum mechanics via Feynman's path integral. As an example, let us calculate a free particle Feynman's kernel $K_F^{(0)}(x_b t_b | x_a t_a)$. For a free particle we have $V(x) = 0$, and Eqs. (6.17) and (6.40) yield

$$K_F^{(0)}(x_b t_b | x_a t_a) = \int\limits_{x(t_a)=x_a}^{x(t_b)=x_b} \mathcal{D}_F x(\tau) \cdot 1 \qquad (6.41)$$

$$= \left(\frac{2\pi i \hbar (t_b - t_a)}{m} \right)^{-1/2} \exp \left\{ \frac{im(x_b - x_a)^2}{2\hbar(t_b - t_a)} \right\}.$$

Feynman's free particle kernel can be expressed in terms of the Fourier transform

$$K_F^{(0)}(x_b t_b | x_a t_a) = \frac{1}{2\pi\hbar} \int\limits_{-\infty}^{\infty} dp \exp \left\{ i\frac{p(x_b - x_a)}{\hbar} - i\frac{p^2(t_b - t_a)}{2m\hbar} \right\}, \qquad (6.42)$$

with the initial condition given by

$$K_F^{(0)}(x_b t_b | x_a t_a)|_{t_b = t_a} = \delta(x_b - x_a).$$

Having $K_F^{(0)}(x_b t_b | x_a t_a)$ either in form (6.41) or (6.42), it is easy to see that the consistency condition holds

$$K_F^{(0)}(x_b t_b | x_a t_a) = \int\limits_{-\infty}^{\infty} dx' K_F^{(0)}(x_b t_b | x' t') K_F^{(0)}(x' t' | x_a t_a).$$

This is a special case of the general quantum-mechanical rule: the amplitudes are multiplied for events occurring in succession in time,

$$K_F(x_b t_b | x_a t_a) = \int\limits_{-\infty}^{\infty} dx' K_F(x_b t_b | x' t') K_F(x' t' | x_a t_a). \qquad (6.43)$$

6.4.1 3D generalization of Feynman's path integral

When $\alpha = 2$ the Hamilton function introduced by Eq. (3.4) has the form

$$H_2(\mathbf{p}, \mathbf{r}) = H_\alpha(\mathbf{p}, \mathbf{r})|_{\alpha=2} = \frac{\mathbf{p}^2}{2m} + V(\mathbf{r}, t), \qquad (6.44)$$

where $\mathbf{p}$ and $\mathbf{r}$ are the 3D vectors and it has been assumed that potential energy term $V(\mathbf{r}, t)$ depends on t.

Then the phase space path integral given by Eq. (6.7) becomes the 3D Feynman's phase space path integral

$$K_F(\mathbf{r}_b t_b | \mathbf{r}_a t_a) = \lim_{N \to \infty} \int d\mathbf{r}_1 ... d\mathbf{r}_{N-1} \frac{1}{(2\pi\hbar)^{3N}} \int d\mathbf{p}_1 ... d\mathbf{p}_N \qquad (6.45)$$

$$\times \exp\left\{ \frac{i}{\hbar} \sum_{j=1}^{N} \mathbf{p}_j (\mathbf{r}_j - \mathbf{r}_{j-1}) \right\} \exp\left\{ -\frac{i}{2m\hbar} \varepsilon \sum_{j=1}^{N} \mathbf{p}_j^2 - \frac{i}{\hbar} \varepsilon \sum_{j=1}^{N} V(\mathbf{r}_j, j\varepsilon) \right\}.$$

Here $\varepsilon = (t_b - t_a)/N$, $\mathbf{r}_j = \mathbf{r}(t_a + i\varepsilon)$, and $\mathbf{r}(t_a + i\varepsilon)|_{i=0} = \mathbf{r}_a$, $\mathbf{r}(t_a + i\varepsilon)|_{i=N} = \mathbf{r}_b$, with $\mathbf{r}_a$ and $\mathbf{r}_b$ being initial and final points of particle paths. Then in the continuum limit $N \to \infty$, $\varepsilon \to 0$ we obtain

$$K_F(\mathbf{r}_b t_b | \mathbf{r}_a t_a) = \int\limits_{\mathbf{r}(t_a)=\mathbf{r}_a}^{\mathbf{r}(t_b)=\mathbf{r}_b} D\mathbf{r}(\tau) \int D\mathbf{p}(\tau) \qquad (6.46)$$

$$\times \exp\left\{ \frac{i}{\hbar} \int\limits_{t_a}^{t_b} d\tau [\mathbf{p}(\tau)\dot{\mathbf{r}}(\tau) - H_2(\mathbf{p}(\tau), \mathbf{r}(\tau), \tau)] \right\},$$

where $\dot{\mathbf{r}}$ denotes the time derivative $d/d\tau$, $H_2(\mathbf{p}(\tau), \mathbf{r}(\tau), \tau)$ is the Hamiltonian given by Eq. (3.3) with the substitutions $\mathbf{p} \to \mathbf{p}(\tau)$, $\mathbf{r} \to \mathbf{r}(\tau)$, $\{\mathbf{p}(\tau), \mathbf{r}(\tau)\}$ is the particle trajectory in 6D phase space, and path integral "measure" $\int\limits_{\mathbf{r}(t_a)=\mathbf{r}_a}^{\mathbf{r}(t_b)=\mathbf{r}_b} D\mathbf{r}(\tau) \int D\mathbf{p}(\tau)...$ is defined by Eq. (6.9).

Quantum kernel $K_F(\mathbf{r}_b t_b | \mathbf{r}_a t_a)$ given by Eq. (6.46) is Feynman's path integral in phase space representation.

To obtain the quantum kernel $K_F(\mathbf{r}_b t_b | \mathbf{r}_a t_a)$ in coordinate representation we have to perform integration over momentums in Eq. (6.46). It can easily be done with the help of the formula

$$\frac{1}{(2\pi\hbar)^3} \int d^3p \exp\left\{ \frac{i}{\hbar}\mathbf{p}(\mathbf{r}_j - \mathbf{r}_{j-1}) - \frac{i}{2m\hbar}\varepsilon\mathbf{p}^2 \right\} \qquad (6.47)$$

$$= \left(\frac{2\pi i\hbar\varepsilon}{m} \right)^{-3N/2} \exp\left\{ i\frac{m(\mathbf{r}_j - \mathbf{r}_{j-1})^2}{2\hbar\varepsilon} \right\}.$$

Then Feynman's quantum kernel $K_F(\mathbf{r}_b t_b | \mathbf{r}_a t_a)$ in 3D coordinate representation can be written as

$$K_F(\mathbf{r}_b t_b | \mathbf{r}_a t_a) = \lim_{N\to\infty} \int d\mathbf{r}_1 ... d\mathbf{r}_{N-1} \left(\frac{2\pi i\hbar\varepsilon}{m} \right)^{-3N/2} \qquad (6.48)$$

$$\times \prod_{j=1}^{N} \exp\left\{ i\frac{m(\mathbf{r}_j - \mathbf{r}_{j-1})^2}{2\hbar\varepsilon} \right\} \exp\left\{ -\frac{i}{\hbar}\varepsilon \sum_{j=1}^{N} V(\mathbf{r}_j, j\varepsilon) \right\},$$

where $V(\mathbf{r}_j, j\varepsilon)$ comes from the potential energy term in Eq. (6.44).

The last equation can be rewritten as

$$K_F(\mathbf{r}_b t_b | \mathbf{r}_a t_a) = \int\limits_{\mathbf{r}(t_a)=\mathbf{r}_a}^{\mathbf{r}(t_b)=\mathbf{r}_b} \mathcal{D}_F \mathbf{r}(\tau) \exp\left\{ -\frac{i}{\hbar} \int\limits_{t_a}^{t_b} d\tau V(\mathbf{r}(\tau), \tau) \right\}, \qquad (6.49)$$

where $\int\limits_{\mathbf{r}(t_a)=\mathbf{r}_a}^{\mathbf{r}(t_b)=\mathbf{r}_b} \mathcal{D}_F \mathbf{r}(\tau)$ stands for 3D Feynman's path integral "measure" given by

$$= \int\limits_{\mathbf{r}(t_a)=\mathbf{r}_a}^{\mathbf{r}(t_b)=\mathbf{r}_b} \mathcal{D}_F \mathbf{r}(\tau) ... \qquad (6.50)$$

$$= \lim_{N\to\infty} \int d\mathbf{r}_1 ... d\mathbf{r}_{N-1} \left(\frac{2\pi i\hbar\varepsilon}{m} \right)^{-3N/2} \prod_{j=1}^{N} \exp\left\{ i\frac{m(\mathbf{r}_j - \mathbf{r}_{j-1})^2}{2\hbar\varepsilon} \right\} ....$$

Thus, Feynman's quantum kernel $K_F(\mathbf{r}_b t_b | \mathbf{r}_a t_a)$ in 3D coordinate representation is defined by Eqs. (6.49) and (6.50).

6.5   Fractional Schrödinger equation from the path integral over Lévy flights

...it is more important to have beauty in one's equations than to have them fit experiment.

P.A.M. Dirac

The kernel $K(x_b t_b | x_a t_a)$ which is defined by Eq. (6.18), describes the evolution of the fractional quantum-mechanical system

$$\psi(x_b, t_b) = \int\limits_{-\infty}^{\infty} dx_a K(x_b t_b | x_a t_a) \psi(x_a, t_a), \qquad t_b \geq t_a, \tag{6.51}$$

where $\psi(x_a, t_a)$ is the wave function of a particle in space point x_a at the time $t = t_a$, and $\psi(x_b, t_b)$ is the fractional wave function of the particle in space point x_b at time $t = t_b$. To obtain the differential equation for the fractional wave function $\psi(x, t)$ we apply Eq. (6.51) in the special case when the time t_b differs only by an infinitesimal interval ε from t_a. By renaming $t_a = t$ and $t_b = t + \varepsilon$ we obtain

$$\psi(x, t + \varepsilon) = \int\limits_{-\infty}^{\infty} dy K(x, t + \varepsilon | y, t) \psi(y, t).$$

Using Feynman's approximation $\int\limits_{t}^{t+\varepsilon} d\tau V(x(\tau), \tau) \simeq \varepsilon V(\frac{x+y}{2}, t)$ and the definition given by Eq. (6.18) we have

$$\psi(x, t + \varepsilon)$$

$$= \int\limits_{-\infty}^{\infty} dy \frac{1}{2\pi\hbar} \int\limits_{-\infty}^{\infty} dp \exp\left\{ i\frac{p(x-y)}{\hbar} - i\frac{D_\alpha \varepsilon |p|^\alpha}{\hbar} - \frac{i}{\hbar}\varepsilon V(\frac{x+y}{2}, t) \right\} \psi(y, t).$$

We may expand the left-hand and the right-hand sides in power series

$$\psi(x, t) + \varepsilon \frac{\partial \psi(x, t)}{\partial t} = \int\limits_{-\infty}^{\infty} dy \frac{1}{2\pi\hbar} \int\limits_{-\infty}^{\infty} dp\, e^{i\frac{p(x-y)}{\hbar}} \left(1 - i\frac{D_\alpha \varepsilon |p|^\alpha}{\hbar} \right) \tag{6.52}$$

$$\times \left(1 - \frac{i}{\hbar} \varepsilon V(\frac{x+y}{2}, t) \right) \psi(y, t).$$

Taking into account the definition (3.14) of quantum Riesz fractional derivative $(\hbar \nabla)^\alpha$ we rewrite Eq. (6.52) as

$$\psi(x, t) + \varepsilon \frac{\partial \psi(x, t)}{\partial t} = \psi(x, t) + i \frac{D_\alpha \varepsilon}{\hbar} (\hbar \nabla)^\alpha \psi(x, t) - \frac{i}{\hbar} \varepsilon V(x, t) \psi(x, t),$$

where we kept the terms up to ε order only.

This will be true to order of ε if $\psi(x, t)$ satisfies the differential equation

$$i\hbar \frac{\partial \psi}{\partial t} = -D_\alpha (\hbar \nabla)^\alpha \psi + V(x, t) \psi. \tag{6.53}$$

This is the fractional Schrödinger equation first derived by Laskin [67], [96] from the Laskin path integral. The space derivative in this equation is of fractional order α, $1 < \alpha \le 2$.

Equation (6.53) may be rewritten in the operator form,

$$i\hbar \frac{\partial \psi}{\partial t} = \widehat{\mathcal{H}}_\alpha \psi, \tag{6.54}$$

where $\widehat{\mathcal{H}}_\alpha$ is the 1D fractional Hamiltonian operator introduced by Eq. (3.19).

Since the kernel $K(x_b t_b | x_a t_a)$ thought of as a function of variables x_b, t_b, is a special wave function (namely, that for a particle which starts at x_a, t_a), we see that $K(x_b t_b | x_a t_a)$ must also satisfy a fractional Schrödinger equation. Thus, for the quantum system described by the fractional Hamiltonian $\widehat{\mathcal{H}}_\alpha$ we have

$$i\hbar \frac{\partial}{\partial t_b} K(x_b t_b | x_a t_a) = -D_\alpha (\hbar \nabla_b)^\alpha K(x_b t_b | x_a t_a) \tag{6.55}$$

$$+ V(x_b) K(x_b t_b | x_a t_a),$$

where $t_b > t_a$ and subscript "b" means that the quantum fractional derivative acts on the variable x_b. The initial condition for this equation is

$$\lim_{t_b \to t_a} K(x_b t_b | x_a t_a) = \delta(x_b - x_a). \tag{6.56}$$

Setting up $x_b = x$, $t_b = t$ and $x_a = 0$, $t_a = 0$, in the above equation and using the fractional Hamiltonian operator $\widehat{\mathcal{H}}_\alpha$ given by Eq. (3.19) we have

$$\left(i\hbar \frac{\partial}{\partial t} - \widehat{\mathcal{H}}_\alpha \right) K(x, t|0, 0) = 0 \tag{6.57}$$

and

$$\lim_{t \to 0} K(x, t|0, 0) = \delta(x), \tag{6.58}$$

where $\delta(x)$ is delta function.

6.5.1   *3D fractional Schrödinger equation from the path integral over Lévy flights*

In the 3D case the evolution of the quantum mechanical system is described by Eq. (6.22). To derive the 3D fractional Schrödinger equation we apply the evolution law (6.22) in the special case when the time t_b differs only by an infinitesimal interval ε from t_a. By renaming $t_a = t$ and $t_b = t + \varepsilon$ we obtain

$$\psi(\mathbf{r}, t + \varepsilon) = \int d^3 r' K(\mathbf{r}, t + \varepsilon | \mathbf{r}', t) \psi(\mathbf{r}', t).$$

Using Feynman's approximation $\int\limits_{t}^{t+\varepsilon} d\tau V(\mathbf{r}(\tau), \tau) \simeq \varepsilon V(\frac{\mathbf{r}+\mathbf{r}'}{2}, t)$ and the definition given by Eq. (6.30) we have

$$\psi(\mathbf{r}, t + \varepsilon) = \int d^3 r' \frac{1}{(2\pi\hbar)^3} \int d^3 p$$

$$\times \exp\left\{ i\frac{\mathbf{p}(\mathbf{r} - \mathbf{r}')}{\hbar} - i\frac{D_\alpha \varepsilon |\mathbf{p}|^\alpha}{\hbar} - \frac{i}{\hbar}\varepsilon V(\frac{\mathbf{r} + \mathbf{r}'}{2}, t) \right\} \psi(\mathbf{r}', t).$$

We may expand the left-hand and the right-hand sides in power series

$$\psi(\mathbf{r}, t) + \varepsilon \frac{\partial \psi(\mathbf{r}, t)}{\partial t}$$

$$= \int d^3 r' \frac{1}{(2\pi\hbar)^3} \int d^3 p\, e^{i\frac{\mathbf{p}(\mathbf{r}-\mathbf{r}')}{\hbar}} \left(1 - i\frac{D_\alpha \varepsilon |\mathbf{p}|^\alpha}{\hbar} \right) \tag{6.59}$$

$$\times \left(1 - \frac{i}{\hbar} \varepsilon V(\frac{\mathbf{r} + \mathbf{r}'}{2}, t) \right) \psi(\mathbf{r}', t).$$

Then, taking into account the definitions of the Fourier transforms

$$\psi(\mathbf{r}, t) = \frac{1}{(2\pi\hbar)^3} \int d^3 p \, e^{i\frac{\mathbf{pr}}{\hbar}} \varphi(\mathbf{p}, t) \qquad (6.60)$$

and

$$\varphi(\mathbf{p}, t) = \int d^3 r \, e^{-i\frac{\mathbf{pr}}{\hbar}} \psi(\mathbf{r}, t), \qquad (6.61)$$

and definition (3.9) of the quantum Riesz fractional derivative $(-\hbar^2\Delta)^{\alpha/2}$ we obtain from Eq. (6.59)

$$\psi(\mathbf{r}, t) + \varepsilon \frac{\partial\psi(\mathbf{r}, t)}{\partial t} = \psi(\mathbf{r}, t) - i\frac{D_\alpha\varepsilon}{\hbar}(-\hbar^2\Delta)^{\alpha/2}\psi(\mathbf{r}, t) - \frac{i}{\hbar}\varepsilon V(\mathbf{r}, t)\psi(\mathbf{r}, t),$$

where we keep the terms up to ε order only.

This will be true to order of ε if $\psi(\mathbf{r}, t)$ satisfies the fractional differential equation

$$i\hbar\frac{\partial\psi(\mathbf{r}, t)}{\partial t} = D_\alpha(-\hbar^2\Delta)^{\alpha/2}\psi(\mathbf{r}, t) + V(\mathbf{r}, t)\psi(\mathbf{r}, t). \qquad (6.62)$$

This is the 3D fractional Schrödinger equation [96] for a quantum particle moving in three-dimensional space. The space derivative in this equation is fractional derivative of order α. Equation (6.62) may be rewritten in the operator form (3.11) by introducing fractional Hamiltonian operator $\widehat{H}_\alpha$ defined by Eq. (3.12).

Since defined by Eq. (6.30) the kernel $K(\mathbf{r}_b t_b | \mathbf{r}_a t_a)$ thought of as a function of variables $\mathbf{r}_b, t_b$, is a special wave function (namely, that for a particle which starts at $\mathbf{r}_a, t_a$), we see that $K(\mathbf{r}_b t_b | \mathbf{r}_a t_a)$ must also satisfy a 3D fractional Schrödinger equation. With the help of Eq. (6.62) we have

$$i\hbar\frac{\partial}{\partial t_b}K(\mathbf{r}_b t_b | \mathbf{r}_a t_a) = D_\alpha(-\hbar^2\Delta_{\mathbf{r}_b})^{\alpha/2}K(\mathbf{r}_b t_b | \mathbf{r}_a t_a) + V(\mathbf{r}_b)K(\mathbf{r}_b t_b | \mathbf{r}_a t_a),$$
$$(6.63)$$

where $t_b > t_a$ subscript "$\mathbf{r}_b$" means that the 3D quantum fractional derivative acts on the variable $\mathbf{r}_b$. The initial condition for this equation is

$$\lim_{t_b \to t_a} K(\mathbf{r}_b t_b | \mathbf{r}_a t_a) = \delta(\mathbf{r}_b - \mathbf{r}_a). \qquad (6.64)$$

Setting up $\mathbf{r}_b = \mathbf{r}$, $t_b = t$ and $\mathbf{r}_a = 0$, $t_a = 0$, in the above equation and using the fractional Hamiltonian operator H_α given by Eq. (3.12) we have

$$\left(i\hbar\frac{\partial}{\partial t} - H_\alpha\right) K(\mathbf{r}, t|0, 0) = 0 \qquad (6.65)$$

and

$$\lim_{t \to 0} K(\mathbf{r}, t|0, 0) = \delta(\mathbf{r}), \qquad (6.66)$$

where $\delta(\mathbf{r})$ is delta function in 3D space.

Chapter 7

A Free Particle Quantum Kernel

7.1 Fundamental properties

For a free particle when $V(x,t) = 0$, we have $H_\alpha(p) = D_\alpha|p|^\alpha$, then Eq. (6.1) results in [67]

$$K^{(0)}(x_b t_b | x_a t_a) = \frac{1}{2\pi\hbar} \int\limits_{-\infty}^{\infty} dp \exp\left\{ i\frac{p(x_b - x_a)}{\hbar} - i\frac{D_\alpha|p|^\alpha(t_b - t_a)}{\hbar} \right\},$$

(7.1)

here $K^{(0)}(x_b t_b | x_a t_a)$ stands for a free particle quantum kernel in the framework of fractional quantum mechanics.

Taking into account Eq. (7.1) it is easy to check the consistency condition

$$K^{(0)}(x_b t_b | x_a t_a) = \int\limits_{-\infty}^{\infty} dx' K^{(0)}(x_b t_b | x' t') K^{(0)}(x' t' | x_a t_a).$$

This is a special case of the general quantum-mechanical rule: for events occurring in succession in time the quantum mechanical amplitudes are multiplied

$$K(x_b t_b | x_a t_a) = \int\limits_{-\infty}^{\infty} dx' K(x_b t_b | x' t') K(x' t' | x_a t_a),$$

(7.2)

where $K(x_b t_b | x_a t_a)$ is given by Eq. (6.18). Introducing notations

$$x = x_b - x_a \qquad \text{and} \qquad t = t_b - t_a,$$

we present Eq. (7.1) as

$$K^{(0)}(x,t) = \frac{1}{2\pi\hbar} \int\limits_{-\infty}^{\infty} dp \exp\left\{ i\frac{px}{\hbar} - i\frac{D_\alpha |p|^\alpha t}{\hbar} \right\}. \qquad (7.3)$$

The fundamental properties of the kernel $K^{(0)}(x,t)$ are:
1. It is the solution to a free particle fractional Schrödinger equation

$$i\hbar\frac{\partial}{\partial t} K^{(0)}(x,t) = D_\alpha(-\hbar^2\Delta)^{\alpha/2} K^{(0)}(x,t), \qquad (7.4)$$

with initial condition

$$K^{(0)}(x,t)|_{t=0} = K^{(0)}(x,0) = \delta(x), \qquad (7.5)$$

where $\delta(x)$ is delta function.
2. It satisfies

$$\int\limits_{-\infty}^{\infty} dx K^{(0)}(x,t) = 1. \qquad (7.6)$$

3. The symmetries hold

$$K^{(0)}(x,t) = K^{(0)}(-x,t) \qquad (7.7)$$

and

$$(K^{(0)}(x,t))^* = K^{(0)}(-x,-t) = K^{(0)}(x,-t), \qquad (7.8)$$

where $(K^{(0)}(x,t))^*$ stands for complex conjugate kernel.
4. When $x = 0$, the kernel $K^{(0)}(x,t)$ is

$$K^{(0)}(0,t) = K^{(0)}(x,t)|_{x=0} = \frac{1}{\alpha\pi\hbar} \left(\frac{\hbar}{iD_\alpha t} \right)^{1/\alpha} \Gamma(\frac{1}{\alpha}), \qquad (7.9)$$

where $\Gamma(1/\alpha)$ is the Gamma function.

7.1.1   Scaling

To make general conclusions on space and time dependencies of a free particle kernel $K^{(0)}(x,t)$, let's study its scaling. We write $K^{(0)}(x,t;D_\alpha)$, where
we keep D_α to remind that besides dependency on x and t the kernel depends on D_α as well. In other words, the scale transformation has to be
applied to x, t and D_α. Hence, we write

$$t = \lambda t', \qquad x = \lambda^\beta x', \qquad D_\alpha = \lambda^\gamma D'_\alpha, \qquad (7.10)$$

$$K^{(0)}(x,t;D_\alpha) = \lambda^\delta K(x',t';D'_\alpha),$$

here β, γ, δ are exponents of the scale transformations. Since the scale
transformations should leave a free particle 1D fractional Schrödinger equation (7.4) invariant and satisfy the condition given by Eq. (7.6) we obtain
the following relationships between scaling exponents,

$$\alpha\beta - \gamma - 1 = 0, \qquad \delta + \beta = 0, \qquad (7.11)$$

which reduce the number of exponents up to 2. Therefore, we have the
two-parameters scale transformation group

$$t = \lambda t', \qquad x = \lambda^\beta x', \qquad D_\alpha = \lambda^{\alpha\beta-1} D'_\alpha, \qquad (7.12)$$

$$K^{(0)}(\lambda^\beta x, \lambda t; \lambda^{\alpha\beta-1} D_\alpha) = \lambda^{-\beta} K^{(0)}(x,t;D_\alpha) \qquad (7.13)$$

where β and λ are arbitrary group parameters.

To get the general scale invariant solutions to Eq. (7.4) we use the
renormalization group framework. As far as the scale invariant solutions to
Eq. (7.4) should satisfy the identity Eq. (7.13) for any arbitrary parameters
β and λ, the solutions depend on a combination of x, and t and D_α to
provide independence on β and λ. Therefore, due to Eqs. (7.12) and (7.13)
the scaling holds

$$K^{(0)}(x,t;D_\alpha) = \frac{1}{x}\mathcal{K}(\frac{x}{\hbar}(\hbar/D_\alpha t)^{1/\alpha}) = \frac{(\hbar/D_\alpha t)^{\frac{1}{\alpha}}}{\hbar}\mathcal{L}(\frac{x}{\hbar}(\hbar/D_\alpha t)^{1/\alpha}), \quad (7.14)$$

where two arbitrary functions $\mathcal{K}$ and $\mathcal{L}$ are determined by the initial conditions, $\mathcal{K}(.) = K^{(0)}(1,.)$ and $\mathcal{L}(.) = K^{(0)}(.,1)$.

Thus, Eq. (7.14) gives us general scale invariant form of a free particle
quantum kernel in the framework of fractional quantum mechanics.

7.2   Fox H-function representation for a free particle kernel

Let's show how a free particle fractional quantum mechanical kernel $K^{(0)}(x_b t_b | x_a t_a)$ defined by Eq. (7.1) can be expressed in terms of the Fox H-function [103]-[106]. First, following the approach proposed by Laskin [93], we obtain the Mellin transform of the quantum mechanical fractional kernel defined by Eq. (7.1). Second, by comparing the inverse Mellin transform with the definition of the Fox function we obtain the desired expression in terms of "known" function, i.e. Fox H-function[1].

Introducing the notations $x \equiv x_b - x_a$, $\tau \equiv t_b - t_a$, we rewrite Eq. (7.1)

$$K^{(0)}(x,\tau) = \frac{1}{2\pi\hbar} \int\limits_{-\infty}^{\infty} dp \exp\left\{ i\frac{px}{\hbar} - i\frac{D_\alpha |p|^\alpha \tau}{\hbar} \right\}. \tag{7.15}$$

Due to the symmetry $K^{(0)}(x,\tau) = K^{(0)}(-x,\tau)$ given by Eq. (7.7) it is sufficient to consider $K_L^{(0)}(x,\tau)$ for $x \geq 0$ only. Further, we will use the following definitions of the Mellin transform

$$\overset{\wedge}{K^{(0)}}(s,\tau) = \int\limits_0^\infty dx x^{s-1} K^{(0)}(x,\tau), \tag{7.16}$$

and inverse Mellin transform

$$K^{(0)}(x,\tau) = \frac{1}{2\pi i} \int\limits_{c-i\infty}^{c+i\infty} ds x^{-s} \overset{\wedge}{K^{(0)}}(s,\tau), \tag{7.17}$$

where the integration path is a straight line from $c - i\infty$ to $c + i\infty$ with $0 < c < 1$.

The Mellin transform of the $K^{(0)}(x,\tau)$ defined in accordance with Eq. (7.16) is

$$\overset{\wedge}{K^{(0)}}(s,\tau) = \frac{1}{2\pi\hbar} \int\limits_0^\infty dx\, x^{s-1} \int\limits_{-\infty}^\infty dp \exp\left\{ i\frac{px}{\hbar} - i\frac{D_\alpha |p|^\alpha \tau}{\hbar} \right\}.$$

---

[1]Note that the H-function bears the name of its discoverer Fox [103] although it has been known since at least 1888, according to [106].

By changing the variables of integration $p \to \left(\frac{\hbar}{iD_\alpha\tau}\right)^{1/\alpha} \varsigma$ and $x \to \left(\frac{\hbar}{iD_\alpha\tau}\right)^{1/\alpha} \xi$, one obtains the integrals in the complex ς and ξ planes. Considering the paths of integration in the ς and ξ planes, it is easy to represent $\hat{K}^{(0)}(s,\tau)$ as follows,

$$\hat{K}^{(0)}(s,\tau)$$

$$= \frac{1}{2\pi}\left(\frac{\hbar}{(\hbar/iD_\alpha\tau)^{1/\alpha}}\right)^{s-1} \int\limits_0^\infty d\xi \xi^{s-1} \int\limits_{-\infty}^\infty d\varsigma \exp\{i\varsigma\xi - |\varsigma|^\alpha\}. \tag{7.18}$$

The integrals over $d\xi$ and $d\varsigma$ can be evaluated by using the equation

$$\int\limits_0^\infty d\xi \xi^{s-1} \int\limits_0^\infty d\varsigma \exp\{i\varsigma\xi - \varsigma^\alpha\} = \frac{4}{s-1}\sin\frac{\pi(s-1)}{2}\Gamma(s)\Gamma(1-\frac{s-1}{\alpha}), \tag{7.19}$$

where $s - 1 < \alpha \le 2$ and $\Gamma(s)$ is the Gamma function.

Inserting Eq. (7.19) into Eq. (7.18) and using the functional relations for the Gamma function, $\Gamma(1-z) = -z\Gamma(-z)$ and $\Gamma(z)\Gamma(1-z) = \pi/\sin\pi z$, yield

$$\hat{K}^{(0)}(s,\tau) = \frac{1}{\alpha}\left(\frac{\hbar}{(\hbar/iD_\alpha\tau)^{1/\alpha}}\right)^{s-1} \frac{\Gamma(s)\Gamma(\frac{1-s}{\alpha})}{\Gamma(\frac{1-s}{2})\Gamma(\frac{1+s}{2})}.$$

The inverse Mellin transform gives a free particle quantum mechanical kernel $K^{(0)}(x,\tau)$

$$K^{(0)}(x,\tau) = \frac{1}{2\pi i}\int\limits_{c-i\infty}^{c+i\infty} ds\, x^{-s} \hat{K}_L^{(0)}(s,\tau)$$

$$= \frac{1}{2\pi i}\frac{1}{\alpha}\int\limits_{c-i\infty}^{c+i\infty} ds \left(\frac{\hbar}{(\hbar/iD_\alpha\tau)^{1/\alpha}}\right)^{s-1} x^{-s} \frac{\Gamma(s)\Gamma(\frac{1-s}{\alpha})}{\Gamma(\frac{1-s}{2})\Gamma(\frac{1+s}{2})},$$

where the integration path is the straight line from $c - i\infty$ to $c + i\infty$ with $0 < c < 1$. By replacing s with $-s$ we obtain

$$K^{(0)}(x,\tau) = \frac{1}{\alpha} \left(\frac{\hbar}{(\hbar/iD_\alpha\tau)^{1/\alpha}} \right)^{-1}$$

$$\times \frac{1}{2\pi i} \int\limits_{-c-i\infty}^{-c+i\infty} ds \left(\frac{1}{\hbar} \left(\frac{\hbar}{iD_\alpha\tau} \right)^{1/\alpha} x \right)^s \frac{\Gamma(-s)\Gamma(\frac{1+s}{\alpha})}{\Gamma(\frac{1+s}{2})\Gamma(\frac{1-s}{2})}.$$

The path of integration may be deformed into one running clockwise around $R_+ - c$. Comparison with the definition of the Fox H-function (see, Eqs. (58) and (59), in [93]) leads to

$$K^{(0)}(x,\tau) = \frac{1}{\alpha} \left(\frac{\hbar}{(\hbar/iD_\alpha\tau)^{1/\alpha}} \right)^{-1} \tag{7.20}$$

$$\times H_{2,2}^{1,1} \left[\frac{1}{\hbar} \left(\frac{\hbar}{iD_\alpha\tau} \right)^{1/\alpha} x \left| \begin{array}{c} (1-1/\alpha, 1/\alpha), (1/2, 1/2) \\ (0,1), (1/2, 1/2) \end{array} \right. \right], \qquad x > 0.$$

Or for any x,

$$K^{(0)}(x,\tau) = \frac{1}{\alpha} \left(\frac{\hbar}{(\hbar/iD_\alpha\tau)^{1/\alpha}} \right)^{-1} \tag{7.21}$$

$$\times H_{2,2}^{1,1} \left[\frac{1}{\hbar} \left(\frac{\hbar}{iD_\alpha\tau} \right)^{1/\alpha} |x| \left| \begin{array}{c} (1-1/\alpha, 1/\alpha), (1/2, 1/2) \\ (0,1), (1/2, 1/2) \end{array} \right. \right].$$

By substituting $x \equiv x_b - x_a$, $\tau \equiv t_b - t_a$ we write

$$K^{(0)}(x_b - x_a, t_b - t_a) = \frac{1}{\alpha} \left(\frac{\hbar}{(\hbar/iD_\alpha(t_b - t_a))^{1/\alpha}} \right)^{-1} \tag{7.22}$$

$$\times H_{2,2}^{1,1} \left[\frac{1}{\hbar} \left(\frac{\hbar}{iD_\alpha\tau} \right)^{1/\alpha} |x_b - x_a| \left| \begin{array}{c} (1-1/\alpha, 1/\alpha), (1/2, 1/2) \\ (0,1), (1/2, 1/2) \end{array} \right. \right].$$

Applying Fox H-function Property 12.2.5 given by Eq. (A.14) in Appendix A, we can express $K^{(0)}(x,\tau)$ as

$$K^{(0)}(x,\tau) = \frac{1}{\alpha|x|} H_{2,2}^{1,1} \left[\frac{1}{\hbar} \left(\frac{\hbar}{iD_\alpha\tau} \right)^{1/\alpha} |x| \left| \begin{array}{c} (1, 1/\alpha), (1, 1/2) \\ (1,1), (1, 1/2) \end{array} \right. \right], \tag{7.23}$$

and with the help of substitution $x \equiv x_b - x_a$, $\tau \equiv t_b - t_a$ we have

$$K^{(0)}(x_b - x_a, t_b - t_a) \tag{7.24}$$

$$= \frac{1}{\alpha|x_b - x_a|} H_{2,2}^{1,1} \left[\frac{1}{\hbar} \left(\frac{\hbar}{iD_\alpha(t_b - t_a)} \right)^{1/\alpha} |x_b - x_a| \left| \begin{array}{c} (1,1/\alpha),(1,1/2) \\ (1,1),(1,1/2) \end{array} \right. \right].$$

It follows from Eqs. (7.21) and (7.23) that there is the identity for $H_{2,2}^{1,1}$-function

$$\left(\frac{\hbar}{(\hbar/iD_\alpha\tau)^{1/\alpha}} \right)^{-1}$$

$$\times H_{2,2}^{1,1} \left[\frac{1}{\hbar} \left(\frac{\hbar}{iD_\alpha\tau} \right)^{1/\alpha} |x| \left| \begin{array}{c} (1 - 1/\alpha, 1/\alpha),(1/2,1/2) \\ (0,1),(1/2,1/2) \end{array} \right. \right] \tag{7.25}$$

$$= \frac{1}{|x|} H_{2,2}^{1,1} \left[\frac{1}{\hbar} \left(\frac{\hbar}{iD_\alpha\tau} \right)^{1/\alpha} |x| \left| \begin{array}{c} (1,1/\alpha),(1,1/2) \\ (1,1),(1,1/2) \end{array} \right. \right],$$

which is in line with the scaling given by Eq. (7.14).

By equating expressions (7.15) and (7.23) we obtain

$$\frac{1}{2\pi\hbar} \int_{-\infty}^{\infty} dp \exp\left\{ i\frac{px}{\hbar} - i\frac{D_\alpha|p|^\alpha\tau}{\hbar} \right\} \tag{7.26}$$

$$= \frac{1}{\alpha|x|} H_{2,2}^{1,1} \left[\frac{1}{\hbar} \left(\frac{\hbar}{iD_\alpha\tau} \right)^{1/\alpha} |x| \left| \begin{array}{c} (1,1/\alpha),(1,1/2) \\ (1,1),(1,1/2) \end{array} \right. \right].$$

With the help of identity (7.25) we can rewrite the above equation in the form

$$\frac{1}{2\pi\hbar} \int_{-\infty}^{\infty} dp \exp\left\{ i\frac{px}{\hbar} - i\frac{D_\alpha|p|^\alpha\tau}{\hbar} \right\} = \frac{1}{\alpha} \left(\frac{\hbar}{(\hbar/iD_\alpha\tau)^{1/\alpha}} \right)^{-1} \tag{7.27}$$

$$\times H_{2,2}^{1,1} \left[\frac{1}{\hbar} \left(\frac{\hbar}{iD_\alpha\tau} \right)^{1/\alpha} |x| \left| \begin{array}{c} (1 - 1/\alpha, 1/\alpha),(1/2,1/2) \\ (0,1),(1/2,1/2) \end{array} \right. \right].$$

Substituting in Eq. (7.23) $x \equiv x_b - x_a$, $\tau \equiv t_b - t_a$, finally yields

$$K^{(0)}(x_b t_b | x_a t_a) = \frac{1}{\alpha |x_b - x_a|}$$

$$\times H_{2,2}^{1,1} \left[\frac{1}{\hbar} \left(\frac{\hbar}{i D_\alpha (t_b - t_a)} \right)^{1/\alpha} |x_b - x_a| \, \middle| \, \begin{array}{c} (1, 1/\alpha), (1, 1/2) \\ (1, 1), (1, 1/2) \end{array} \right]. \quad (7.28)$$

With the help of Property 12.2.5 (see Appendix A, Eq. (A.14)) it can be rewritten as

$$K^{(0)}(x_b - x_a, t_b - t_a) = \frac{1}{\alpha} \left(\frac{\hbar}{(\hbar / i D_\alpha (t_b - t_a))^{1/\alpha}} \right)^{-1} \quad (7.29)$$

$$\times H_{2,2}^{1,1} \left[\frac{1}{\hbar} \left(\frac{\hbar}{i D_\alpha \tau} \right)^{1/\alpha} |x_b - x_a| \, \middle| \, \begin{array}{c} (1 - 1/\alpha, 1/\alpha), (1/2, 1/2) \\ (0, 1), (1/2, 1/2) \end{array} \right].$$

Thus, we found two representations for the 1D free particle quantum mechanical kernel $K^{(0)}(x_b t_b | x_a t_a)$ in the framework of fractional quantum mechanics.

The equations (7.28) and (7.29) for the 1D free particle fractional quantum mechanical kernel are in agreement with scaling law given by Eq. (7.14).

Let us show that Eq. (7.23) includes the well-known Feynman's quantum mechanical kernel as a particular case at $\alpha = 2$, see Eq. (3-3) in [12]. Putting $\alpha = 2$ in Eq. (7.23), we write

$$K^{(0)}(x, \tau)|_{\alpha=2} = \frac{1}{2|x|} H_{2,2}^{1,1} \left[\frac{1}{\hbar} \left(\frac{\hbar}{i D_2 \tau} \right)^{1/2} |x| \, \middle| \, \begin{array}{c} (1, 1/2), (1, 1/2) \\ (1, 1), (1, 1/2) \end{array} \right]. \quad (7.30)$$

Applying the series expansion defined by Eq. (A.9) to the $H_{2,2}^{1,1}$-function and substituting $k \to 2l$ yield

$$K^{(0)}(x, \tau)|_{\alpha=2} = \frac{1}{2\hbar} \left(\frac{\hbar}{i D_2 \tau} \right)^{1/2} \sum_{l=0}^{\infty} \left(-\frac{1}{\hbar} \left(\frac{\hbar}{i D_2 \tau} \right)^{1/2} \right)^{2l} \quad (7.31)$$

$$\times \frac{|x|^{2l}}{(2l)!} \frac{1}{\Gamma(\frac{1}{2} - l)}.$$

Taking into account identity $\Gamma(\frac{1}{2}+z)\Gamma(\frac{1}{2}-z) = \pi/\cos\pi z$, and applying the Gauss multiplication formula

$$\Gamma(2l) = \sqrt{\frac{2^{4l-1}}{2\pi}}\Gamma(l)\Gamma(l+\frac{1}{2}), \qquad (7.32)$$

we find that

$$(2l)!\Gamma(\frac{1}{2} - l) = \frac{\sqrt{\pi}}{(-1)^l}(2)^{2l}l!. \qquad (7.33)$$

With the help of Eq. (7.33) kernel $K^{(0)}(x,\tau)|_{\alpha=2}$ can be rewritten as

$$K^{(0)}(x,\tau)|_{\alpha=2} = \frac{1}{2\sqrt{\pi}\hbar}\left(\frac{\hbar}{iD_2\tau}\right)^{1/2}\sum_{l=0}^{\infty}\left(-\frac{1}{\hbar}\left(\frac{\hbar}{iD_2\tau}\right)^{1/2}\right)^{2l}$$

$$\times \frac{(-1)^l|x|^{2l}}{2^{2l}l!} \qquad (7.34)$$

$$= \frac{1}{2\sqrt{\pi}\hbar}\left(\frac{\hbar}{iD_2\tau}\right)^{1/2}\exp\left\{-\frac{1}{4}\frac{|x|^2}{\hbar iD_2\tau}\right\}.$$

Since $D_2 = 1/2m$, we come to the Feynman's kernel

$$K^{(0)}(x,\tau)|_{\alpha=2} \equiv K_F^{(0)}(x,\tau) = \sqrt{\frac{m}{2\pi i\hbar\tau}}\exp\left\{\frac{im|x|^2}{2\hbar\tau}\right\}, \qquad (7.35)$$

or (see Eq. (3-3), [12])

$$K^{(0)}(x_b - x_a, t_b - t_a)|_{\alpha=2} \equiv K_F^{(0)}(x_b - x_a, t_b - t_a) \qquad (7.36)$$

$$= \left(\frac{2\pi i\hbar(t_b - t_a)}{m}\right)^{-1/2}\exp\left\{\frac{im|x_b - x_a|^2}{2\hbar(t_b - t_a)}\right\},$$

which coincides with Eq. (6.41).

Thus, it is shown how Feynman's free particle kernel can be derived from the general equation (7.23).

By equating the expressions (7.30) and (7.35) we come to the identity with involvement of $H_{2,2}^{1,1}$-function

$$\frac{1}{2|x|}H_{2,2}^{1,1}\left[\frac{1}{\hbar}\left(\frac{2m\hbar}{i\tau}\right)^{1/2}|x|\,\middle|\,\begin{array}{c}(1,1/2),(1,1/2)\\(1,1),(1,1/2)\end{array}\right] \qquad (7.37)$$

$$= \sqrt{\frac{m}{2\pi i \hbar \tau}} \exp\left\{ \frac{im|x|^2}{2\hbar\tau} \right\}.$$

This equation can be presented in the form

$$H_{2,2}^{1,1}\left[w \left| \begin{array}{l} (1,1/2),(1,1/2) \\ (1,1),(1,1/2) \end{array} \right. \right] \tag{7.38}$$

$$= \frac{w}{\sqrt{\pi}} \exp\left\{ -\frac{w^2}{4} \right\},$$

where w has been introduced as

$$w = \left(\frac{2m}{i\hbar\tau} \right)^{1/2} |x|. \tag{7.39}$$

Using H-function Property 12.2.2 given by Eq. (A.11) from Appendix A, we can rewrite Eq. (7.38) in terms of $H_{1,1}^{1,0}$-function

$$H_{1,1}^{1,0}\left[w \left| \begin{array}{l} (1,1/2) \\ (1,1) \end{array} \right. \right] \tag{7.40}$$

$$= \frac{w}{\sqrt{\pi}} \exp\left\{ -\frac{w^2}{4} \right\}.$$

Considering Eq. (7.21) in the case when $\alpha = 2$, and equating it to Eq. (7.35) we come to another identity with involvement of $H_{2,2}^{1,1}$-function

$$H_{2,2}^{1,1}\left[\frac{1}{\hbar} \left(\frac{2m\hbar}{i\tau} \right)^{1/2} |x| \left| \begin{array}{l} (1/2,1/2),(1/2,1/2) \\ (0,1),(1/2,1/2) \end{array} \right. \right] \tag{7.41}$$

$$= \frac{1}{\sqrt{\pi}} \exp\left\{ \frac{im|x|^2}{2\hbar\tau} \right\},$$

which can be rewritten in the form

$$H_{2,2}^{1,1}\left[w \left| \begin{array}{l} (1/2,1/2),(1/2,1/2) \\ (0,1),(1/2,1/2) \end{array} \right. \right] \tag{7.42}$$

$$= \frac{1}{\sqrt{\pi}} \exp\left\{ -\frac{w^2}{4} \right\},$$

with w given by Eq. (7.39).

The last equation can be expressed in terms of $H_{1,1}^{1,0}$-function

$$H_{1,1}^{1,0}\left[w\,\middle|\,\begin{matrix}(1/2,1/2)\\(0,1)\end{matrix}\right] \tag{7.43}$$

$$= \frac{1}{\sqrt{\pi}}\exp\left\{-\frac{w^2}{4}\right\},$$

if we apply H-function Property 12.2.2 given by Eq. (A.11) from Appendix A.

It follows from Eqs. (7.38) and (7.42) that the identity holds

$$\frac{1}{|w|}H_{1,1}^{1,0}\left[w\,\middle|\,\begin{matrix}(1,1/2)\\(1,1)\end{matrix}\right] = H_{1,1}^{1,0}\left[w\,\middle|\,\begin{matrix}(1/2,1/2)\\(0,1)\end{matrix}\right], \tag{7.44}$$

which is in line with H-function Property 12.2.5 given by Eq. (A.14).

7.2.1 A free particle kernel: 3D case

The generalization of Eq. (7.15) to the 3D case has the form

$$K^{(0)}(\mathbf{r}_b t_b|\mathbf{r}_a t_a) \tag{7.45}$$

$$= \frac{1}{(2\pi\hbar)^3}\int d^3p\exp\left\{i\frac{\mathbf{p}(\mathbf{r}_b-\mathbf{r}_a)}{\hbar}-i\frac{D_\alpha|\mathbf{p}|^\alpha(t_b-t_a)}{\hbar}\right\},$$

where $\mathbf{r}$ and $\mathbf{p}$ are the 3D vectors.

In terms of the Lévy probability distribution function defined by (6.28) the kernel $K^{(0)}(\mathbf{r}_b t_b|\mathbf{r}_a t_a)$ can be expressed as

$$K^{(0)}(\mathbf{r}_b t_b|\mathbf{r}_a t_a) = \frac{1}{\hbar^3}\left(\frac{\hbar}{iD_\alpha(t_b-t_a)}\right)^{3/\alpha} \tag{7.46}$$

$$\times L_\alpha\left(\frac{1}{\hbar}\left(\frac{\hbar}{iD_\alpha(t_b-t_a)}\right)^{1/\alpha}|\mathbf{r}_b-\mathbf{r}_a|\right).$$

To express the kernel $K^{(0)}(\mathbf{r}_b t_b|\mathbf{r}_a t_a)$ in terms of Fox's H-function we write

$$K^{(0)}(\mathbf{r}_b t_b | \mathbf{r}_a t_a) = \frac{1}{(2\pi\hbar)^3} \int\limits_0^\infty dp\, p^2 \int\limits_0^\pi d\vartheta \sin\vartheta \int\limits_0^{2\pi} d\varphi$$

$$\times \exp\left\{ i\frac{p|\mathbf{r}_b - \mathbf{r}_a|\cos\vartheta}{\hbar} - i\frac{D_\alpha |\mathbf{p}|^\alpha (t_b - t_a)}{\hbar} \right\}.$$

By performing integration over $d\vartheta$ and $d\varphi$ we come to

$$K^{(0)}(\mathbf{r}_b t_b | \mathbf{r}_a t_a) \tag{7.47}$$

$$= \frac{1}{2\pi^2 \hbar^2 |\mathbf{r}_b - \mathbf{r}_a|} \int\limits_0^\infty dp\, p \sin(\frac{p|\mathbf{r}_b - \mathbf{r}_a|}{\hbar}) \exp\left\{ -i\frac{D_\alpha |\mathbf{p}|^\alpha (t_b - t_a)}{\hbar} \right\}.$$

Noting that

$$p\sin(\frac{p|\mathbf{r}_b - \mathbf{r}_a|}{\hbar}) = -\hbar\frac{\partial}{\partial|\mathbf{r}_b - \mathbf{r}_a|}\cos(\frac{p|\mathbf{r}_b - \mathbf{r}_a|}{\hbar}), \tag{7.48}$$

we obtain

$$K^{(0)}(\mathbf{r}_b t_b | \mathbf{r}_a t_a) = -\frac{1}{2\pi|\mathbf{r}_b - \mathbf{r}_a|}\frac{\partial}{\partial x}K^{(0)}(x; t_b - t_a)|_{x=|\mathbf{r}_b - \mathbf{r}_a|}, \tag{7.49}$$

here the kernel $K^{(0)}(x; t_b - t_a)$ is 1D kernel introduced by Eq. (7.15) and presented in terms of H-function by Eq. (7.23).

Let us note that Eq. (7.49) is a special case of a general relation that holds between the D-dimensional and $(D + 2)$-dimensional Fourier transforms of any isotropic function. An important consequence of Eq. (7.49) is that it allows us to evaluate the 3D kernel knowing the expression (7.23) for a 1D quantum kernel. Thus, the problem is to calculate the derivative of the 1D kernel in Eq. (7.49). With the help of Eq. (A.19) from Appendix A we find

$$K^{(0)}(\mathbf{r}_b t_b | \mathbf{r}_a t_a) = -\frac{1}{2\pi\alpha}\frac{1}{|\mathbf{r}_b - \mathbf{r}_a|^3} \tag{7.50}$$

$$\times H_{3,3}^{1,2}\left[\frac{1}{\hbar}\left(\frac{\hbar}{iD_\alpha(t_b - t_a)} \right)^{1/\alpha} |\mathbf{r}_b - \mathbf{r}_a| \left| \begin{array}{c} (1,1),(1,1/\alpha),(1,1/2) \\ (1,1),(1,1/2),(2,1) \end{array} \right. \right].$$

This is fractional quantum mechanical kernel for a free particle in 3D space. We see that in comparison with 1D case the 3D fractional quantum mechanical kernel is represented in terms of Fox's $H_{3,3}^{1,2}$-function.

By applying Property 12.2.5 for Fox's H-function we express $K^{(0)}(\mathbf{r}_b t_b|\mathbf{r}_a t_a)$ in the form

$$K^{(0)}(\mathbf{r}_b t_b|\mathbf{r}_a t_a) = -\frac{1}{2\pi\alpha\hbar}\frac{1}{|\mathbf{r}_b - \mathbf{r}_a|^2}\left(\frac{\hbar}{iD_\alpha(t_b - t_a)}\right)^{1/\alpha} \tag{7.51}$$

$$\times H_{3,3}^{1,2}\left[\frac{1}{\hbar}\left(\frac{\hbar}{iD_\alpha(t_b - t_a)}\right)^{1/\alpha}|\mathbf{r}_b - \mathbf{r}_a|\ \middle|\ \begin{array}{l}(0,1),(1-1/\alpha,1/\alpha),(1/2,1/2)\\(0,1),(1/2,1/2),(1,1)\end{array}\right].$$

It is interesting to find out how the Feynman a free particle 3D quantum mechanical kernel $K_F^{(0)}(\mathbf{r}_b t_b|\mathbf{r}_a t_a)$ can be obtained from the general equation (7.51) at $\alpha = 2$. Setting in Eq. (7.51) $\alpha = 2$ and applying the series expansion (A.9) for the $H_{3,3}^{1,2}$-function we come to

$$K^{(0)}(\mathbf{r}_b t_b|\mathbf{r}_a t_a)|_{\alpha=2} = K_F^{(0)}(\mathbf{r}_b t_b|\mathbf{r}_a t_a) = -\frac{1}{4\pi|\mathbf{r}_b - \mathbf{r}_a|^3}$$

$$\times \sum_{k=0}^{\infty}\frac{\Gamma(1+k)}{\Gamma(k)\Gamma(\frac{1-k}{2})}\frac{(-1)^k}{k!}\left(\frac{1}{\hbar}\left(\frac{\hbar}{iD_2(t_b - t_a)}\right)^{1/2}|\mathbf{r}_b - \mathbf{r}_a|\right)^{1+k} \tag{7.52}$$

$$= -\frac{1}{4\pi\hbar|\mathbf{r}_b - \mathbf{r}_a|^2}\left(\frac{\hbar}{iD_2(t_b - t_a)}\right)^{1/2}$$

$$\times \sum_{k=0}^{\infty}\frac{k}{\Gamma(\frac{1-k}{2})}\frac{(-1)^k}{k!}\left(\frac{1}{\hbar}\left(\frac{\hbar}{iD_2(t_b - t_a)}\right)^{1/2}|\mathbf{r}_b - \mathbf{r}_a|\right)^k.$$

Substitution $k \to 2l$ yields

$$K_F^{(0)}(\mathbf{r}_b t_b|\mathbf{r}_a t_a) \tag{7.53}$$

$$= -\frac{1}{4\pi\hbar|\mathbf{r}_b - \mathbf{r}_a|^2}\left(\frac{\hbar}{iD_2(t_b - t_a)}\right)^{1/2}$$

$$\times \sum_{l=0}^{\infty}\frac{(2l)}{\Gamma(\frac{1}{2} - l)}\frac{1}{(2l)!}\left(-\frac{1}{\hbar}\left(\frac{\hbar}{iD_2(t_b - t_a)}\right)^{1/2}|\mathbf{r}_b - \mathbf{r}_a|\right)^{2l}.$$

Using the identity

$$\sum_{l=0}^{\infty}(2l)\frac{(W)^{2l}}{\Gamma(\frac{1}{2}-l)(2l)!} = W\frac{\partial}{\partial W}\sum_{k=0}^{\infty}\frac{(W)^{2l}}{\Gamma(\frac{1}{2}-l)(2l)!},$$

and introducing the notation

$$W = \frac{1}{\hbar}\left(\frac{\hbar}{iD_2(t_b-t_a)}\right)^{1/2}|\mathbf{r}_b-\mathbf{r}_a|, \qquad (7.54)$$

we obtain from Eq. (7.53)

$$K_F^{(0)}(\mathbf{r}_bt_b|\mathbf{r}_at_a)$$

$$= -\frac{1}{4\pi\hbar|\mathbf{r}_b-\mathbf{r}_a|^2}\left(\frac{\hbar}{iD_2(t_b-t_a)}\right)^{1/2}W\frac{\partial}{\partial W}\sum_{l=0}^{\infty}\frac{1}{\Gamma(\frac{1}{2}-l)}\frac{(-W)^{2l}}{(2l)!}.$$

With the help of Eq. (7.33) the sum in the last equation can be expressed as

$$\sum_{l=0}^{\infty}\frac{1}{\Gamma(\frac{1}{2}-l)}\frac{(-W)^{2l}}{(2l)!} = \frac{1}{\sqrt{\pi}}\exp\left\{-\frac{W^2}{4}\right\}.$$

Then taking into account Eq. (7.54) we find

$$K_F^{(0)}(\mathbf{r}_bt_b|\mathbf{r}_at_a) = \left(\frac{1}{4\pi\hbar iD_2(t_b-t_a)}\right)^{3/2}\exp\left\{-\frac{|\mathbf{r}_b-\mathbf{r}_a|^2}{4\hbar iD_2(t_b-t_a)}\right\}. \quad (7.55)$$

Since $D_2 = 1/2m$, where m is a particle mass, we rediscover Feynman's 3D quantum mechanical kernel of a free quantum particle with the mass m (see, Eq. (1.382) for the case $D = 3$, in [107]),

$$K_F^{(0)}(\mathbf{r}_bt_b|\mathbf{r}_at_a) = \left(\frac{m}{2\pi i\hbar(t_b-t_a)}\right)^{3/2}\exp\left\{\frac{im|\mathbf{r}_b-\mathbf{r}_a|^2}{2\hbar(t_b-t_a)}\right\}. \quad (7.56)$$

Thus, we see that Eq. (7.51) includes Feynman's 3D kernel $K_F^{(0)}(\mathbf{r}_bt_b|\mathbf{r}_at_a)$ as a special case at $\alpha = 2$.

7.2.2   A free particle kernel: D-dimensional generalization

To generalize the above developments to D-dimensional space $(D \geq 2)$ we write for the D-dimensional a free particle quantum mechanical kernel

$$K^{(0)}(\mathbf{r}_b t_b | \mathbf{r}_a t_a) \qquad (7.57)$$

$$= \frac{1}{(2\pi\hbar)^{\mathrm{D}}} \int d^{\mathrm{D}} p \exp \left\{ i\frac{\mathbf{p}(\mathbf{r}_b - \mathbf{r}_a)}{\hbar} - i\frac{D_\alpha |\mathbf{p}|^\alpha (t_b - t_a)}{\hbar} \right\},$$

where $\mathbf{r}$ and $\mathbf{p}$ are the D-dimensional vectors

$$\mathbf{r} = (r_1, r_2, ..., r_{\mathrm{D}}) \quad \text{and} \quad \mathbf{p} = (p_1, p_2, ..., p_{\mathrm{D}}).$$

Introducing the spherical coordinate system in the D-dimensional momentum space

$$(p_1, p_2, ..., p_{\mathrm{D}}) \rightarrow (p, \varphi, \theta_1, ..., \theta_{\mathrm{D}-2}),$$

where $p = |\mathbf{p}|$, gives us

$$d^{\mathrm{D}} p = p^{\mathrm{D}-1} dp d\varphi \sin\theta_1 d\theta_1 \sin^2\theta_2 d\theta_2 ... \sin^{\mathrm{D}-2}\theta_{\mathrm{D}-2} d\theta_{\mathrm{D}-2},$$

with the variables p, φ and θ_k assuming values over the intervals

$$0 \leq p < \infty, \qquad 0 \leq \varphi \leq 2\pi, \qquad 0 \leq \theta_k \leq \pi, \qquad k = 1, ..., \mathrm{D} - 2.$$

Thus, in the spherical momentum system Eq. (7.57) reads (for $D \geq 4$)

$$K^{(0)}(\mathbf{r}_b t_b | \mathbf{r}_a t_a) = \frac{1}{(2\pi\hbar)^{\mathrm{D}}} \int\limits_0^\infty dp\, p^{\mathrm{D}-1} \exp \left\{ -i\frac{D_\alpha p^\alpha (t_b - t_a)}{\hbar} \right\}$$

$$\times \int\limits_0^{2\pi} d\varphi \left(\prod_{k=1}^{\mathrm{D}-3} \int\limits_0^\pi d\theta_k \sin^k \theta_k \right) \qquad (7.58)$$

$$\times \int\limits_0^\pi d\theta_{\mathrm{D}-2} \sin^{\mathrm{D}-2}\theta_{\mathrm{D}-2} \exp \left\{ i\frac{p|\mathbf{r}_b - \mathbf{r}_a| \cos\theta_{\mathrm{D}-2}}{\hbar} \right\}.$$

The integral over $d\varphi$ results in

$$\int\limits_0^{2\pi} d\varphi = 2\pi. \tag{7.59}$$

To evaluate the integrals over $d\theta_1...d\theta_{D-3}$ we use the formula

$$\int\limits_0^{\pi} d\theta_k \sin^k \theta_k = \sqrt{\pi}\frac{\Gamma(\frac{k+1}{2})}{\Gamma(\frac{k+2}{2})}, \tag{7.60}$$

and obtain

$$\left(\prod_{k=1}^{D-3}\int\limits_0^{\pi} d\theta_k \sin^{k-1}\theta_k\right) = \prod_{k=1}^{D-3}\left(\sqrt{\pi}\frac{\Gamma(\frac{k+1}{2})}{\Gamma(\frac{k+2}{2})}\right) = \frac{\pi^{\frac{D-3}{2}}}{\Gamma(\frac{D-1}{2})}. \tag{7.61}$$

The integration over $d\theta_{D-2}$ results in[2]

$$\int\limits_0^{\pi} d\theta_{D-2}(\sin\theta_{D-2})^{D-2}\exp\left\{i\frac{p|\mathbf{r}_b - \mathbf{r}_a|\cos\theta_{D-2}}{\hbar}\right\} \tag{7.62}$$

$$= 2^{2-\frac{D}{2}}\pi\frac{\Gamma(D-2)}{\Gamma(\frac{D-2}{2})}\left(\frac{\hbar}{p|\mathbf{r}_b - \mathbf{r}_a|}\right)^{\frac{D-2}{2}}J_{\frac{D-2}{2}}\left(\frac{p|\mathbf{r}_b - \mathbf{r}_a|}{\hbar}\right).$$

Gathering together Eqs. (7.59) - (7.62) and taking into account Eq. (7.32) yield for Eq. (7.58)

$$K^{(0)}(\mathbf{r}_b t_b|\mathbf{r}_a t_a) = \frac{1}{(2\pi\hbar)^{\frac{D}{2}}}\frac{1}{\hbar(|\mathbf{r}_b - \mathbf{r}_a|)^{\frac{D-2}{2}}} \tag{7.63}$$

$$\times\int\limits_0^{\infty} dp p^{\frac{D}{2}}J_{\frac{D-2}{2}}(\frac{p|\mathbf{r}_b - \mathbf{r}_a|}{\hbar})\exp\left\{-i\frac{D_\alpha p^\alpha(t_b - t_a)}{\hbar}\right\},$$

---

[2]We used the table integral (see, Eq. (3.387) in [86])

$$\int\limits_0^{\pi} d\theta \sin^m\theta\exp\{iA\cos\theta\}$$

$$= \int\limits_{-1}^{1} dx(1-x^2)^{\frac{m-1}{2}}e^{iAx} = 2^{1-\frac{m}{2}}\frac{\Gamma(m)}{\Gamma(\frac{m}{2})}A^{-\frac{m}{2}}J_{\frac{m}{2}}(A).$$

where $J_\nu(z)$ is the Bessel function of the first kind of order ν defined by Eq. (8.402) in [86]

$$J_\nu(z) = \left(\frac{z}{2}\right)^\nu \sum_{k=0}^{\infty} (-1)^k \frac{z^{2k}}{2^{2k} k! \Gamma(\nu + k + 1)}. \tag{7.64}$$

From definition (7.64) we have

$$J_0(z) = J_\nu(z)|_{\nu=0} = \sum_{k=0}^{\infty} (-1)^k \frac{z^{2k}}{2^{2k}(k!)^2} \tag{7.65}$$

and

$$J_{1/2}(z) = J_\nu(z)|_{\nu=1/2} = \sqrt{\frac{z}{2}} \sum_{k=0}^{\infty} (-1)^k \frac{z^{2k}}{k! \Gamma(k + 3/2)} \left(\frac{z}{2}\right)^{2k} \tag{7.66}$$

$$= \sqrt{\frac{2}{\pi z}} \sum_{k=0}^{\infty} \frac{(-1)^k z^{2k+1}}{(2k+1)!} = \sqrt{\frac{2}{\pi z}} \sin z,$$

where we took into account the identity for Gamma function $\Gamma(k + 3/2)$

$$\Gamma(k + 3/2) = \frac{\sqrt{\pi}(2k+1)!}{k! 2^{2k+1}}.$$

With the help of the above equations it can be easily verified that Eq. (7.63) includes expressions for a free particle kernel in space dimensions $D = 2$ and $D = 3$ as two particular cases. Indeed, it is easy to see, for example, that in the case $D = 3$, substituting $J_{1/2}(z)$ into Eq. (7.63) gives us the 3D kernel defined by Eq. (7.47).

Chapter 8

Transforms of a Free Particle Kernel

Such is the advantage of a well-constructed language that its sim-
plified notation often becomes the source of profound theories.

Pierre-Simon Laplace

8.1 Laplace transform

The Laplace transform $\widetilde{K}^{(0)}(x,s)$ of a 1D free particle kernel is defined as

$$\widetilde{K}^{(0)}(x,s) = \int_0^\infty d\tau e^{-s\tau} K^{(0)}(x,\tau), \qquad (8.1)$$

where $K^{(0)}(x,\tau)$ is given by Eq. (7.23).

By using Eq. (7.23) and applying the series expansion (A.9) for the function $H_{2,2}^{1,1}$ we obtain

$$
\widetilde{K}^{(0)}(x,s) = \frac{1}{\alpha x} \sum_{k=0}^\infty \frac{\Gamma(\frac{1+k}{\alpha})}{\Gamma(\frac{1+k}{2})\Gamma(\frac{1-k}{2})} \frac{(-1)^k}{k!}
$$

$$
\times \left(\frac{1}{\hbar} \left(\frac{\hbar}{iD_\alpha} \right)^{1/\alpha} |x| \right)^{1+k} \int_0^\infty d\tau e^{-s\tau} \tau^{-\frac{1+k}{\alpha}} \qquad (8.2)
$$

$$
= \frac{1}{\alpha x s} \sum_{k=0}^\infty \frac{\Gamma(\frac{1+k}{\alpha})\Gamma(1 - \frac{1+k}{\alpha})}{\Gamma(\frac{1+k}{2})\Gamma(\frac{1-k}{2})} \frac{(-1)^k}{k!} \left(\frac{1}{\hbar} \left(\frac{\hbar s}{iD_\alpha} \right)^{1/\alpha} |x| \right)^{1+k},
$$

where it was taken into account that

$$\int\limits_0^\infty d\tau e^{-s\tau}\tau^{-\frac{1+k}{\alpha}} = s^{\frac{1+k}{\alpha}-1}\Gamma\left(1 - \frac{1+k}{\alpha}\right).$$

Applying the definition of Fox's function $H_{3,2}^{1,2}$ we can finally rewrite the Laplace transform $\widetilde{K}^{(0)}(x,s)$ in terms of H-function

$$\widetilde{K}^{(0)}(x,s) = \frac{1}{\alpha x s} \tag{8.3}$$

$$\times H_{3,2}^{1,2}\left[\frac{1}{\hbar}\left(\frac{\hbar s}{iD_\alpha}\right)^{1/\alpha}|x| \,\middle|\, \begin{array}{c}(1,1/\alpha),(0,-1/\alpha),(1,1/2)\\(1,1),(1,1/2)\end{array}\right].$$

Putting $\alpha = 2$ in Eq. (8.3) and using the series expansion for the $H_{3,2}^{1,2}$-function we obtain the Laplace transform of a free particle kernel for standard quantum mechanics

$$\widetilde{K}^{(0)}(x,s)|_{\alpha=2} = \sqrt{\frac{m}{2si\hbar}}\exp\left\{-\sqrt{\frac{2ms}{i\hbar}}|x|\right\}, \tag{8.4}$$

here we used $D_2 = 1/2m$, and m is particle mass.

8.2 The energy-time transformation

In fractional quantum mechanics an important role is played by the Fourier transform of the kernel $K^{(0)}(x_2 - x_1, \tau)$ in the time variable, which is the fractional fixed-energy amplitude $k^{(0)}(x_2, x_1; E)$

$$k^{(0)}(x_2, x_1; E) = \int\limits_0^\infty d\tau e^{(i/\hbar)E\tau} K^{(0)}(x_2 - x_1, \tau), \tag{8.5}$$

where $K^{(0)}(x_2 - x_1, \tau)$ is a free particle kernel given by Eq. (7.24).

To make the integral convergent, we have to move the energy into the upper complex half-plane by an infinitesimal amount ϵ. Then the fractional fixed-energy amplitude becomes

$$k^{(0)}(x_2, x_1; E) = \sum_{n=1}^\infty \phi_n(x_2)\phi_n^*(x_1)\frac{i\hbar}{E - E_n + i\epsilon}, \tag{8.6}$$

where $\phi_n(x)$ is the eigenfunction of the 1D fractional Schrödinger equation with eigenvalue E_n.

The small $i\epsilon$-shift in the energy E in Eq. (8.6) may be thought of as attached to each of the energies E_n which are thus shifted by an infinitesimal value below the real energy axis. When doing the Fourier integral (8.5), the exponential $e^{-(i/\hbar)E\tau}$ always makes it possible to close the integration contour along the energy axis by an infinite semicircle in the complex energy plane, which lies in the upper half-plane for $\tau < 0$ and in the lower half-plane for $\tau > 0$. The $i\epsilon$-shift guarantees that for $\tau < 0$, there is no pole inside the closed contour, making the propagator vanish. For $\tau > 0$, on the other hand, poles in the lower half-plane give a spectral representation of the fractional kernel (8.6) via Cauchy's residue theorem.

It follows that the fixed-energy amplitude $k^{(0)}(x_2, x_1; E)$ and the kernel $K^{(0)}(x_2 - x_1, \tau)$ are related to each other by the inverse energy Fourier transform

$$K^{(0)}(x_2 - x_1, \tau) = \frac{1}{2\pi\hbar} \int\limits_{-\infty}^{\infty} dE\, e^{-(i/\hbar)E\tau} k^{(0)}(x_2, x_1; E). \qquad (8.7)$$

To obtain the equation for $k^{(0)}(x_2, x_1; E)$ in terms of H-function, let's substitute $K^{(0)}(x_2 - x_1, \tau)$ given by

$$K^{(0)}(x_2 - x_1, \tau)$$

$$= \frac{1}{\alpha|x_2 - x_1|} H_{2,2}^{1,1}\left[\frac{1}{\hbar}\left(\frac{\hbar}{iD_\alpha\tau}\right)^{1/\alpha} |x_2 - x_1| \,\middle|\, \begin{matrix} (1, 1/\alpha), (1, 1/2) \\ (1, 1), (1, 1/2) \end{matrix} \right]$$

into Eq. (8.5), and then evaluate the integral over $d\tau$. The result is

$$k^{(0)}(x_2, x_1; E) \qquad (8.8)$$

$$= \frac{i\hbar}{\alpha|x_2 - x_1|E} H_{2,3}^{2,1}\left[\frac{1}{\hbar}\left(-\frac{E}{D_\alpha}\right)^{1/\alpha} |x_2 - x_1| \,\middle|\, \begin{matrix} (1, 1/\alpha), (1, 1/2) \\ (1, 1/\alpha), (1, 1), (1, 1/2) \end{matrix} \right].$$

The above equations can be easily generalized to the 3D case. Let us calculate a free particle fixed-energy amplitude $k_L^{(0)}(\mathbf{r}_2, \mathbf{r}_1; E)$ defined as follows

$$k^{(0)}(\mathbf{r}_2, \mathbf{r}_1; E) = \int_0^\infty d\tau e^{(i/\hbar)E\tau} K^{(0)}(\mathbf{r}_2 - \mathbf{r}_1, \tau). \qquad (8.9)$$

Then the inverse energy Fourier transform expresses $K^{(0)}(\mathbf{r}_2 - \mathbf{r}_1, \tau)$ in terms of $k^{(0)}(\mathbf{r}_2, \mathbf{r}_1; E)$

$$K^{(0)}(\mathbf{r}_2 - \mathbf{r}_1, \tau) = \frac{1}{2\pi\hbar} \int_0^\infty dE e^{(i/\hbar)E\tau} k^{(0)}(\mathbf{r}_2, \mathbf{r}_1; E). \qquad (8.10)$$

With the help of Eq. (7.51) we have

$$k^{(0)}(\mathbf{r}_2, \mathbf{r}_1; E) = -\int_0^\infty d\tau e^{(i/\hbar)E\tau} \frac{1}{2\pi\alpha} \frac{1}{|\mathbf{r}_2 - \mathbf{r}_1|^3} \qquad (8.11)$$

$$\times H_{3,3}^{1,2} \left[\frac{1}{\hbar} \left(\frac{\hbar}{iD_\alpha\tau} \right)^{1/\alpha} |\mathbf{r}_2 - \mathbf{r}_1| \left| \begin{array}{c} (1,1), (1,1/\alpha), (1,1/2) \\ (1,1), (1,1/2), (2,1) \end{array} \right. \right].$$

Using Property 12.2.8 of H-function (see Eq. (A.17) in Appendix A) we find

$$k^{(0)}(\mathbf{r}_2, \mathbf{r}_1; E) = -\frac{\hbar}{2\pi\alpha i E |\mathbf{r}_2 - \mathbf{r}_1|^3} \qquad (8.12)$$

$$\times H_{3,4}^{2,2} \left[\frac{1}{\hbar} \left(-\frac{E}{D_\alpha} \right)^{1/\alpha} |\mathbf{r}_2 - \mathbf{r}_1| \left| \begin{array}{c} (1,1), (1,1/\alpha), (1,1/2) \\ (1,1/\alpha), (1,1), (1,1/2), (2,1) \end{array} \right. \right].$$

Thus the expression for the fixed-energy amplitude $k_L^{(0)}(\mathbf{r}_2, \mathbf{r}_1; E)$ involves the $H_{3,4}^{2,2}$ Fox's H-function. The inverse energy Fourier transform defined by Eq. (8.10) and Eq. (8.12) allow us to obtain the following alternative representation for the fractional kernel $K^{(0)}(\mathbf{r}_2 - \mathbf{r}_1, \tau)$

$$K^{(0)}(\mathbf{r}_2 - \mathbf{r}_1, \tau) = -\frac{1}{2\pi\hbar} \int_{-\infty}^\infty dE \, e^{-(i/\hbar)E\tau} \frac{\hbar}{2\pi\alpha i E |\mathbf{r}_2 - \mathbf{r}_1|^3} \qquad (8.13)$$

$$\times H_{3,4}^{2,2} \left[\frac{1}{\hbar} \left(-\frac{E}{D_\alpha} \right)^{1/\alpha} |\mathbf{r}_2 - \mathbf{r}_1| \left| \begin{array}{c} (1,1), (1,1/\alpha), (1,1/2) \\ (1,1/\alpha), (1,1), (1,1/2), (2,1) \end{array} \right. \right].$$

If we put $\alpha = 2$ then Eq. (8.12) gives us the standard quantum mechanical fixed-energy amplitude $k_{\mathrm{QM}}^{(0)}(\mathbf{r}_2, \mathbf{r}_1; E) = k^{(0)}(\mathbf{r}_2, \mathbf{r}_1; E)|_{\alpha=2}$. Indeed, by setting $\alpha = 2$ into Eq. (8.12) and taking into account that in accordance with Property 12.2.2 of H-function and Eq. (A.8) we have

$$H_{3,4}^{2,2}\left[\kappa|\mathbf{r}_2 - \mathbf{r}_1|\ \middle|\ \begin{matrix} (1,1),(1,1/2),(1,1/2) \\ (1,1/2),(1,1),(1,1/2),(2,1) \end{matrix}\right]$$

$$= H_{0,1}^{1,0}\left[\kappa|\mathbf{r}_2 - \mathbf{r}_1|\ \middle|\ \begin{matrix} - \\ (2,1) \end{matrix}\right]$$

$$= (\kappa|\mathbf{r}_2 - \mathbf{r}_1|)^2 \exp\left(-\kappa|\mathbf{r}_2 - \mathbf{r}_1|\right),$$

where the notation

$$\kappa = \frac{1}{\hbar}\sqrt{-2mE} \tag{8.14}$$

has been introduced.

Hence, we obtain in the case when $\alpha = 2$

$$k_{\mathrm{QM}}^{(0)}(\mathbf{r}_2, \mathbf{r}_1; E) = k^{(0)}(\mathbf{r}_2, \mathbf{r}_1; E)|_{\alpha=2}$$

$$= \frac{-im}{2\pi\hbar\kappa^2|\mathbf{r}_2 - \mathbf{r}_1|^3} \times (\kappa|\mathbf{r}_2 - \mathbf{r}_1|)^2 \exp\left(-\kappa|\mathbf{r}_2 - \mathbf{r}_1|\right) \tag{8.15}$$

$$= -\frac{im}{2\pi\hbar|\mathbf{r}_2 - \mathbf{r}_1|} \times \exp\left(-\kappa|\mathbf{r}_2 - \mathbf{r}_1|\right),$$

which is Eq. (1.390) from Chapter 1, [108].

8.3 A free particle Green function

In terms of the fractional fixed-energy amplitude $k^{(0)}(x_2, x_1; E)$ introduced by Eq. (8.5), 1D free particle Green function $G^{(0)}(x_2, x_1; E)$ is defined as

$$G^{(0)}(x_2, x_1; E) = \frac{1}{i\hbar}k^{(0)}(x_2, x_1; E). \tag{8.16}$$

We can rewrite this equation to present a free particle Green function in terms of a free particle fractional quantum mechanical kernel $K^{(0)}(x_2 - x_1, \tau)$ given by Eq. (8.7)

$$G^{(0)}(x_2, x_1; E) = \frac{1}{i\hbar} \int\limits_0^\infty d\tau e^{(i/\hbar)E\tau} K^{(0)}(x_2 - x_1, \tau). \qquad (8.17)$$

Substituting $k^{(0)}(x_2, x_1; E)$ given by Eq. (8.8) into Eq. (8.16) yields the expression for a free particle Green function $G^{(0)}(r_2, r_1; E)$ in terms of $H_{2,3}^{2,1}$-function first found in [109]

$$G^{(0)}(x_2, x_1; E) = \frac{1}{\alpha E |x_2 - x_1|} \qquad (8.18)$$

$$\times H_{2,3}^{2,1} \left[\frac{1}{\hbar} \left(-\frac{E}{D_\alpha} \right)^{1/\alpha} |x_2 - x_1| \,\middle|\, \begin{matrix} (1, 1/\alpha), (1, 1/2) \\ (1, 1/\alpha), (1, 1), (1, 1/2) \end{matrix} \right].$$

This free particle Green function $G^{(0)}(r_2, r_1; \tau)$ is related to $G^{(0)}(r_2, r_1; E)$ by the energy-time Fourier transform

$$G^{(0)}(r_2, r_1; \tau) = \frac{1}{2\pi\hbar} \int\limits_{-\infty}^\infty dE \, e^{-(i/\hbar)E\tau} G^{(0)}(r_2, r_1; E). \qquad (8.19)$$

Taking into account the path integral representation (6.5) for a free fractional quantum mechanical kernel we can write

$$G^{(0)}(r_2, r_1; \tau) = \frac{1}{i\hbar} \int\limits_0^\infty d\tau e^{(i/\hbar)E\tau}$$

$$\times \int\limits_{r(t_a)=r_a}^{r(t_b)=r_b} \mathrm{D}x(\tau) \int \mathrm{D}p(\tau) \exp\{iS_\alpha(p, x)/\hbar\} \qquad (8.20)$$

$$= \int\limits_{r(0)=r_a}^{r(\tau)=r_b} \mathrm{D}x(\tau) \int \mathrm{D}p(\tau) \exp\left\{ \frac{i}{\hbar} \int\limits_0^\tau d\tau [p(\tau)\dot{x}(\tau) - H_\alpha(p(\tau), x(\tau), \tau)] \right\},$$

which is the path integral representation for a free particle Green function in 1D case in the framework of fractional quantum mechanics.

In 3D case the Green function for a free particle can be expressed in terms of a fractional fixed-energy amplitude. It follows from Eqs. (8.9) and (8.17) that

$$G^{(0)}(\mathbf{r}_2, \mathbf{r}_1; E) = \frac{1}{i\hbar} k^{(0)}(\mathbf{r}_2, \mathbf{r}_1; E). \qquad (8.21)$$

In terms of the fractional fixed-energy amplitude $k^{(0)}(\mathbf{r}_2, \mathbf{r}_1; E)$ introduced by Eq. (8.9), a free particle Green function $G^{(0)}(\mathbf{r}_2, \mathbf{r}_1; E)$ in the framework of fractional quantum mechanics is defined as

$$G^{(0)}(\mathbf{r}_2, \mathbf{r}_1; E) = \frac{1}{i\hbar} \int\limits_0^\infty d\tau e^{(i/\hbar)E\tau} K^{(0)}(\mathbf{r}_2 - \mathbf{r}_1, \tau), \qquad (8.22)$$

where $K^{(0)}(\mathbf{r}_2 - \mathbf{r}_1, \tau)$ is a free particle fractional quantum mechanical kernel given by Eq. (7.45).

A free particle Green function $G^{(0)}(\mathbf{r}_2, \mathbf{r}_1; \tau)$ is related to $G^{(0)}(\mathbf{r}_2, \mathbf{r}_1; E)$ by the energy-time Fourier transform

$$G^{(0)}(\mathbf{r}_2, \mathbf{r}_1; \tau) = \frac{1}{2\pi\hbar} \int\limits_{-\infty}^\infty dE\, e^{-(i/\hbar)E\tau} G^{(0)}(\mathbf{r}_2, \mathbf{r}_1; E). \qquad (8.23)$$

Taking into account the path integral representation (6.10) for a free particle fractional quantum mechanical kernel we can write

$$G^{(0)}(\mathbf{r}_2, \mathbf{r}_1; E) = \frac{1}{i\hbar} \int\limits_0^\infty d\tau e^{(i/\hbar)E\tau} \qquad (8.24)$$

$$\times \int\limits_{\mathbf{r}(0)=\mathbf{r}_1}^{\mathbf{r}(\tau)=\mathbf{r}_2} D\mathbf{r}(\tau') \int D\mathbf{p}(\tau') \exp\left\{ \frac{i}{\hbar} \int\limits_0^\tau d\tau'[\mathbf{p}(\tau')\dot{\mathbf{r}}(\tau') - D_\alpha|\mathbf{p}(\tau')|^\alpha] \right\},$$

which is the path integral representation for a free particle Green function in the 3D case in the framework of fractional quantum mechanics.

The fractional quantum mechanical kernel $K^{(0)}(\mathbf{r}_2 - \mathbf{r}_1, \tau)$ is expressed in terms of the Green function $G^{(0)}(\mathbf{r}_2, \mathbf{r}_1; E)$ as

$$K^{(0)}(\mathbf{r}_2 - \mathbf{r}_1, \tau) = \frac{i}{2\pi} \int\limits_{-\infty}^\infty dE e^{-(i/\hbar)E\tau} G^{(0)}(\mathbf{r}_2, \mathbf{r}_1; E). \qquad (8.25)$$

Using the 3D generalization of Eq. (8.6) we can define a free particle Green function by the following equation

$$G^{(0)}(\mathbf{r}_2, \mathbf{r}_1; E) = \sum_{n=1}^{\infty} \phi_n(\mathbf{r}_2)\phi_n^*(\mathbf{r}_1)\frac{1}{E - E_n + i\epsilon}, \qquad (8.26)$$

where $\phi_n(\mathbf{r})$ is the eigenfunction of a free particle time independent fractional Schrödinger equation (4.4) with eigenvalue E_n, in the 3D case.

The Green $G(\mathbf{r}_2, \mathbf{r}_1; E)$ function of a particle moving in potential field $V(\mathbf{r})$ is expressed in terms of fractional kernel $K(\mathbf{r}_2 - \mathbf{r}_1, \tau)$

$$G(\mathbf{r}_2, \mathbf{r}_1; E) = \frac{1}{i\hbar} \int_0^{\infty} d\tau e^{(i/\hbar)E\tau} K(\mathbf{r}_2 - \mathbf{r}_1, \tau), \qquad (8.27)$$

here $K(\mathbf{r}_2 - \mathbf{r}_1, \tau)$ is defined by Eq. (6.7). We come to the definition of the Green function in the framework of fractional quantum mechanics in terms of path integral

$$G(\mathbf{r}_2, \mathbf{r}_1; E) = \frac{1}{i\hbar} \int_0^{\infty} d\tau e^{(i/\hbar)E\tau} \int_{\mathbf{r}(t_a)=\mathbf{r}_a}^{\mathbf{r}(t_b)=\mathbf{r}_b} D\mathbf{r}(\tau') \int D\mathbf{p}(\tau') \qquad (8.28)$$

$$\times \exp\left\{ \frac{i}{\hbar} \int_{t_a}^{t_b} d\tau' [\mathbf{p}(\tau')\dot{\mathbf{r}}(\tau') - H_\alpha(\mathbf{p}(\tau'), \mathbf{r}(\tau'), \tau')] \right\},$$

where $\dot{\mathbf{r}}$ denotes the time derivative, $H_\alpha(\mathbf{p}(\tau'), \mathbf{r}(\tau'), \tau')$ is the fractional Hamiltonian given by Eq. (3.3) with the substitutions $\mathbf{p} \to \mathbf{p}(\tau')$, $\mathbf{r} \to \mathbf{r}(\tau')$, and $\{\mathbf{p}(\tau'), \mathbf{r}(\tau')\}$ is a particle trajectory in phase space.

The fractional quantum mechanical kernel $K(\mathbf{r}_2 - \mathbf{r}_1, \tau)$ is expressed in terms of the Green function $G^{(0)}(\mathbf{r}_2, \mathbf{r}_1; E)$ as

$$K(\mathbf{r}_2 - \mathbf{r}_1, \tau) = \frac{i}{2\pi} \int_{-\infty}^{\infty} dE e^{-(i/\hbar)E\tau} G(\mathbf{r}_2, \mathbf{r}_1; E). \qquad (8.29)$$

The Green function for a free particle can be expressed in terms of fractional fixed-energy amplitude. It follows from Eqs. (8.9) and (8.22) that

$$G^{(0)}(\mathbf{r}_2, \mathbf{r}_1; E) = \frac{1}{i\hbar} k^{(0)}(\mathbf{r}_2, \mathbf{r}_1; E). \qquad (8.30)$$

Hence, we can use Eqs. (8.12) and (8.13) to present a free particle Green function in terms of H-function

$$G^{(0)}(\mathbf{r}_2, \mathbf{r}_1; E) = -\frac{1}{2\pi\alpha E|\mathbf{r}_2 - \mathbf{r}_1|^3} \tag{8.31}$$

$$\times H^{2,2}_{3,4} \left[\frac{1}{\hbar} \left(-\frac{E}{D_\alpha} \right)^{1/\alpha} |\mathbf{r}_2 - \mathbf{r}_1| \left| \begin{matrix} (1,1),(1,1/\alpha),(1,1/2) \\ (1,1/\alpha),(1,1),(1,1/2),(2,1) \end{matrix} \right. \right].$$

In the limit case when $\alpha = 2$, Eq. (8.13) gives us the standard quantum mechanical Green function of a free particle, $G^{(0)}_{\mathrm{QM}}(\mathbf{r}_2, \mathbf{r}_1; E) = G^{(0)}(\mathbf{r}_2, \mathbf{r}_1; E)|_{\alpha=2}$. Indeed, having Eq. (8.15) we write

$$G^{(0)}_{\mathrm{QM}}(\mathbf{r}_2, \mathbf{r}_1; E) = -\frac{m}{2\pi\hbar^2|\mathbf{r}_2 - \mathbf{r}_1|} \times \exp\left(-\kappa|\mathbf{r}_2 - \mathbf{r}_1| \right), \tag{8.32}$$

where κ has been introduced by Eq. (8.14).

Substituting Eq. (8.32) into Eq. (8.23) and integrating over dE yield

$$G^{(0)}_{\mathrm{QM}}(\mathbf{r}_2, \mathbf{r}_1; \tau) = \frac{1}{i\hbar} \left(\frac{m}{2\pi i\hbar\tau} \right)^{3/2} \exp\left\{ \frac{im|\mathbf{r}_b - \mathbf{r}_a|^2}{2\hbar\tau} \right\} \tag{8.33}$$

or

$$G^{(0)}_{\mathrm{QM}}(\mathbf{r}_2, \mathbf{r}_1; \tau) = \frac{1}{i\hbar} K^{(0)}_F(\mathbf{r}_b, \mathbf{r}_a, \tau), \qquad \tau \geq 0, \tag{8.34}$$

where $K^{(0)}_F(\mathbf{r}_b, \mathbf{r}_a, \tau)$ is Feynman's free particle kernel defined by Eq. (7.56) with substitution $t_b - t_a = \tau$.

8.4 Fractional kernel in momentum representation

To obtain the 1D fractional quantum-mechanical kernel in momentum representation we start from Eq. (6.51). Taking into account the definitions of the Fourier transforms given by Eqs. (3.15) and (3.16) let us present Eq. (6.51) in momentum representation,

$$\varphi(p_b, t_b) = \frac{1}{2\pi\hbar} \int_{-\infty}^{\infty} dp_a \mathcal{K}(p_b t_b | p_a t_a) \varphi(p_a, t_a), \qquad t_b \geq t_a, \tag{8.35}$$

where $\varphi(p_a, t_a)$ is the wave function of a particle with momentum p_a at the time $t = t_a$, and $\varphi(p_b, t_b)$ is the wave function of the particle with momentum p_b at time t_b. This equation describes the evolution of a wave function in momentum space. The kernel in momentum representation is expressed in terms of the kernel in coordinate representation by the following equation

$$\mathcal{K}(p_b t_b | p_a t_a) = \int\limits_{-\infty}^{\infty} dx_b \int\limits_{-\infty}^{\infty} dx_a e^{-\frac{i}{\hbar} p_b x_b + \frac{i}{\hbar} p_a x_a} K(x_b t_b | x_a t_a), \qquad (8.36)$$

where $t_b \geq t_a$.

As an example, let us obtain a free particle kernel $\mathcal{K}^{(0)}(p_b t_b | p_a t_a)$ in momentum representation. By substituting into Eq. (8.36) a free particle kernel in coordinate representation $K^{(0)}(x_b t_b | x_a t_a)$ given by Eq. (7.1), we find

$$\mathcal{K}^{(0)}(p_b t_b | p_a t_a) = (2\pi\hbar)\delta(p_a - p_b)\exp\{-\frac{i}{\hbar}D_\alpha |p_a|^\alpha (t_b - t_a)\}, \qquad (8.37)$$

when $t_b > t_a$ and $\mathcal{K}^{(0)}(p_b t_b | p_a t_a) = 0$, when $t_b < t_a$.

The presence of the delta function $\delta(p_a - p_b)$ shows that the momentum of a free quantum particle does not change, that is the kernel $\mathcal{K}^{(0)}(p_b t_b | p_a t_a)$ supports the moment conservation law.

It is easy to see that $\mathcal{K}^{(0)}(p_b t_b | p_a t_a)$ satisfies

$$\mathcal{K}^{(0)}(p_b t_b | p_a t_a) = \int\limits_{-\infty}^{\infty} dp' \mathcal{K}^{(0)}(p_b t_b | p't) \mathcal{K}^{(0)}(p't | p_a t_a), \qquad t_b > t > t_a.$$

This is a special case of the general quantum-mechanical rule in momentum representation

$$\mathcal{K}(p_b t_b | p_a t_a) = \int\limits_{-\infty}^{\infty} dp' \mathcal{K}(p_b t_b | p't) \mathcal{K}(p't | p_a t_a) \qquad t_b > t > t_a. \qquad (8.38)$$

The above consideration can be generalized to the 3D case. Substituting the wave function $\psi(\mathbf{r}, t)$ with the wave function $\varphi(\mathbf{p}, t)$ into Eq. (6.22) (see, Eq. (6.51)) $\varphi(\mathbf{p}, t)$ yields

$$\varphi(\mathbf{p}_b, t_b) = \frac{1}{(2\pi\hbar)^3} \int d^3 p_a \mathcal{K}(\mathbf{p}_b t_b | \mathbf{p}_a t_a)\varphi(\mathbf{p}_a, t_a), \qquad (8.39)$$

where the kernel in the momentum representation $\mathcal{K}(\mathbf{p}_b t_b|\mathbf{p}_a t_a)$ is defined in terms of the kernel in coordinate representation $K(\mathbf{r}_b t_b|\mathbf{r}_a t_a)$ as follows

$$\mathcal{K}(\mathbf{p}_b t_b|\mathbf{p}_a t_a) = \int d^3 r_a d^3 r_b e^{-\frac{i}{\hbar}\mathbf{p}_b \mathbf{r}_b + \frac{i}{\hbar}\mathbf{p}_a \mathbf{r}_a} K(\mathbf{r}_b t_b|\mathbf{r}_a t_a). \tag{8.40}$$

For example, for a free particle in the momentum representation $\mathcal{K}^{(0)}(\mathbf{p}_b t_b|\mathbf{p}_a t_a)$ we have

$$\mathcal{K}^{(0)}(\mathbf{p}_b t_b|\mathbf{p}_a t_a) = \int d^3 r_a d^3 r_b e^{-\frac{i}{\hbar}\mathbf{p}_b \mathbf{r}_b + \frac{i}{\hbar}\mathbf{p}_a \mathbf{r}_a} K^{(0)}(\mathbf{r}_b t_b|\mathbf{r}_a t_a) \tag{8.41}$$

$$= (2\pi\hbar)^3 \delta(\mathbf{p}_a - \mathbf{p}_b) \exp\{-\frac{i}{\hbar}D_\alpha |\mathbf{p}_a|^\alpha (t_b - t_a)\}, \quad \text{for} \quad t_b > t_a,$$

and $\mathcal{K}^{(0)}(\mathbf{p}_b t_b|\mathbf{p}_a t_a) = 0$ for $t_b < t_a$.

The presence of the delta function $\delta(\mathbf{p}_a - \mathbf{p}_b)$ shows that the momentum of a free quantum particle does not change. It follows from Eq. (8.41) that the consistency condition holds

$$\mathcal{K}^{(0)}(\mathbf{p}_b t_b|\mathbf{p}_a t_a) = \int d^3 p' \mathcal{K}^{(0)}(\mathbf{p}_b t_b|\mathbf{p}' t)\mathcal{K}^{(0)}(\mathbf{p}' t|\mathbf{p}_a t_a), \quad t_b > t > t_a.$$

This is a special case of the general quantum-mechanical rule: amplitudes for events occurring in succession in time multiply

$$\mathcal{K}(\mathbf{p}_b t_b|\mathbf{p}_a t_a) = \int d^3 p' \mathcal{K}(\mathbf{p}_b t_b|\mathbf{p}' t)\mathcal{K}(\mathbf{p}' t|\mathbf{p}_a t_a), \quad t_b > t > t_a. \tag{8.42}$$

From Eq. (8.41) we see that a free particle kernel in momentum space is expressed in terms of an exponential function, while in coordinate representation (7.50) it has a more complicated representation in terms of Fox's H-function.

Chapter 9

Fractional Oscillator

9.1 Quarkonium and quantum fractional oscillator

As a physical application of fractional quantum mechanics we consider a new approach to study the quark–antiquark bound states $q\bar{q}$ treated within the non-relativistic potential picture [67]. Note that the non-relativistic approach can only be justified for heavy quark systems (for example, charmonium $c\bar{c}$ or bottonium $b\bar{b}$). The term quarkonium is used to denote any $q\bar{q}$ bound state system [110] in analogy to positronium in the e^+e^- system. The non-relativistic potential approach remains the most successful and simplest way to calculate and predict energy levels and decay rates of quarkonium.

From the stand point of the "potential" view, we can assume that the confining potential energy of two quarks localized at the space points $\mathbf{r}_i$ and $\mathbf{r}_j$, is

$$V(|\mathbf{r}_i - \mathbf{r}_j|) = q_i q_j |\mathbf{r}_i - \mathbf{r}_j|^\beta, \tag{9.1}$$

where q_i and q_j are the color charges of i and j quarks respectively, and the index $\beta > 0$. Equation (9.1) coincides with the QCD requirements: (i) At short distances the quarks and gluons appear to be weakly coupled; (ii) At large distances the effective coupling becomes strong, resulting in the phenomena of quark confinement[1].

Considering a multiparticle system of N quarks yields the following equation for the potential energy $U(\mathbf{r}_1, ..., \mathbf{r}_N)$ of the system

$$U(\mathbf{r}_1, ..., \mathbf{r}_N) = \sum_{1 \leq i < j \leq N} q_i q_j |\mathbf{r}_i - \mathbf{r}_j|^\beta. \tag{9.2}$$

---

[1]The term quark confinement describes the observation that quarks do not occur isolated in nature, but only in hadronic bound states as the colorless objects such as baryons and mesons.

In order to illustrate the main idea, we consider the simplest case, when color q_i charge can only be in two states q or $-q$, and the colorless condition $\sum_{i=1}^{N} q_i = 0$ takes place. Using the general statistical mechanics approach (see Definition 3.2.1 and Proposition 3.2.2 in [111]) we can conclude that the many-particle system with the potential energy (9.2) will be thermodynamically stable only for $0 < \beta \leq 2$.

In order to study the problem of quarkonium it seems reasonable to consider the non-relativistic fractional quantum mechanical model with the fractional Hamiltonian operator $H_{\alpha,\beta}$ defined as

$$H_{\alpha,\beta} = D_\alpha(-\hbar^2\Delta)^{\alpha/2} + q^2|\mathbf{r}|^\beta, \quad 1 < \alpha \leq 2, \quad 1 < \beta \leq 2, \qquad (9.3)$$

where $\mathbf{r}$ is the 3D vector, $\Delta = \partial^2/\partial\mathbf{r}^2$ is the Laplacian, and the operator $(-\hbar^2\Delta)^{\alpha/2}$ is defined by Eq. (3.9).

Thus, we come to the fractional Schrödinger equation for the wave function $\psi(\mathbf{r}, t)$ of quantum fractional oscillator,

$$i\hbar\frac{\partial\psi(\mathbf{r}, t)}{\partial t} = D_\alpha(-\hbar^2\Delta)^{\alpha/2}\psi(\mathbf{r}, t) + q^2|\mathbf{r}|^\beta\psi(\mathbf{r}, t). \qquad (9.4)$$

Searching for the solution in form

$$\psi(\mathbf{r}, t) = e^{-iEt/\hbar}\phi(\mathbf{r}), \qquad (9.5)$$

with E being energy of oscillator, we come to the time-independent fractional Schrödinger equation of the form,

$$D_\alpha(-\hbar^2\Delta)^{\alpha/2}\phi(\mathbf{r}) + q^2|\mathbf{r}|^\beta\phi(\mathbf{r}) = E\phi(\mathbf{r}). \qquad (9.6)$$

For the special case, when $\alpha = \beta$, the Hamiltonian operator (9.3) has the form

$$H_\alpha = D_\alpha(-\hbar^2\Delta)^{\alpha/2} + q^2|\mathbf{r}|^\alpha, \quad 1 < \alpha \leq 2, \qquad (9.7)$$

and Eq. (9.4) becomes

$$i\hbar\frac{\partial\psi(\mathbf{r}, t)}{\partial t} = D_\alpha(-\hbar^2\Delta)^{\alpha/2}\psi(\mathbf{r}, t) + q^2|\mathbf{r}|^\alpha\psi(\mathbf{r}, t). \qquad (9.8)$$

It is easy to see that Hamiltonian H_α (9.7) is the fractional generalization of the 3D harmonic oscillator Hamiltonian of standard quantum mechanics. Following [67] and [96], we will call this model *quantum fractional oscillator*.

In the 1D case the fractional Schrödinger equation for the wave function $\psi(r,t)$ of quantum fractional oscillator has the form

$$i\hbar\frac{\partial\psi(r,t)}{\partial t} = -D_\alpha(\hbar\nabla)^\alpha\psi(r,t) + q^2|r|^\beta\psi(r,t), \qquad (9.9)$$

where $1 < \alpha \le 2$, $1 < \beta \le 2$ and $\nabla = \partial/\partial x$.

In 1D case the fractional Hamiltonian operator $H_{\alpha,\beta}$ is defined as

$$H_{\alpha,\beta} = -D_\alpha(\hbar\nabla)^\alpha\psi(r,t) + q^2|r|^\beta, \qquad (9.10)$$

The time independent fractional Schrödinger equation for 1D quantum fractional oscillator is

$$-D_\alpha(\hbar\nabla)^\alpha\phi(r) + q^2|r|^\beta\phi(r) = E\phi(r), \qquad (9.11)$$

$$1 < \alpha \le 2, \quad 1 < \beta \le 2,$$

where the wave function $\psi(r,t)$ is related to the time independent wave function $\phi(r)$ by

$$\psi(r,t) = e^{-iEt/\hbar}\phi(r), \qquad (9.12)$$

here E is the energy of quantum fractional oscillator.

9.1.1 *Fractional oscillator in momentum representation*

Using the definitions of the quantum Riesz fractional derivative (3.9) and the Fourier transforms of the wave functions (3.10) one can rewrite the fractional Schrödinger equation (9.4) in momentum representation

$$i\hbar\frac{\partial\varphi(\mathbf{p},t)}{\partial t} = q^2(-\hbar^2\Delta_\mathbf{p})^{\beta/2}\varphi(\mathbf{p},t) + D_\alpha|\mathbf{p}|^\alpha\varphi(\mathbf{p},t), \qquad (9.13)$$

$$1 < \alpha \le 2, \quad 1 < \beta \le 2,$$

where $\varphi(\mathbf{p},t)$ is the wave function in momentum representation

$$\varphi(\mathbf{p},t) = \int d^3r e^{-i\frac{\mathbf{pr}}{\hbar}}\psi(\mathbf{r},t) \qquad (9.14)$$

and $(-\hbar^2\Delta_\mathbf{p})$ is the quantum Riesz fractional derivative in momentum representation introduced by

$$(-\hbar^2\Delta_\mathbf{p})^{\beta/2}\varphi(\mathbf{p},t) = \int d^3r e^{-i\frac{\mathbf{pr}}{\hbar}}|\mathbf{r}|^\beta\psi(\mathbf{r},t), \qquad (9.15)$$

with $\Delta_\mathbf{p} = \partial^2/\partial\mathbf{p}^2$ being the Laplacian in momentum representation. Searching for solution to Eq. (9.13) in the form

$$\varphi(\mathbf{p},t) = e^{-iEt/\hbar}\chi(\mathbf{p}),$$

where E is the energy of quantum fractional oscillator, we come to the time-independent fractional Schrödinger equation in momentum representation,

$$q^2(-\hbar^2\Delta_\mathbf{p})^{\beta/2}\chi(\mathbf{p}) + D_\alpha|\mathbf{p}|^\alpha\chi(\mathbf{p}) = E\chi(\mathbf{p}). \qquad (9.16)$$

For the special case, when $\alpha = \beta$, Eq. (9.13) becomes

$$i\hbar\frac{\partial\varphi(\mathbf{p},t)}{\partial t} = q^2(-\hbar^2\Delta_\mathbf{p})^{\alpha/2}\varphi(\mathbf{p},t) + D_\alpha|\mathbf{p}|^\alpha\varphi(\mathbf{p},t), \qquad (9.17)$$

$$1 < \alpha \leq 2.$$

It is easy to see that this equation is a fractional generalization of the standard quantum mechanical 3D Schrödinger equation for the oscillator in momentum representation.

9.2 Spectrum of 1D quantum fractional oscillator in semi-classical approximation

9.2.1 *Coordinate representation*

In the 1D case time independent fractional Schrödinger equation for quantum fractional oscillator is given by expression

$$-D_\alpha(\hbar\nabla)^\alpha\phi(x) + q^2|x|^\beta\phi(x) = E\phi(x), \qquad (9.18)$$

$$1 < \alpha \leq 2, \quad 1 < \beta \leq 2,$$

where $\nabla = \partial/\partial x$.

The energy levels of the 1D quantum fractional oscillator with the Hamiltonian $H_\alpha = D_\alpha |p|^\alpha + q^2 |x|^\beta$ can be found in semiclassical approximation [96]. We set the total energy equal to E, so that

$$E = D_\alpha |p|^\alpha + q^2 |x|^\beta, \qquad (9.19)$$

whence $|p| = \left(\frac{1}{D_\alpha}(E - q^2|x|^\beta)\right)^{1/\alpha}$. Thus, at the turning points where $p = 0$, classical motion is possible in the range $|x| \leq (E/q^2)^{1/\beta}$.

A routine use of the Bohr–Sommerfeld quantization rule [94] yields

$$2\pi\hbar(n + \frac{1}{2}) = \oint p\,dx = 4\int\limits_0^{x_m} p\,dx = 4\int\limits_0^{x_m} D_\alpha^{-1/\alpha}(E - q^2|x|^\beta)^{1\alpha}dx, \quad (9.20)$$

where the notation $\oint$ means the integral over one complete period of the classical motion, and $x_m = (E/q^2)^{1/\beta}$ is the turning point of classical motion. To evaluate the integral on the right-hand of Eq. (9.20) we introduce a new variable $y = x(E/q^2)^{-1/\beta}$. Then we have

$$\int\limits_0^{x_m} D_\alpha^{-1/\alpha}(E - q^2|x|^\beta)^{1\alpha}dx = \frac{1}{D_\alpha^{1/\alpha}q^{2/\beta}}E^{\frac{1}{\alpha}+\frac{1}{\beta}}\int\limits_0^1 dy(1 - y^\beta)^{1/\alpha}.$$

The integral over dy can be expressed in terms of the B-function. Indeed, substitution $z = y^\beta$ yields[2]

$$\int\limits_0^1 dy(1 - y^\beta)^{1/\alpha} = \frac{1}{\beta}\int\limits_0^1 dz\, z^{\frac{1}{\beta}-1}(1 - z)^{\frac{1}{\alpha}} = \frac{1}{\beta}B(\frac{1}{\beta}, \frac{1}{\alpha} + 1). \qquad (9.22)$$

With the help of Eq. (9.22) we rewrite Eq. (9.20) as

$$\pi\hbar\left(n + \frac{1}{2}\right) = \frac{2}{D_\alpha^{1/\alpha}q^{2/\beta}}E^{\frac{1}{\alpha}+\frac{1}{\beta}}\frac{1}{\beta}B\left(\frac{1}{\beta}, \frac{1}{\alpha} + 1\right).$$

---

[2]The B-function is defined by

$$B(u, v) = \int\limits_0^1 dy\, y^{u-1}(1 - y)^{v-1}. \qquad (9.21)$$

The above equation gives the energy levels of stationary states for the 1D quantum fractional oscillator [96],

$$E_n = \left(\frac{\pi \hbar \beta D_\alpha^{1/\alpha} q^{2/\beta}}{2B(\frac{1}{\beta}, \frac{1}{\alpha} + 1)} \right)^{\frac{\alpha\beta}{\alpha+\beta}} \left(n + \frac{1}{2} \right)^{\frac{\alpha\beta}{\alpha+\beta}}, \qquad (9.23)$$

$$1 < \alpha \leq 2, \quad 1 < \beta \leq 2.$$

This equation is a generalization of the well-known equation for the energy levels of the standard quantum harmonic oscillator (see, for example, [94]), and is transformed into it at $\alpha = 2$, $\beta = 2$. Indeed, when $\alpha = 2$ and $\beta = 2$, we have $D_2 = 1/2m$, $q^2 = m\omega^2/2$, where m is the oscillator mass and its frequency is ω. In this case equation (9.23) results[3]

$$E_n = \hbar\omega \left(n + \frac{1}{2} \right). \qquad (9.24)$$

We note that at the condition

$$\frac{1}{\alpha} + \frac{1}{\beta} = 1, \qquad (9.25)$$

Eq. (9.23) gives equidistant energy levels. When $1 < \alpha \leq 2$ and $1 < \beta \leq 2$ the condition given by Eq. (9.25) takes place for $\alpha = 2$ and $\beta = 2$ only. It means that only the standard quantum harmonic oscillator has an equidistant energy spectrum.

9.2.2 *Momentum representation*

In the 1D case the fractional Schrödinger equation (9.13) reads

$$i\hbar \frac{\partial \varphi(p,t)}{\partial t} = q^2(-\hbar^2 \Delta_p)^{\beta/2} \varphi(p,t) + D_\alpha |p|^\alpha \varphi(p,t), \qquad (9.26)$$

where $\Delta_p = \partial^2/\partial p^2$ and $1 < \alpha \leq 2$, $1 < \beta \leq 2$.

---

[3]The value for the B-function $B(1/2, 3/2) = \pi/2$ has been used.

We set the total energy equal to E, so that

$$E = q^2|x|^\beta + D_\alpha|p|^\alpha, \tag{9.27}$$

where the dependence $|x|$ on $|p|$ is given by

$$|x| = \left(\frac{1}{q^2}(E - D_\alpha|p|^\alpha)\right)^{1/\beta}. \tag{9.28}$$

Thus, at the turning points $x = 0$, classical motion is possible in the range $|p_m| \leq (E/D_\alpha)^{1/\alpha}$. The Bohr–Sommerfeld quantization rule reads

$$2\pi\hbar(n + \frac{1}{2}) = \oint xdp = 4\int_0^{p_m} xdp = 4\int_0^{p_m} \frac{1}{q^{2/\beta}}(E - q^2|p|^\alpha)^{1\beta}dp. \tag{9.29}$$

The energy levels of stationary states for the 1D quantum fractional oscillator in momentum representation can be obtained from Eq. (9.23) by applying the transformations (9.20), that is

$$D_\alpha \leftrightarrow q^2, \qquad \alpha \leftrightarrow \beta. \tag{9.30}$$

It gives us

$$E_n = \left(\frac{\pi\hbar\alpha D_\alpha^{1/\alpha}q^{2/\beta}}{2B(\frac{1}{\alpha},\frac{1}{\beta}+1)}\right)^{\frac{\alpha\beta}{\alpha+\beta}}(n + \frac{1}{2})^{\frac{\alpha\beta}{\alpha+\beta}}, \tag{9.31}$$

$$1 < \alpha \leq 2, \quad 1 < \beta \leq 2.$$

We note that at the condition

$$\frac{1}{\alpha} + \frac{1}{\beta} = 1,$$

Eq. (9.31) gives the equidistant energy levels.

9.3 Dimensionless fractional Schrödinger equation for fractional oscillator

Aiming to present the 1D fractional Schrödinger equation (9.18) in dimensionless form, we introduce dimensionless coordinate ξ and scale parameter λ

$$x = \lambda \xi, \tag{9.32}$$

where units of λ is cm.

The scale transformation (9.32) initiates the scaling of the Riesz fractional derivative

$$(-\Delta_x)^{\alpha/2} = \lambda^{-\alpha}(-\Delta_\xi)^{\alpha/2}, \qquad 1 < \alpha \leq 2, \tag{9.33}$$

where $\Delta_x = \partial^2/\partial x^2$ and $\Delta_\xi = \partial^2/\partial \xi^2$.

Substituting Eqs. (9.32) and (9.33) into Eq. (9.18) yields

$$D_\alpha \hbar^\alpha \lambda^{-\alpha}(-\Delta_\xi)^{\alpha/2}\overline{\phi}(\xi) + q^2 \lambda^\beta |\xi|^\beta \overline{\phi}(\xi) = E\overline{\phi}(\xi), \tag{9.34}$$

here, wave function $\overline{\phi}(\xi)$ is related to wave function $\phi(r)$ from Eq. (9.18) by

$$\overline{\phi}(\xi) = \lambda^{1/2}\phi(r). \tag{9.35}$$

Assuming that $\phi(x)$ is normalized as

$$\int\limits_{-\infty}^{\infty} dx |\phi(x)|^2 = 1,$$

we conclude, that $\overline{\phi}(\xi)$ is normalized as

$$\int\limits_{-\infty}^{\infty} d\xi |\overline{\phi}(\xi)|^2 = 1. \tag{9.36}$$

Choosing λ as

$$\lambda = \left(\frac{D_\alpha \hbar^\alpha}{q^2}\right)^{1/(\alpha+\beta)}, \tag{9.37}$$

and introducing dimensionless energy ϵ defined by

$$\epsilon = \frac{E}{(D_\alpha^\beta \hbar^{\alpha\beta} q^{2\alpha})^{1/(\alpha+\beta)}}, \tag{9.38}$$

yield

$$(-\Delta_\xi)^{\alpha/2} \overline{\phi}(\xi) + |\xi|^\beta \overline{\phi}(\xi) = \epsilon \overline{\phi}(\xi). \tag{9.39}$$

To remind us that besides the dependency on ξ, the wave function $\overline{\phi}(\xi)$ also depends on α and β, we rename

$$\overline{\phi}(\xi) \to \overline{\phi}_{\alpha,\beta}(\xi). \tag{9.40}$$

Renaming allows us to present Eq. (9.39) in the form

$$(-\Delta_\xi)^{\alpha/2} \overline{\phi}_{\alpha,\beta}(\xi) + |\xi|^\beta \overline{\phi}_{\alpha,\beta}(\xi) = \epsilon \overline{\phi}_{\alpha,\beta}(\xi), \tag{9.41}$$

$$1 < \alpha \leq 2, \quad 1 < \beta \leq 2.$$

This is a dimensionless fractional Schrödinger equation for the fractional oscillator. From the stand point of fractional calculus, Eq. (9.41) can be considered as a fractional eigenvalue problem with involvement of the fractional Riesz derivative.

9.4 Symmetries of quantum fractional oscillator

It is easy to see that there is a set of transformations to go from Eq. (9.4) to Eq. (9.13) and vice versa

$$\mathbf{r} \leftrightarrow \mathbf{p}, \quad \psi(\mathbf{r}, t) \leftrightarrow \varphi(\mathbf{p}, t), \quad \Delta \leftrightarrow \Delta_{\mathbf{p}}, \tag{9.42}$$

$$D_\alpha \to q^2, \quad q^2 \to D_\beta, \quad \alpha \leftrightarrow \beta.$$

By applying this symmetry we can go from coordinate representation to momentum representation in all formulas for the fractional oscillator.

The symmetry

$$D_\alpha \to q^2, \quad q^2 \to D_\beta, \quad \alpha \leftrightarrow \beta \tag{9.43}$$

leaves ϵ defined by Eq. (9.38) invariant, while changing the wave function

$$\overline{\phi}_{\alpha,\beta}(\xi) \to \overline{\chi}_{\alpha,\beta}(\xi).$$

In other words, the symmetry (9.43) transforms the wave function $\overline{\phi}_{\alpha,\beta}(\xi)$ introduced by Eq. (9.40) into the wave function $\overline{\chi}_{\alpha,\beta}(\xi)$. The transformed wave function $\overline{\chi}_{\alpha,\beta}(\xi)$ satisfies the dimensionless fractional Schrödinger equation

$$(-\Delta_\xi)^{\beta/2}\overline{\chi}_{\alpha,\beta}(\xi) + |\xi|^\alpha\overline{\chi}_{\alpha,\beta}(\xi) = \epsilon\overline{\chi}_{\alpha,\beta}(\xi), \qquad (9.44)$$

$$1 < \alpha \le 2, \quad 1 < \beta \le 2,$$

and the normalization condition (9.36).

By comparing Eq. (9.41) and Eq. (9.44) we conclude that the solution to eigenvalue problem (9.44) is

$$\overline{\chi}_{\alpha,\beta}(\xi) = \overline{\phi}_{\alpha,\beta}(\xi; \epsilon)|_{\alpha\leftrightarrow\beta} = \overline{\phi}_{\beta,\alpha}(\xi; \epsilon). \qquad (9.45)$$

Let us note that applying the symmetry (9.43) to Eq. (9.44) brings us back to Eq. (9.41).

Chapter 10

Some Analytically Solvable Models of Fractional Quantum Mechanics

Thus, the task is, not so much to see what no one has yet seen; but to think what nobody has yet thought, about that which everybody sees.

Erwin Schrödinger

10.1 A free particle

10.1.1 *Scaling properties of the 1D fractional Schrödinger equation*

To make general conclusions regarding solutions of the 1D fractional Schrödinger equation for a free particle, let's study the scaling for the wave function $\psi(x, t; D_\alpha)$, where we keep D_α to remind that besides dependency on x and t the wave function depends on D_α as well. Taking into account Eq. (7.10) we write

$$t = \lambda t', \qquad x = \lambda^\beta x', \qquad D_\alpha = \lambda^\gamma D'_\alpha, \tag{10.1}$$

$$\psi(x, t; D_\alpha) = \lambda^\delta \psi(x', t'; D'_\alpha),$$

where β, γ, δ are exponents of the scale transformations which should leave a free particle 1D fractional Schrödinger equation invariant

$$i\hbar \frac{\partial \psi(x, t; D_\alpha)}{\partial t} = -D_\alpha (\hbar \nabla)^\alpha \psi(x, t; D_\alpha), \tag{10.2}$$

and save the normalization condition $\int\limits_{-\infty}^{\infty} dx |\psi(x, t; D_\alpha)|^2 = 1$. It results in the relationships between scaling exponents,

$$\alpha\beta - \gamma - 1 = 0, \qquad \delta + \beta/2 = 0, \tag{10.3}$$

141

and reduces the number of exponents up to 2. Therefore, we have the two-parameter scale transformation group

$$t = \lambda t', \qquad x = \lambda^\beta x', \qquad D_\alpha = \lambda^{\alpha\beta - 1} D'_\alpha, \tag{10.4}$$

$$\psi(\lambda^\beta x, \lambda t; \lambda^{\alpha\beta - 1} D_\alpha) = \lambda^{-\beta/2} \psi(x, t; D_\alpha), \tag{10.5}$$

where β and λ are arbitrary group parameters.

When the initial condition $\psi(x, t = 0)$ is invariant under the scaling group introduced by Eqs. (10.4) and (10.5) then the solution to Eq. (10.2) remains the group invariant. As an example of the invariant initial condition one may keep in mind $\psi(x, t = 0) = \delta(x)$, which gives us the Green function of the 1D fractional Schrödinger equation.

To get the general scale invariant solutions of a free particle 1D fractional Schrödinger equation we may use the renormalization group framework. As far as the scale invariant solutions to Eq. (10.2) should satisfy the identity Eq. (10.5) for any arbitrary parameters β and λ, the solutions can depend on a combination of x and t to provide the independence on β and λ. It allows us to conclude that the scaling holds

$$\psi(x, t) = \frac{1}{x} \Phi(\frac{x}{\hbar} (\hbar/D_\alpha t)^{1/\alpha}) = \frac{(\hbar/D_\alpha t)^{\frac{1}{\alpha}}}{\hbar} \Psi(\frac{x}{\hbar} (\hbar/D_\alpha t)^{1/\alpha}), \tag{10.6}$$

where two arbitrary functions Φ and Ψ are determined by the initial conditions, $\Phi(.) = \psi(1, .)$ and $\Psi(.) = \psi(., 1)$.

Thus, Eq. (10.6) gives us general scale invariant form of a free particle wave function in the framework of fractional quantum mechanics.

10.1.2   *Exact 1D solution*

Following [95], [109] let's solve the 1D fractional Schrödinger equation for a free particle (10.2) with initial condition $\psi_0(x)$

$$\psi(x, t = 0) = \psi_0(x). \tag{10.7}$$

Applying the Fourier transforms defined by Eqs. (3.15) and using the quantum Riesz fractional derivative given by Eq. (3.14) yield for the wave function $\varphi(p, t)$ in the momentum representation,

$$i\hbar \frac{\partial \varphi(p, t)}{\partial t} = D_\alpha |p|^\alpha \varphi(p, t), \tag{10.8}$$

with the initial condition $\varphi_0(p)$ given by

$$\varphi_0(p) = \varphi(p, t = 0) = \int\limits_{-\infty}^{\infty} dx e^{-ipx/\hbar} \psi_0(x). \tag{10.9}$$

Equation (10.8) is the 1D fractional Schrödinger equation for a free particle in momentum representation.

The solution of the problem introduced by Eqs. (10.8) and (10.9) is

$$\varphi(p, t) = \exp\{-i\frac{D_\alpha |p|^\alpha t}{\hbar}\}\varphi_0(p). \tag{10.10}$$

Therefore, the solution to the 1D fractional Schrödinger equation Eq. (10.2) with initial condition given by Eq. (10.7) can be presented as

$$\psi(x, t) = \frac{1}{2\pi\hbar} \int\limits_{-\infty}^{\infty} dx' \int\limits_{-\infty}^{\infty} dp \exp\{i\frac{p(x - x')}{\hbar} - i\frac{D_\alpha |p|^\alpha t}{\hbar}\}\psi_0(x'). \tag{10.11}$$

To perform the integral over dp let us first note that this equation can be rewritten as

$$\psi(x, t) = \frac{1}{\pi\hbar} \int\limits_{-\infty}^{\infty} dx' \int\limits_{0}^{\infty} dp \cos(\frac{p(x - x')}{\hbar}) \exp\{-i\frac{D_\alpha |p|^\alpha t}{\hbar}\}\psi_0(x'). \tag{10.12}$$

Second, $\exp(-iD_\alpha |p|^\alpha t/\hbar)$ is expressed in terms of the Fox H-function in the following way (see Eq. (A.21))

$$\exp\{-i\frac{D_\alpha |p|^\alpha t}{\hbar}\} = H_{0,1}^{1,0}\left[\frac{iD_\alpha t}{\hbar}|p|^\alpha \left|\begin{array}{c} - \\ (0, 1) \end{array}\right.\right]. \tag{10.13}$$

Further, using the cosine transform of H-function defined by Eq. (A.20), we obtain

$$\psi(x, t) = \int\limits_{-\infty}^{\infty} dx' \frac{1}{|x - x'|} \tag{10.14}$$

$$\times H_{2,2}^{1,1}\left[\frac{1}{\hbar\alpha}\left(\frac{\hbar}{iD_\alpha t}\right)|x - x'|^\alpha \left|\begin{array}{c} (1, 1), (1, \alpha/2) \\ (1, \alpha), (1, \alpha/2) \end{array}\right.\right]\psi_0(x').$$

Alternatively, because of the H-function's Property 12.2.4 (see, Eq. (A.13) in Appendix A), $\psi(x,t)$ can be presented as

$$\psi(x,t) = \int_{-\infty}^{\infty} dx' \frac{1}{\alpha|x-x'|}$$ (10.15)

$$\times H_{2,2}^{1,1}\left[\frac{1}{\hbar}\left(\frac{\hbar}{iD_\alpha t}\right)^{1/\alpha}|x-x'|\,\middle|\,\begin{matrix}(1,1/\alpha),(1,1/2)\\(1,1),(1,1/2)\end{matrix}\right]\psi_0(x').$$

Hence, the integral over dp in Eq. (10.11) is expressed in terms of the $H_{2,2}^{1,1}$-function. If we choose the initial condition $\psi_0(x) = \delta(x)$, then Eq. (10.15) gives us the quantum mechanical kernel $K^{(0)}(x,t|0,0)$ for the 1D free particle

$$K^{(0)}(x,t|0,0) = \frac{1}{\alpha|x|}$$ (10.16)

$$\times H_{2,2}^{1,1}\left[\frac{1}{\hbar}\left(\frac{\hbar}{iD_\alpha t}\right)^{1/\alpha}|x|\,\middle|\,\begin{matrix}(1,1/\alpha),(1,1/2)\\(1,1),(1,1/2)\end{matrix}\right].$$

Applying H-function's Property 12.2.5 given by Eq. (A.14), see Appendix A, we can rewrite $K^{(0)}(x,t|0,0)$ in the following form

$$K^{(0)}(x,t|0,0) = \frac{1}{\alpha\hbar}\left(\frac{\hbar}{iD_\alpha t}\right)^{1/\alpha}$$ (10.17)

$$\times H_{2,2}^{1,1}\left[\frac{1}{\hbar}\left(\frac{\hbar}{iD_\alpha t}\right)^{1/\alpha}|x|\,\middle|\,\begin{matrix}(1-1/\alpha,1/\alpha),(1/2,1/2)\\(0,1),(1/2,1/2)\end{matrix}\right].$$

Therefore, it allows us to rewrite Eq. (10.15) as

$$\psi(x,t) = \frac{1}{\alpha\hbar}\left(\frac{\hbar}{iD_\alpha t}\right)^{1/\alpha}$$ (10.18)

$$\times \int_{-\infty}^{\infty} dx' H_{2,2}^{1,1}\left[\frac{1}{\hbar}\left(\frac{\hbar}{iD_\alpha t}\right)^{1/\alpha}|x-x'|\,\middle|\,\begin{matrix}(1-1/\alpha,1/\alpha),(1/2,1/2)\\(0,1),(1/2,1/2)\end{matrix}\right]\psi_0(x'),$$

which is an alternative form of the solution to the 1D fractional Schrödinger equation.

10.1.3 *Exact 3D solution*

Quantum dynamics of a free particle in the 3D space is governed by the following equation,

$$i\hbar \frac{\partial \psi(\mathbf{r}, t)}{\partial t} = D_\alpha (-\hbar^2 \Delta)^{\alpha/2} \psi(\mathbf{r}, t), \qquad (10.19)$$

with initial condition

$$\psi(\mathbf{r}, t = 0) = \psi_0(\mathbf{r}).$$

Using the 3D Fourier transforms defined by Eq. (12.19) and the definition of the 3D quantum fractional Riesz derivative given by Eq. (3.9) yield for the wave function $\varphi(\mathbf{p}, t)$ in the momentum representation,

$$i\hbar \frac{\partial \varphi(\mathbf{p}, t))}{\partial t} = D_\alpha |\mathbf{p}|^\alpha \varphi(\mathbf{p}, t), \qquad (10.20)$$

with the initial condition $\varphi_0(\mathbf{p})$ given by

$$\varphi_0(\mathbf{p}) = \varphi(\mathbf{p}, t = 0) = \int d^3 r e^{-i\frac{\mathbf{p}\mathbf{r}}{\hbar}} \psi_0(\mathbf{r}). \qquad (10.21)$$

Going back to Eq. (10.19) we can see that the solution $\psi(\mathbf{r}, t)$ has a form

$$\psi(\mathbf{r}, t) = \frac{1}{(2\pi\hbar)^3} \int d^3 r' \int d^3 p \exp\{i\frac{\mathbf{p}(\mathbf{r} - \mathbf{r}')}{\hbar} - i\frac{D_\alpha |\mathbf{p}|^\alpha t}{\hbar}\} \psi_0(\mathbf{r}').$$

The integral over $d^3 p$ can be expressed in terms of $H_{3,3}^{1,2}$-function, see, for instance, Eqs. (33) and (34) in [93]. The solution to the problem defined by Eq. (10.19) is

$$\psi(\mathbf{r}, t) = -\frac{1}{2\pi\alpha} \int d^3 r' \frac{1}{|\mathbf{r} - \mathbf{r}'|^3} \qquad (10.22)$$

$$\times H_{3,3}^{1,2} \left[\frac{1}{\hbar} \left(\frac{\hbar}{iD_\alpha t} \right)^{1/\alpha} |\mathbf{r} - \mathbf{r}'| \,\middle|\, \begin{matrix} (1,1), (1,1/\alpha), (1,1/2) \\ (1,1), (1,1/2), (2,1) \end{matrix} \right] \psi_0(\mathbf{r}').$$

Substituting $\psi_0(\mathbf{r}) = \delta_0(\mathbf{r})$ into Eq. (10.22) gives us the quantum mechanical kernel $K^{(0)}(\mathbf{r}, t|0, 0)$ for a free particle

$$K^{(0)}(\mathbf{r}, t|0, 0) \tag{10.23}$$

$$= -\frac{1}{2\pi\alpha} \frac{1}{|\mathbf{r}|^3} H_{3,3}^{1,2} \left[\frac{1}{\hbar} \left(\frac{\hbar}{iD_\alpha t} \right)^{1/\alpha} |\mathbf{r}| \left| \begin{array}{c} (1,1), (1, 1/\alpha), (1, 1/2) \\ (1,1), (1, 1/2), (2, 1) \end{array} \right. \right].$$

This is the equation for a free particle 3D kernel in the framework of fractional quantum mechanics. We see that in comparison with the 1D case, the 3D quantum kernel is expressed in terms of the $H_{3,3}^{1,2}$ Fox H-function.

In the case $\alpha = 2$, we come to the well-known equation for the Feynman 3D quantum kernel $K_F^{(0)}(\mathbf{r}, t|0, 0)$,

$$K^{(0)}(\mathbf{r}, t|0, 0)|_{\alpha=2} \equiv K_F^{(0)}(\mathbf{r}, t|0, 0) = \left(\frac{m}{2\pi i\hbar t} \right)^{3/2} \exp\left\{ \frac{im|\mathbf{r}|^2}{2\hbar t} \right\}. \tag{10.24}$$

10.2 Infinite potential well

A particle in a one-dimensional well moves in a potential field $V(x)$ which is zero for $-a \leq x \leq a$, and which is infinite elsewhere,

$$V(x) = \infty, \qquad x < -a, \qquad \text{(i)}$$

$$V(x) = 0, \qquad -a \leq x \leq a, \qquad \text{(ii)} \tag{10.25}$$

$$V(x) = \infty, \qquad x > a. \qquad \text{(iii)}$$

It is evident *a priori* that the spectrum will be discrete. We are interested in the solutions of the fractional Schrödinger equation (12.31) that describe the stationary states with well-defined energies. Such a stationary state with an energy E is described by a wave function $\psi(x, t)$, which can be written as $\psi(x, t) = \exp\{-iEt/\hbar\}\phi(x)$, where $\phi(x)$ is now time independent. In regions (i) and (iii), (see Eq. (10.25)) we have to substitute ∞ for $V(x)$ into Eq. (12.31), and it is easy to see that here the fractional Schrödinger equation can be satisfied only if we take $\phi(x) = 0$. In the middle region (ii), the time-independent fractional Schrödinger equation is

$$-D_\alpha(\hbar\nabla)^\alpha \phi(x) = E\phi(x). \tag{10.26}$$

We can treat this equation as a fractional eigenvalue problem [95]. Within region (ii), the eigenfunctions are determined by Eq. (10.26). Outside of the region (ii), $x < -a$ and $x > a$, the eigenfunctions are zero. We want the wave function $\phi(x)$ to be continuous everywhere, thus we impose the boundary conditions

$$\phi(-a) = \phi(a) = 0 \qquad (10.27)$$

for the solutions to the fractional differential equation (10.26).

Hence, we have the eigenvalue problem

$$-D_\alpha(\hbar\nabla)^\alpha \phi(x) = E\phi(x), \qquad \phi(-a) = \phi(a) = 0. \qquad (10.28)$$

It follows from Eq. (3.42) that the general solution to this eigenvalue problem is superposition of plane waves

$$\phi(x) = Ae^{ikx} + Be^{-ikx}, \qquad (10.29)$$

where the following notation has been introduced

$$k = \frac{1}{\hbar}\left(\frac{E}{D_\alpha}\right)^{1/\alpha}, \qquad 1 < \alpha \leq 2, \qquad (10.30)$$

and constants A and B have to be chosen to satisfy the boundary conditions (10.27). From Eqs. (10.29) and (10.27) we obtain

$$\phi(a) = Ae^{ika} + Be^{-ika} = 0 \qquad (10.31)$$

and

$$\phi(-a) = Ae^{-ika} + Be^{ika} = 0. \qquad (10.32)$$

It follows from Eq. (10.31) that

$$A = -Be^{-2ika}. \qquad (10.33)$$

Inserting the last equation back into Eq. (10.31) brings us

$$\sin(2ka) = 0. \qquad (10.34)$$

To satisfy Eq. (10.34) we need to restrict values of k by

$$2ka = n\pi$$

or

$$k = \frac{n\pi}{2a}, \tag{10.35}$$

where n is a positive integer.

By substituting Eqs. (10.33) and (10.35) into Eq. (10.29) we obtain

$$\phi_n(x) = B\{\exp(-i\frac{n\pi}{2a}x) - (-1)^n \exp(i\frac{n\pi}{2a}x)\}, \tag{10.36}$$

with the constant B defined from the normalization condition

$$\int\limits_{-a}^{a} dx |\phi_n(x)|^2 = 1. \tag{10.37}$$

It is easy to see that to satisfy this normalization condition the constant B in Eq. (10.36) has to be

$$B = \frac{1}{2\sqrt{a}}. \tag{10.38}$$

Then the normalized solution to the problem (10.28) is

$$\phi_n(x) = \frac{1}{2\sqrt{a}}\{\exp(-i\frac{n\pi}{2a}x) - (-1)^n \exp(i\frac{n\pi}{2a}x)\}. \tag{10.39}$$

The potential $V(x)$ defined by Eq. (10.25) is symmetric potential

$$V(x) = V(-x).$$

Hence, one can search for the solutions $\phi_n(x)$, which have definite parity. Indeed, in the case when $n = 2m + 1$, where $m = 0, 1, 2, \ldots$ we obtain from Eq. (10.39)

$$\phi_m^{\text{even}}(x) = \phi_n(x)|_{n=2m+1} = \frac{1}{\sqrt{a}} \cos(\frac{(2m+1)\pi}{2a}x), \tag{10.40}$$

$$m = 0, 1, 2, \ldots,$$

where the wave function $\phi_m^{\text{even}}(x)$ is even under reflection $x \to -x$ solution to the problem (10.28), $\phi_m^{\text{even}}(x) = \phi_m^{\text{even}}(-x)$.

In the case when $n = 2m$, where $m = 1, 2, \ldots$ we obtain Eq. (10.39)

$$\phi_m^{\text{odd}}(x) = \phi_n(x)_{n=2m} = -\frac{i}{\sqrt{a}}\sin(\frac{m\pi}{a}x), \qquad m = 1, 2, \ldots, \qquad (10.41)$$

where wave function $\phi_m^{\text{odd}}(x)$ is odd under reflection $x \to -x$ solution to the problem (10.28), $\phi_m^{\text{odd}}(x) = -\phi_m^{\text{odd}}(-x)$.

Solutions $\phi_m^{\text{even}}(x)$ and $\phi_m^{\text{odd}}(x)$ satisfy the orthonormality property that is

$$\int\limits_{-a}^{a} dx \phi_m^{\text{even}}(x)(\phi_{m'}^{\text{even}}(x))^* = \int\limits_{-a}^{a} dx \phi_m^{\text{odd}}(x)(\phi_{m'}^{\text{odd}}(x))^* = \delta_{mm'},$$

where $\delta_{mm'}$ is the Kronecker symbol and $(\phi_{m'}^{\text{even}}(x))^*$ is complex conjugate wave function.

Solutions $\phi_m^{\text{even}}(x)$ and $\phi_m^{\text{odd}}(x)$ also satisfy

$$\int\limits_{-a}^{a} dx \phi_m^{\text{even}}(x)\phi_{m'}^{\text{odd}}(x) = 0, \qquad m \neq m'.$$

The eigenvalues of the fractional problem given by Eqs. (10.26) and (10.27) are energy levels of the particle in the infinite symmetric potential well (10.25). With the help of Eqs. (10.30) and (10.35) we find [95]

$$E_n = D_\alpha \left(\frac{\pi\hbar}{2a}\right)^\alpha n^\alpha, \qquad n = 1, 2, 3, \ldots, \qquad 1 < \alpha \leq 2. \qquad (10.42)$$

It is obvious that in the Gaussian case ($\alpha = 2$), Eq. (10.42) is transformed into the standard quantum mechanical equation (for example, see Eq. (20.7), [94]) for the energy levels of a particle in the infinite potential well.

The time-dependent solutions $\psi_n(x, t)$ for a particle in a one-dimensional infinite symmetric potential well are

$$\psi_n(x, t) = \frac{1}{2\sqrt{a}}\{\exp(-i\frac{n\pi}{2a}x) - (-1)^n \exp(i\frac{n\pi}{2a}x)\} \qquad (10.43)$$

$$\times \exp\left\{-iD_\alpha \left(\frac{\pi\hbar}{2a}\right)^\alpha n^\alpha t/\hbar\right\}, \qquad n = 1, 2, 3, \ldots.$$

In the infinite potential well the ground state is represented by the wave function $\psi_{\text{ground}}(x,t)$ at $n = 1$, and it has the form

$$\psi_{\text{ground}}(x,t) = \psi_n(x,t)|_{n=1}$$

$$= \frac{1}{\sqrt{a}} \cos\{\frac{\pi x}{2a}\} \exp\left\{ -iD_\alpha \left(\frac{\pi \hbar}{2a}\right)^\alpha t/\hbar \right\} \qquad (10.44)$$

$$= \phi_{\text{ground}}(x) \exp\{-iE_{\text{ground}}t/\hbar\},$$

where time independent wave function of ground state $\phi_{\text{ground}}(x)$ is given by

$$\phi_{\text{ground}}(x) \equiv \phi_0^{\text{even}}(x) = \frac{1}{\sqrt{a}} \cos\{\frac{\pi x}{2a}\}, \qquad (10.45)$$

with $\phi_0^{\text{even}}(x)$ defined by Eq. (10.40) and the energy of the ground state is

$$E_{\text{ground}} = D_\alpha \left(\frac{\pi \hbar}{2a}\right)^\alpha. \qquad (10.46)$$

It follows from Eq. (10.43) that the even time independent solution to the problem (10.28) is

$$\psi_m^{\text{even}}(x,t) = \frac{1}{\sqrt{a}} \cos(\frac{(2m+1)\pi}{2a}x) \qquad (10.47)$$

$$\times \exp\{-iD_\alpha \left(\frac{\pi \hbar}{2a}\right)^\alpha (2m+1)^\alpha t/\hbar\}, \qquad m = 0,1,2,...,$$

and the odd time independent solution to the problem (10.28) is

$$\psi_m^{\text{odd}}(x,t) = -\frac{i}{\sqrt{a}} \sin(\frac{m\pi}{a}x) \exp\left\{ -iD_\alpha \left(\frac{\pi \hbar}{a}\right)^\alpha m^\alpha t/\hbar \right\}, \qquad (10.48)$$

$$m = 1,2,....$$

10.2.1   Consistency of the solution for infinite potential well

Following the papers [112] and [113] let's show that, the solution $\phi_{\text{ground}}(x)$ given by Eq. (10.45) satisfies the fractional differential equation (10.26),

$$-D_\alpha(\hbar\nabla)^\alpha\phi_{\text{ground}}(x) = E_{\text{ground}}\phi_{\text{ground}}(x), \quad n = 1, 2, 3, ... \quad (10.49)$$

and the boundary conditions

$$\phi_{\text{ground}}(a) = \phi_{\text{ground}}(-a) = 0. \quad (10.50)$$

The energy E_{ground} is given by Eq. (10.46).

By definition of quantum Riesz fractional derivatives (3.14) we have

$$-D_\alpha(\hbar\nabla)^\alpha\phi_{\text{ground}}(x) = \frac{D_\alpha}{2\pi\hbar} \int\limits_{-\infty}^{\infty} dp\, e^{i\frac{px}{\hbar}}|p|^\alpha\varphi_{\text{ground}}(p), \quad (10.51)$$

where $\varphi_{\text{ground}}(p)$ is the Fourier transform of the ground state wave function $\phi_{\text{ground}}(x)$,

$$\varphi_{\text{ground}}(p) = \int\limits_{-\infty}^{\infty} dx\, e^{-i\frac{px}{\hbar}}\phi_{\text{ground}}(x).$$

Further, substituting $\phi_{\text{ground}}(x)$ given by Eq. (10.45) yields [112]

$$\varphi_{\text{ground}}(p) = \frac{1}{\sqrt{a}} \int\limits_{-a}^{a} dx\, e^{-i\frac{px}{\hbar}} \cos\{\frac{\pi x}{2a}\} \quad (10.52)$$

$$= -\frac{\pi}{\sqrt{a}} \left(\frac{\hbar^2}{a}\right) \frac{\cos(ap/\hbar)}{p^2 - (\pi\hbar/2a)^2}.$$

It follows from Eqs. (10.49) and (10.51) that $\phi_{\text{ground}}(x)$ reads

$$\phi_{\text{ground}}(x) = \frac{D_\alpha}{2\pi\hbar E_{\text{ground}}} \int\limits_{-\infty}^{\infty} dp\, e^{i\frac{px}{\hbar}}|p|^\alpha\varphi_{\text{ground}}(p). \quad (10.53)$$

Now we are aiming to show that equation (10.53) holds. Substituting $\varphi_{\text{ground}}(p)$ given by Eq. (10.52) into the right-hand side of Eq. (10.53) yields

$$\phi_{\text{ground}}(x) = -\frac{D_\alpha}{2E_{\text{ground}}} \frac{\pi}{\sqrt{a}} \left(\frac{\hbar}{a}\right) \int\limits_{-\infty}^{\infty} dp\, e^{i\frac{px}{\hbar}}|p|^\alpha \frac{\cos(ap/\hbar)}{p^2 - (\pi\hbar/2a)^2} \quad (10.54)$$

$$= -\frac{D_\alpha}{2E_{\text{ground}}} \frac{\pi}{\sqrt{a}} \left(\frac{\hbar}{a}\right) \left(\frac{2a}{\pi\hbar}\right)^2 \int\limits_{-\infty}^{\infty} dp\, e^{i\frac{px}{\hbar}} \frac{|p|^\alpha \cos(ap/\hbar)}{(2ap/\pi\hbar)^2 - 1}.$$

To evaluate the integral we introduce a new integration variable

$$q = \frac{2a}{\pi\hbar} p.$$

Then Eq. (10.54) reads

$$\phi_{\text{ground}}(x) = -\frac{D_\alpha}{\pi E_{\text{ground}}} \frac{1}{\sqrt{a}} \left(\frac{\pi\hbar}{2a}\right)^\alpha \int\limits_{-\infty}^{\infty} dq\, |q|^\alpha \frac{\cos(\pi q/2)}{q^2 - 1} e^{i\pi q x/2a}. \quad (10.55)$$

The integral

$$I = \int\limits_{-\infty}^{\infty} dq\, |q|^\alpha \frac{\cos(\pi q/2)}{q^2 - 1} e^{i\pi q x/2a} \quad (10.56)$$

has poles on the real axis at $q = \pm 1$, and it can be evaluated via analytic continuation as a Cauchy principal value integral [114].

Substitution

$$\cos(\pi q/2) = \frac{1}{2}(e^{i\pi q/2} + e^{-i\pi q/2})$$

allows us to present the integral I as

$$I = I_1 + I_2, \quad (10.57)$$

where

$$I_1 = \frac{1}{2} \int\limits_{-\infty}^{\infty} dq\, |q|^\alpha \frac{e^{i(x/a+1)\pi q/2}}{(q+1)(q-1)} \quad (10.58)$$

and

$$I_2 = \frac{1}{2} \int\limits_{-\infty}^{\infty} dq\, |q|^\alpha \frac{e^{i(x/a-1)\pi q/2}}{(q+1)(q-1)}. \quad (10.59)$$

In the first integral I_1, we close the contour in the upper half q-plane with a semicircular of radius R and then go around the poles on the real axis in the upper q-plane, with small semicircular of radius ρ. With the help of Jordan's lemma we obtain for I_1 in the limit $R \to \infty$ and $\rho \to 0$

$$I_1 = i\pi \left(\frac{i}{2} \cos \frac{\pi x}{2a} \right), \tag{10.60}$$

where the value of the integral I_1 has to be understood as a Cauchy principal value [114].

Similarly, for the second integral I_2, we close the contour in the lower q-plane and circle around the poles in the lower half q-plane to obtain

$$I_2 = i\pi \left(\frac{i}{2} \cos \frac{\pi x}{2a} \right), \tag{10.61}$$

where the value of the integral I_2 has to be understood as a Cauchy principal value [114].

By adding I_1 and I_2 we come up with the value of the integral I because of Eq. (10.57)

$$I = -\pi \cos \frac{\pi x}{2a}. \tag{10.62}$$

Substituting the value of the integral I into Eq. (10.55) yields

$$\phi_{\text{ground}}(x) = \frac{D_\alpha}{E_{\text{ground}}} \frac{1}{\sqrt{a}} \left(\frac{\pi \hbar}{2a} \right)^\alpha \cos \frac{\pi x}{2a}. \tag{10.63}$$

Noting that E_{ground} is given by Eq. (10.46) we obtain $\phi_{\text{ground}}(x)$ in the form given by Eq. (10.41).

Thus, it has been shown that $\phi_{\text{ground}}(x)$ is indeed the solution to eigenvalue problem presented by Eq. (10.28).

A similar consideration can be provided to prove that $\phi_m^{\text{even}}(x)$ and $\phi_m^{\text{odd}}(x)$, given by Eqs. (10.40) and (10.41) respectively, are in fact the solutions to eigenvalue problem (10.28).

10.3   Bound state in δ-potential well

For one dimensional attractive δ-potential well, $V(x) = -V_0 \delta(x)$, $(V_0 > 0)$, where $\delta(x)$ is the Dirac delta function, the time-independent fractional Schrödinger equation (4.9) is

$$-D_\alpha(\hbar\nabla)^\alpha\phi(x) - V_0\delta(x)\phi(x) = E\phi(x), \qquad 1 < \alpha \leq 2. \qquad (10.64)$$

Searching for the bound state in δ-potential well, we consider this fractional quantum mechanical problem for negative energies $E < 0$. Dong and Xu [115] were the first who found the bound energy and wave function in δ-potential well in the framework of fractional quantum mechanics. In our consideration we will follow [115] and [116].

To solve the problem defined by Eq. (10.64) we go to momentum representation according to Eqs. (3.25) - (3.27). Thus, we have

$$D_\alpha|p|^\alpha\overline{\phi}(p) - \frac{V_0}{2\pi\hbar}\int\limits_{-\infty}^{\infty} dq\overline{\phi}(q) = E\overline{\phi}(p), \qquad 1 < \alpha \leq 2, \qquad (10.65)$$

where the wave function $\overline{\phi}(p)$ in momentum representation and the wave function in coordinate representation $\phi(x)$ are related to each other by

$$\phi(x) = \frac{1}{2\pi\hbar}\int\limits_{-\infty}^{\infty} dp e^{ipx/\hbar}\overline{\phi}(p), \qquad \overline{\phi}(p) = \int\limits_{-\infty}^{\infty} dx e^{-ipx/\hbar}\phi(x). \qquad (10.66)$$

Let us define the parameter λ, $\lambda > 0$,

$$\lambda^\alpha = -\frac{E}{D_\alpha}, \qquad E < 0. \qquad (10.67)$$

Then we can rewrite Eq. (10.65) in the form

$$\overline{\phi}(p) = \frac{\gamma U}{|p|^\alpha + \lambda^\alpha}, \qquad (10.68)$$

where the following notations have been introduced

$$\gamma = \frac{V_0}{2\pi\hbar D_\alpha} \qquad (10.69)$$

and

$$U = \int\limits_{-\infty}^{\infty} dq\overline{\phi}(q) = 2\pi\hbar\phi(x)|_{x=0} = 2\pi\hbar\phi(0). \qquad (10.70)$$

Substituting $\overline{\phi}(p)$ given by Eq. (10.68) into Eq. (10.70) yields

$$1 = \gamma \int\limits_{-\infty}^{\infty} \frac{dp}{|p|^{\alpha} + \lambda^{\alpha}} = 2\gamma\lambda^{1-\alpha} \int\limits_{0}^{\infty} \frac{dq}{|q|^{\alpha} + 1}. \tag{10.71}$$

To calculate the integral $\int\limits_{0}^{\infty} dq/(|q|^{\alpha} + 1)$ we use formula (see, formula 3.241.2, page 322 [86])

$$\int\limits_{0}^{\infty} \frac{x^{\mu-1}dx}{1 + x^{\nu}} = \frac{\pi}{\nu}\mathrm{cosec}\frac{\mu\pi}{\nu}, \qquad [\mathrm{Re}\,\nu > \mathrm{Re}\,\mu > 0],$$

where $\mathrm{cosec}\,x = 1/\sin x$.
Hence, we have

$$\int\limits_{0}^{\infty} \frac{dq}{|q|^{\alpha} + 1} = \frac{\pi}{\alpha}\mathrm{cosec}\frac{\pi}{\alpha}$$

and Eq. (10.71) becomes

$$1 = 2\gamma\lambda^{1-\alpha}\frac{\pi}{\alpha}\mathrm{cosec}\frac{\pi}{\alpha}.$$

Thus, we obtain

$$\lambda = (\frac{2\gamma\pi}{\alpha}\mathrm{cosec}\frac{\pi}{\alpha})^{1/(\alpha-1)}. \tag{10.72}$$

Using the definition given by Eq. (10.67) we find the bound energy in δ-potential well

$$E = -\left(\frac{V_0\mathrm{cosec}(\pi/\alpha)}{\alpha\hbar D_{\alpha}^{1/\alpha}}\right)^{\alpha/(\alpha-1)}. \tag{10.73}$$

By substituting Eq. (10.68) into Eq. (10.66) we find the wave function $\phi(x)$ of the bound state

$$\phi(x) = \frac{\gamma U}{2\pi\hbar} \int\limits_{-\infty}^{\infty} dp \frac{e^{ipx/\hbar}}{|p|^{\alpha} + \lambda^{\alpha}}. \tag{10.74}$$

The integral $\int\limits_{-\infty}^{\infty} dp e^{ipx/\hbar}/(|p|^\alpha + \lambda^\alpha)$ has been calculated in Appendix C, and the result is given by Eq. (C.14). Using this result we obtain $\phi(x)$

$$\phi(x) = \frac{\gamma U}{\lambda^\alpha |x|} H_{2,3}^{2,1} \left[(\hbar^{-1}\lambda)^\alpha |x|^\alpha \, \middle| \, \begin{matrix} (1,1), (1,\alpha/2) \\ (1,\alpha), (1,1), (1,\alpha/2) \end{matrix} \right]. \tag{10.75}$$

The wave function has to satisfy the normalization condition

$$\int\limits_{-\infty}^{\infty} dx |\phi(x)|^2 = 1, \tag{10.76}$$

which allows us to calculate factor γU in Eq. (10.75). Indeed, by substituting Eq. (10.74) into Eq. (10.76) we have

$$\int\limits_{-\infty}^{\infty} dx |\phi(x)|^2 = \left(\frac{\gamma U}{2\pi\hbar}\right)^2 \int\limits_{-\infty}^{\infty} dx \int\limits_{-\infty}^{\infty} dp_1 \frac{e^{ip_1 x/\hbar}}{|p_1|^\alpha + \lambda^\alpha} \int\limits_{-\infty}^{\infty} dp_2 \frac{e^{-ip_2 x/\hbar}}{|p_2|^\alpha + \lambda^\alpha}$$

$$= \frac{(\gamma U)^2}{2\pi\hbar} \int\limits_{-\infty}^{\infty} dp_1 \int\limits_{-\infty}^{\infty} dp_2 \delta(p_1 - p_2) \frac{1}{|p_1|^\alpha + \lambda^\alpha} \frac{1}{|p_2|^\alpha + \lambda^\alpha}$$

$$= \frac{(\gamma U)^2}{2\pi\hbar} \int\limits_{-\infty}^{\infty} dp \frac{1}{(|p_1|^\alpha + \lambda^\alpha)^2} = 1,$$

that is

$$\gamma U = \sqrt{2\pi\hbar} \left(\int\limits_{-\infty}^{\infty} dp \frac{1}{(|p_1|^\alpha + \lambda^\alpha)^2} \right)^{-1/2}. \tag{10.77}$$

To calculate the integral $\int\limits_{-\infty}^{\infty} dp/(|p|^\alpha + \lambda^\alpha)^2$ we use formula (see, formula 3.241.4, page 322 [86])

$$\int\limits_{0}^{\infty} \frac{x^{\mu-1} dx}{(p + qx^\nu)^{n+1}} = \frac{1}{\nu p^{n+1}} \left(\frac{p}{q}\right)^{\mu/\nu} \frac{\Gamma(\frac{\mu}{\nu})\Gamma(1 + n - \frac{\mu}{\nu})}{\Gamma(1 + n)},$$

here

$$[0 < \frac{\mu}{\nu} < n + 1, \qquad p \neq 0, \quad q \neq 0].$$

Then the integral $\displaystyle\int_{-\infty}^{\infty} dp/(|p|^\alpha + \lambda^\alpha)^2$ is

$$\int_{-\infty}^{\infty} dp \frac{1}{(|p|^\alpha + \lambda^\alpha)^2} = \frac{2\lambda}{\alpha\lambda^{2\alpha}} \frac{\Gamma(\frac{1}{\alpha})\Gamma(2 - \frac{1}{\alpha})}{\Gamma(2)} \tag{10.78}$$

$$= \frac{2\pi}{\alpha} \lambda^{1-2\alpha} \left(\frac{\alpha - 1}{\alpha} \right) \operatorname{cosec}(\pi/\alpha),$$

where the identities

$$\Gamma(\frac{1}{\alpha})\Gamma(1 - \frac{1}{\alpha}) = \frac{\pi}{\sin(\pi/\alpha)}, \qquad \Gamma(2 - \frac{1}{\alpha}) = (1 - \frac{1}{\alpha})\Gamma(1 - \frac{1}{\alpha})$$

have been used.

Substituting the result (10.78) of calculation of the integral $\displaystyle\int_{-\infty}^{\infty} dp/(|p|^\alpha + \lambda^\alpha)^2$ into Eq. (10.77) yields

$$\gamma U = \sqrt{2\pi\hbar} \left(\frac{2\pi}{\alpha} \lambda^{1-2\alpha} \left(\frac{\alpha - 1}{\alpha} \right) \operatorname{cosec}(\pi/\alpha) \right)^{-1/2} \tag{10.79}$$

$$= \alpha\lambda^\alpha \sqrt{\frac{\hbar}{\lambda(\alpha - 1)}} \sin(\pi/\alpha).$$

Then, the wave function $\phi(x)$ given by Eq. (10.75) reads

$$\phi(x) = \frac{C_\alpha}{|x|} H_{2,3}^{2,1} \left[(\hbar^{-1}\lambda)^\alpha |x|^\alpha \left| \begin{array}{l} (1,1), (1,\alpha/2) \\ (1,\alpha), (1,1), (1,\alpha/2) \end{array} \right. \right], \tag{10.80}$$

where $1 < \alpha \leq 2$, C_α constant is defined by

$$C_\alpha = \alpha\sqrt{\frac{\hbar}{\lambda(\alpha - 1)}} \sin(\pi/\alpha), \tag{10.81}$$

and λ is introduced by Eq. (10.67).

We conclude that the solution to fractional quantum mechanics problem - finding bound energy and wave function of the bound state in δ-potential well, is given by Eqs. (10.73), (10.80) and (10.81).

In the case of standard quantum mechanics, when $\alpha = 2$ and $D_2 = 1/2m$, Eqs. (10.73), (10.80) and (10.81) are transformed into the well-known formulas [117] for the bound energy and bound state wave function in δ-potential well. Indeed, when $\alpha = 2$ the bound energy given by Eq. (10.73) has the form

$$\mathcal{E} = E|_{\alpha=2} = -\frac{mV_0^2}{2\hbar^2}. \tag{10.82}$$

When $\alpha = 2$, the constant C_α defined by Eq. (10.81) is

$$C_2 = C_\alpha|_{\alpha=2} = 2\sqrt{\frac{\hbar}{\lambda}} = 2\sqrt{\hbar/\sqrt{-2m\mathcal{E}}}, \tag{10.83}$$

with λ given by Eq. (10.67) at $\alpha = 2$

$$\lambda = \sqrt{-2m\mathcal{E}}. \tag{10.84}$$

When $\alpha = 2$, wave function $\phi(x)$ of the bound state in δ-potential well becomes (see, Eq. (C.16) in Appendix C)

$$\phi(x)|_{\alpha=2} = \frac{C_2}{|x|} H_{2,3}^{2,1}\left[(\hbar^{-1}\lambda)^2|x|^2 \,\middle|\, \begin{matrix} (1,1),(1,1) \\ (1,2),(1,1),(1,1) \end{matrix} \right] \tag{10.85}$$

$$= \frac{\lambda C_2}{2\hbar} \exp\{-\frac{\lambda|x|}{\hbar}\} = \sqrt{\frac{\lambda}{\hbar}} \exp\{-\frac{\lambda|x|}{\hbar}\},$$

which can be expressed in the form

$$\phi(x)|_{\alpha=2} = \left(\sqrt{-2m\mathcal{E}}/\hbar\right)^{1/2} \exp\{-\frac{\sqrt{-2m\mathcal{E}}}{\hbar}|x|\}, \tag{10.86}$$

if we take into account Eq. (10.84).

10.4 Linear potential field

For a particle in a linear potential field (for example, see [94], page 74), the potential function $V(x)$ can be written as:

$$V(x) = \begin{cases} Fx, & x \geq 0, \\ \infty, & x < 0, \end{cases} \tag{10.87}$$

where $F > 0$. Therefore, the fractional Schrödinger equation Eq. (12.31) becomes

$$-D_\alpha(\hbar\nabla)^\alpha \phi(x) + Fx\phi(x) = E\phi(x), \qquad 1 < \alpha \leq 2, \qquad x \geq 0. \tag{10.88}$$

The continuity and bounded conditions of the wave function let us conclude that $\phi(x) = 0$, $x < 0$. Besides, $\phi(x)$ must satisfy the boundary conditions

$$\begin{aligned} \phi(x) &= 0, & x &= 0, \\ \phi(x) &= 0, & x &\to \infty. \end{aligned} \tag{10.89}$$

Then, wave function $\phi_n(x)$ of the quantum state with energy E_n, $n = 1, 2, 3, \dots$ is [115]

$$\phi_n(x) = \frac{2\pi A_\alpha}{\alpha + 1} \tag{10.90}$$

$$\times H^{1,1}_{2,2}\left[(x - \frac{E_n}{F})\frac{1}{\hbar}\left(\frac{D_\alpha}{(\alpha+1)F\hbar}\right)^{-\frac{1}{\alpha+1}} \left| \begin{matrix} (1 - \frac{1}{\alpha+1}, \frac{1}{\alpha+1}), (\frac{\alpha+2}{2(\alpha+1)}, \frac{\alpha}{2(\alpha+1)}) \\ (0, 1), (\frac{\alpha+2}{2(\alpha+1)}, \frac{\alpha}{2(\alpha+1)}) \end{matrix} \right. \right],$$

where the normalization constant A_α is given by

$$A_\alpha = \frac{1}{2\pi\hbar}\left(\frac{D_\alpha}{(\alpha+1)F\hbar}\right)^{-1/(\alpha+1)}. \tag{10.91}$$

The energy spectra E_n has the form

$$E_n = \lambda_n F\hbar\left(\frac{D_\alpha}{(\alpha+1)F\hbar}\right)^{1/(\alpha+1)}, \tag{10.92}$$

$$1 < \alpha \leq 2, \qquad n = 1, 2, 3, \dots,$$

where λ_n are the solutions of the equation [115]

$$H_{2,2}^{1,1}\left[-\lambda_n \left| \begin{array}{c} (1-\frac{1}{\alpha+1}, \frac{1}{\alpha+1}), (\frac{\alpha+2}{2(\alpha+1)}, \frac{\alpha}{2(\alpha+1)}) \\ (0,1), (\frac{\alpha+2}{2(\alpha+1)}, \frac{\alpha}{2(\alpha+1)}) \end{array} \right. \right] = 0. \qquad (10.93)$$

When $\alpha = 2$, Eqs. (10.90) and (10.92) turn into well-known equations of standard quantum mechanics [94], [115].

Other solvable physical models of fractional quantum mechanics include the 1D Coulomb potential [115], a finite square potential well, dynamics in the field of 1D lattice, penetration through a δ-potential barrier, the Dirac comb [99], the bound state problem, and penetration through double δ-potential barrier [118].

10.5   Quantum kernel for a free particle in the box

Now we are going to consider the impact of integration over Lévy flight paths on a quantum kernel for a free particle in a 1D box of length $2a$ confined by infinitely high walls at $x = -a$ and $x = a$. With the help of Eq. (6.23) the kernel $K_{\text{box}}^{(0)}(x_b t | x_a 0)$ of a free particle in the infinite symmetric potential well (10.25) can be presented as

$$K_{\text{box}}^{(0)}(x_b t | x_a 0) = \sum_{n=1}^{\infty} \phi_n(x_b) \phi_n^*(x_a) e^{-(i/\hbar)E_n t}, \quad t > 0, \qquad (10.94)$$

where the time-independent wave functions $\phi_n(x)$ are given by Eq. (10.39) for a particle in a one-dimensional infinite symmetric potential well (10.25). The energy levels E_n are defined by Eq. (10.46). Substituting $\phi_n(x)$ and E_n into Eq. (10.94) yields

$$K_{\text{box}}^{(0)}(x_b t | x_a 0)$$

$$= \frac{1}{4a} \sum_{n=1}^{\infty} [\exp\{i\frac{\pi n}{2a}(-x_b + x_a)\} - (-1)^n \exp\{-i\frac{\pi n}{2a}(x_b + x_a)\}$$

$$- (-1)^n \exp\{i\frac{\pi n}{2a}(x_b + x_a)\} + \exp\{i\frac{\pi n}{2a}(x_b - x_a)\}] \qquad (10.95)$$

$$\times \exp\left(-\frac{itD_\alpha}{\hbar}\left(\frac{\pi\hbar}{2a}\right)^\alpha n^\alpha\right), \quad t > 0,$$

or

$$K_{\text{box}}^{(0)}(x_b t | x_a 0) \tag{10.96}$$

$$= \frac{1}{2a} \sum_{n=1}^{\infty} [\cos\{\frac{\pi n}{2a}(x_b - x_a)\} - (-1)^n \cos\{\frac{\pi n}{2a}(x_b + x_a)\}]$$

$$\times \exp\left(-\frac{itD_\alpha}{\hbar}\left(\frac{\pi\hbar}{2a}\right)^\alpha n^\alpha\right).$$

The kernel $K_{\text{box}}^{(0)}(x_b t | x_a 0)$ satisfies the boundary conditions

$$K_{\text{box}}^{(0)}(x_b = a, t | x_a, 0) = 0,$$
$$K_{\text{box}}^{(0)}(x_b, t | x_a = -a, 0) = 0, \tag{10.97}$$

enforced by the two infinite walls at $x = -a$ and $x = a$ at all times.

Further, it is easy to see that the following chain of transformations holds

$$K_{\text{box}}^{(0)}(x_b t | x_a 0) = \frac{1}{2a} \sum_{n=1}^{\infty} [\cos\{\frac{\pi n}{2a}(x_b - x_a)\} - (-1)^n \cos\{\frac{\pi n}{2a}(x_b + x_a)\}]$$

$$\times \exp\left(-\frac{itD_\alpha}{\hbar}\left(\frac{\pi\hbar}{2a}\right)^\alpha n^\alpha\right)$$

$$= \frac{1}{4a} \sum_{l=-\infty}^{\infty} \int_{-\infty}^{\infty} dp\, \delta(p - \frac{\pi\hbar}{2a}l) \left\{ \exp[\frac{ip(x_b - x_a)}{\hbar}] - (-1)^l \exp[\frac{ip(x_b + x_a)}{\hbar}] \right\} \tag{10.98}$$

$$\times \exp\left(-\frac{itD_\alpha |p|^\alpha}{\hbar}\right)$$

$$= \frac{1}{2\pi\hbar} \sum_{l=-\infty}^{\infty} \int_{-\infty}^{\infty} dp \left\{ \exp[\frac{ip(x_b - x_a + 4la)}{\hbar}] - (-1)^l \exp[\frac{ip(x_b + x_a + 4la)}{\hbar}] \right\}$$

$$\times \exp\left(-\frac{itD_\alpha |p|^\alpha}{\hbar}\right),$$

where the Poisson summation formula[1] has been taken into consideration.

If we take into account the definition of a free particle kernel $K^{(0)}(x_b t | x_a 0)$ given by Eq. (12.20) then the kernel for a free particle in the box with infinitely high walls $K_{\text{box}}(x_b t | x_a 0)$ becomes

$$K_{\text{box}}^{(0)}(x_b t | x_a 0) \tag{10.100}$$

$$= \sum_{l=-\infty}^{\infty} \left\{ K^{(0)}(x_b + 4la, t | x_a 0) - (-1)^l K^{(0)}(x_b + 4la, t | - x_a 0) \right\}.$$

In terms of Fox's $H_{2,2}^{1,1}$-function $K_{\text{box}}^{(0)}(x_b t | x_a 0)$ is

$$K_{\text{box}}^{(0)}(x_b t | x_a 0) = \frac{1}{\alpha} \left(\frac{\hbar}{(\hbar / i D_\alpha t)^{1/\alpha}} \right)^{-1}$$

$$\times \sum_{l=-\infty}^{\infty} \left\{ H_{2,2}^{1,1} \left[\frac{1}{\hbar} \left(\frac{\hbar}{i D_\alpha t} \right)^{1/\alpha} |x_b - x_a + 4la| \left| \begin{array}{c} (1 - 1/\alpha, 1/\alpha), (1/2, 1/2) \\ (0, 1), (1/2, 1/2) \end{array} \right. \right] \right.$$

$$\tag{10.101}$$

$$\left. - (-1)^l H_{2,2}^{1,1} \left[\frac{1}{\hbar} \left(\frac{\hbar}{i D_\alpha t} \right)^{1/\alpha} |x_b + x_a + 4la| \left| \begin{array}{c} (1 - 1/\alpha, 1/\alpha), (1/2, 1/2) \\ (0, 1), (1/2, 1/2) \end{array} \right. \right] \right\}.$$

Alternatively, $K_{\text{box}}^{(0)}(x_b t | x_a 0)$ can be presented in terms of Fox's $H_{2,2}^{1,1}$-function as

$$K_{\text{box}}^{(0)}(x_b t | x_a 0) = \frac{1}{\alpha} \frac{1}{|x_b - x_a|}$$

$$\times \sum_{l=-\infty}^{\infty} \left\{ H_{2,2}^{1,1} \left[\frac{1}{\hbar} \left(\frac{\hbar}{i D_\alpha t} \right)^{1/\alpha} |x_b - x_a + 4la| \left| \begin{array}{c} (1, 1/\alpha), (1, 1/2) \\ (1, 1), (1, 1/2) \end{array} \right. \right] \right.$$

$$\tag{10.102}$$

$$\left. - (-1)^l H_{2,2}^{1,1} \left[\frac{1}{\hbar} \left(\frac{\hbar}{i D_\alpha t} \right)^{1/\alpha} |x_b + x_a + 4la| \left| \begin{array}{c} (1, 1/\alpha), (1, 1/2) \\ (1, 1), (1, 1/2) \end{array} \right. \right] \right\},$$

---

[1]

$$\frac{\pi \hbar}{2a} \sum_{l=-\infty}^{\infty} \delta(p - \frac{\pi \hbar}{2a} l) = \sum_{l=-\infty}^{\infty} \exp\{i \frac{4pa}{\hbar} l\}, \tag{10.99}$$

where δ is the delta function.

where Property 12.2.5 (see Eq. (A.14) in Appendix A) of Fox's H-function has been used.

We see that for the kernel $K_{\text{box}}(x_b t | x_a 0)$ we have two different representations given by Eq. (10.96) and Eq. (10.101). By equating them we come to a new identity with involvement of Fox's $H_{2,2}^{1,1}$-function

$$\frac{1}{2a} \sum_{n=1}^{\infty} [\cos\{\frac{\pi n}{2a}(x_b - x_a)\} - (-1)^n \cos\{\frac{\pi n}{2a}(x_b + x_a)\}] \qquad (10.103)$$

$$\times \exp\left(-\frac{itD_\alpha}{\hbar} \left(\frac{\pi \hbar}{2a}\right)^\alpha n^\alpha\right)$$

$$= \frac{1}{\alpha} \left(\frac{\hbar}{(\hbar/iD_\alpha t)^{1/\alpha}}\right)^{-1}$$

$$\times \sum_{l=-\infty}^{\infty} \left\{ H_{2,2}^{1,1}\left[\frac{1}{\hbar}\left(\frac{\hbar}{iD_\alpha t}\right)^{1/\alpha} |x_b - x_a + 4la| \, \middle| \, \begin{matrix} (1 - 1/\alpha, 1/\alpha), (1/2, 1/2) \\ (0, 1), (1/2, 1/2) \end{matrix} \right] \right.$$

$$\left. -(-1)^l H_{2,2}^{1,1}\left[\frac{1}{\hbar}\left(\frac{\hbar}{iD_\alpha t}\right)^{1/\alpha} |x_b + x_a + 4la| \, \middle| \, \begin{matrix} (1 - 1/\alpha, 1/\alpha), (1/2, 1/2) \\ (0, 1), (1/2, 1/2) \end{matrix} \right] \right\}.$$

In the case when $\alpha = 2$ this equation gives us the solution for a quantum kernel of a free particle in the box $K_{\text{box}}^{(0)}(x_b t | x_a 0)|_{\alpha=2}$ in the framework of standard quantum mechanics. Indeed, using the identity (7.37) and substituting $D_2 = 1/2m$, where m is particle mass, we obtain

$$K_{\text{box}}^{(0)}(x_b t | x_a 0)|_{\alpha=2} = \sqrt{\frac{m}{2\pi i \hbar t}} \qquad (10.104)$$

$$\times \sum_{l=-\infty}^{\infty} \left\{ \exp(\frac{im|x_b - x_a + 4la|^2}{2\hbar t}) - (-1)^l \exp(\frac{im|x_b + x_a + 4la|^2}{2\hbar t}) \right\}.$$

Thus, we obtained the standard quantum mechanics solution to the quantum kernel of a free particle in a symmetric 1D box of length $2a$ confined by infinitely high walls at $x = -a$ and $x = a$.

10.6   Fractional Bohr atom

It could be that I've perhaps found out a little bit about the structure of atoms. ... If I'm right, it would not be an indication of the nature of a possibility [marginal note in the original: "i.e., impossibility"] (like J. J. Thomson's theory) but perhaps a little piece of reality.

N. Bohr (1912)[2]

In 1913 Bohr proposed a model for atoms and molecules by synthesizing Planck's quantum hypothesis with classical mechanics. When the atomic number Z is small, his model provides good accuracy for the ground-state energy. When Z is large, his model is not as accurate in comparison with the experimental data but still provides a good trend agreeing with the experimental values of the ground-state energy of atoms.

Bohr developed his approach for modeling atoms and molecules by synthesizing Planck's quantum hypothesis with classical mechanics [120], [121]. Bohr tried to explain the hydrogen spectral lines with a radical planetary model of electrons orbiting around a nucleus. He made a set of assumptions to quantify his model, leading to the existence a discrete set of stable, stationary orbits for electrons in the atom:

1. *The dynamical equilibrium of the stationary electronic orbits in an atom is achieved by balancing the electrostatic Coulomb forces of attraction between nuclei and electrons against the centrifugal effect and the interelectronic repelling treated in the framework of classical mechanics;*

2. *Stationary states satisfy the quantization condition that the ratio of the total kinetic energy of the electron to its orbital frequency be an integer multiple of $\hbar$. For circular orbits, this signifies that the angular momentum of the electron is restricted by integer multiple of $\hbar$. Therefore, electrons in the stationary orbits do not radiate in spite of their acceleration due to orbital motion.*

3. *Energy is emitted by an atom only when an electron makes a jump i.e. noncontinuous transition between two stationary orbits, and the frequency of such radiation emission is determined by $E/\hbar$, where E is the energy difference between higher and lower energy orbits where the transition occurs.*

---

[2]A letter from Niels Bohr to his brother, Harald, dated 19 June 1912 (see, page 238 in [119]).

And vice versa, an atom absorbs radiation by having its electrons make a transition from lower to higher energy orbit.

The success of the Bohr theory [120], [121] with hydrogen-like atoms gave great impetus to further research on the "Bohr atom". In spite of some extraordinary achievements by Bohr and others, it was clear that the theory was provisional.

The fractional generalization of the "*Bohr atom*" is based on the following fractional Schrödinger equation

$$i\hbar\frac{\partial\psi(\mathbf{r},t)}{\partial t} = D_\alpha(-\hbar^2\Delta)^{\alpha/2}\psi(\mathbf{r},t) - \frac{Ze^2}{|\mathbf{r}|}\psi(\mathbf{r},t), \qquad (10.105)$$

where e is the electron charge, Ze is the nuclear charge of the hydrogen-like atom, $\mathbf{r}$ is the 3D vector, $\Delta = \partial^2/\partial\mathbf{r}^2$ is the Laplacian, and the operator $(-\hbar^2\Delta)^{\alpha/2}$ is the 3D quantum Riesz fractional derivative defined by Eq. (3.9).

The fractional Hamiltonian of the considered quantum-mechanical system has the form

$$H(\mathbf{p},\mathbf{r}) = D_\alpha|\mathbf{p}|^\alpha - \frac{Ze^2}{|\mathbf{r}|}. \qquad (10.106)$$

The eigenvalue problem for a fractional hydrogen-like atom is [95]

$$D_\alpha(-\hbar^2\Delta)^{\alpha/2}\phi(\mathbf{r}) - \frac{Ze^2}{|\mathbf{r}|}\phi(\mathbf{r}) = E\phi(\mathbf{r}),$$

where $\phi(\mathbf{r})$ is related to $\psi(\mathbf{r},t)$ by the equation

$$\psi(\mathbf{r},t) = \exp\{-i\frac{Et}{\hbar}\}\phi(\mathbf{r}).$$

The total energy E is

$$E = E_{kin} + V,$$

where E_{kin} is the kinetic energy

$$E_{kin} = D_\alpha|\mathbf{p}|^\alpha,$$

and V is the potential energy

$$V = -\frac{Ze^2}{|\mathbf{r}|}.$$

It is well known that if the potential energy is a homogeneous function of coordinates, and the motion takes place in a finite region of space, there exists a simple relation between the time average values of the kinetic and potential energies, known as the *virial theorem* (see, page 23, [122]). It follows from the virial theorem that between average kinetic energy and average potential energy of the system with Hamiltonian (10.106) there exists the relation

$$\alpha \overline{E}_{kin} = -\overline{V}, \qquad (10.107)$$

where the average value $\overline{f}$ of any function of time is defined as

$$\overline{f} = \lim_{T \to \infty} \frac{1}{T} \int\limits_{0}^{\infty} dt f(t).$$

In order to evaluate the energy spectrum of the fractional hydrogen-like atom let us remind the *Niels Bohr postulates* [120]:

1. The electron moves in orbits restricted by the requirement that the angular momentum be an integral multiple of $\hbar$, that is, for circular orbits of radius a_n, the electron momentum is restricted by

$$pa_n = n\hbar, \qquad (n = 1, 2, 3, \ldots), \qquad (10.108)$$

and furthermore the electrons in these orbits do not radiate in spite of their acceleration. They were said to be in stationary states.

2. Electrons can make discontinuous transitions from one allowed orbit corresponding to $n = n_1$ to another corresponding to $n = n_2$, and the change in energy will appear as radiation with frequency

$$\omega = \frac{E_{n_2} - E_{n_1}}{\hbar}. \qquad (10.109)$$

An atom may absorb radiation by having its electrons make a transition to a higher energy orbit.

Using the first Bohr's postulate and Eq. (10.108) we obtain

$$\alpha D_\alpha \left(\frac{n\hbar}{a_n} \right)^\alpha = \frac{Ze^2}{a_n},$$

from which it follows that the equation for the radius of the fractional Bohr orbits is

$$a_n = a_0 n^{\frac{\alpha}{\alpha - 1}}, \qquad (10.110)$$

here a_0 is the fractional Bohr radius (the radius of the lowest, $n = 1$ Bohr orbit) defined as,

$$a_0 = \left(\frac{\alpha D_\alpha \hbar^\alpha}{Ze^2}\right)^{\frac{1}{\alpha-1}}. \tag{10.111}$$

By using Eq. (10.107) we find for the total average energy $\overline{E}$

$$\overline{E} = (1 - \alpha)\overline{E}_{kin}.$$

Thus, for the energy levels of the fractional hydrogen-like atom we have

$$E_n = (1 - \alpha)E_0 n^{-\frac{\alpha}{\alpha-1}}, \qquad 1 < \alpha \leq 2, \tag{10.112}$$

where E_0 is the binding energy of the electron in the lowest Bohr orbit $n = 1$, that is, the energy required to put it in a state with $E = 0$ corresponding to $n = \infty$,

$$E_0 = \left(\frac{(Ze^2)^\alpha}{\alpha^\alpha D_\alpha \hbar^\alpha}\right)^{\frac{1}{\alpha-1}}. \tag{10.113}$$

The quantity $(\alpha - 1)E_0$ in Eq. (10.112) can be considered as a fractional generalization of the Rydberg constant $\mathrm{Ry} = me^4/2\hbar^2$.

According to the second Bohr postulate the frequency of the radiation ω associated with the transition, say, for example from m to n, $m \to n$, is,

$$\omega = \frac{(1 - \alpha)E_0}{\hbar} \cdot \left[\frac{1}{n^{\frac{\alpha}{\alpha-1}}} - \frac{1}{m^{\frac{\alpha}{\alpha-1}}}\right]. \tag{10.114}$$

The new equations (10.110)-(10.114) give us the fractional generalization of the "*Bohr atom*" theory. In a special Gaussian case (standard quantum mechanics) Eqs. (10.110)-(10.114) allow one to reproduce the well-known results of the Bohr theory [120], [121]. The existence of Eqs. (10.110)-(10.114) is a result of deviation of fractal dimension $d_{\text{fractal}}^{(L\acute{e}vy)}$ of the Lévy-like quantum mechanical path from 2, that is $d_{\text{fractal}}^{(L\acute{e}vy)} = \alpha < 2$.

Chapter 11

Fractional Nonlinear Quantum Dynamics

The nonlinear Schrödinger equation is an attractive and fast developing area of studies. Not pretending to give a complete list of publications on the topic, let us mention some of them, such as [123]-[129].

Our intent is to show how nonlinear fractional Schrödinger equation and fractional Ginzburg–Landau equation emerge from quantum dynamics with long-range interaction.

11.1 Exciton-phonon quantum dynamics with long-range interaction

Dynamic lattice models are widely used to study a broad set of physical phenomena and systems. In the early 1970s a novel mechanism for the localization and transport of vibrational energy in certain types of molecular chains was proposed by A.S. Davydov [130]. He pioneered the concept of the solitary excitons or the Davydov's soliton [131]. Our focus is analytical developments on quantum 1D exciton-phonon dynamics with power-law long-range exciton-exciton interaction $J_{n,m} = J/|n - m|^s$, $(s > 0)$ for excitons located at the lattice sites n and m. In addition to the well-known interactions with integer values of s, some complex media can be described by fractional values of s (see, for example, references in [132]). Using the ideas first developed by Laskin and Zaslavsky [133], we elaborate the Davydov model for the exciton-phonon system with a long-range power-law exciton-exciton interaction. We will show that 1D lattice exciton-phonon dynamics in the long-wave limit can be effectively presented by the system of two coupled equations for exciton and phonon sub-systems. The dynamic equation describing the exciton sub-system is the fractional differential equation, which is a manifestation of nonlocality of interaction, originating from the long-range interaction term. The dynamic equation describing the phonon sub-system is the differential equation.

From this system of two coupled equations we obtain few fundamental theoretical frameworks to study nonlinear quantum dynamics with long-range interaction. They are: non-linear fractional Schrödinger equation, non-linear Hilbert–Schrödinger equation, fractional generalization of Zakharov system, fractional Ginzburg–Landau equation [134].

11.2   Long-range exciton-exciton interaction

To model 1D quantum lattice dynamics let us consider a linear, rigid arrangement of sites with one molecule at each lattice site. Then, Davydov's Hamiltonian [135] reads

$$\widehat{H} = \widehat{H}_{\mathrm{ex}} + \widehat{H}_{\mathrm{ph}} + \widehat{H}_{\mathrm{int}}, \tag{11.1}$$

where $\widehat{H}_{\mathrm{ex}}$ is exciton Hamilton operator, which describes dynamics of intramolecular excitations or simply excitons, $\widehat{H}_{\mathrm{ph}}$ is phonon Hamiltonian operator, which describes displacements of molecules from their equilibrium states or, in other words, the lattice vibrations, and H_{int} is the exciton-phonon operator, which describes interaction of molecular excitations (excitons) with their displacements (lattice vibrations).

The exciton Hamiltonian H_{ex} is

$$\widehat{H}_{\mathrm{ex}} = \varepsilon \sum_{n=-\infty}^{\infty} b_n^+ b_n - \mathcal{J} \sum_{n=-\infty}^{\infty} (b_n^+ b_{n+1} + b_n^+ b_{n-1}), \tag{11.2}$$

here b_n^+ is creation and b_n is annihilation operators of an excitation on a molecule on site n, parameter ε is exciton energy and $\mathcal{J}$ is interaction constant.

Operators b_n^+ and b_n satisfy the commutation relations $[b_n, b_m^+] = \delta_{n,m}$, $[b_n, b_m] = 0$, $[b_n^+, b_m^+] = 0$.

Alternatively, the Hamiltonian H_{ex} can be written as

$$\widehat{H}_{\mathrm{ex}} = \varepsilon \sum_{n=-\infty}^{\infty} b_n^+ b_n - \sum_{n,m=-\infty}^{\infty} \mathcal{J}_{n,m} b_n^+ b_m, \tag{11.3}$$

if we introduce exciton transfer matrix $\mathcal{J}_{n,m}$, which describes exciton-exciton interaction,

$$\mathcal{J}_{n,m} = \mathcal{J}(\delta_{(n+1),m} + \delta_{(n-1),m}), \tag{11.4}$$

where the Kronecker symbols $\delta_{(n+1),m}$ mean that only the nearest-neighbor sites have been considered. In other words, the interaction term $\mathcal{J}_{n,m} b_n^+ b_m$ in Eq. (11.3) is responsible for transfer from site n to the nearest-neighbor sites $n \pm 1$.

The phonon Hamiltonian H_{ph} is

$$\widehat{H}_{\mathrm{ph}} = \sum_{n=-\infty}^{\infty} \left(\frac{\widehat{p}_n^2}{2m} + \frac{w}{2} (\widehat{u}_{n+1} - \widehat{u}_n)^2 \right), \qquad (11.5)$$

where w is the elasticity constant of the 1D lattice and $\widehat{u}_n$ is the operator of displacement of a molecule from its equilibrium position on the site n, $\widehat{p}_n$ is the momentum operator of a molecule on site n and m is molecular mass.

The exciton-phonon Hamiltonian $\widehat{H}_{\mathrm{int}}$, which describes the coupling between internal molecular excitations with their displacements, has the form

$$\widehat{H}_{\mathrm{int}} = \frac{\chi}{2} \sum_{n=-\infty}^{\infty} (\widehat{u}_{n+1} - \widehat{u}_n) b_n^+ b_n, \qquad (11.6)$$

with the coupling constant χ.

To extend Davydov's model and go beyond the nearest-neighbor interaction we introduce, follow Laskin [134], the power-law interaction between excitons on sites n and m. Thus, to study long-range exciton-exciton interaction we come up with exciton transfer matrix $\mathcal{J}_{n,m}$ given by

$$\mathcal{J}_{n-m} = \frac{\mathcal{J}}{|n-m|^s}, \qquad n \neq m, \qquad (11.7)$$

where $\mathcal{J}$ is the interaction constant, parameter s covers different physical models; the nearest-neighbor approximation $(s = \infty)$, the dipole-dipole interaction $(s = 3)$, the Coulomb potential $(s = 1)$. Our main interest will be in fractional values of s that can appear for more sophisticated interaction potentials attributed to complex media.

Aiming to obtain a system of coupled dynamic equations for the exciton-photon system under consideration, we introduce Davydov's ansatz.

11.2.1 *Davydov's ansatz*

To study system described by Eq. (11.1) with exciton transfer matrix given by Eq. (11.7) we introduce quantum state vector following [135] [136], see, also [137]

$$|\phi(t) \rangle = |\Psi(t) > |\Phi(t) >, \qquad (11.8)$$

where quantum vectors $|\Psi(t)>$ and $|\Phi(t)>$ are defined by

$$|\Psi(t)> = \sum_n \psi_n(t) b_n^+ |0>_{\text{ex}} \qquad (11.9)$$

and

$$|\Phi(t)> = \exp\left\{-\frac{i}{\hbar} \sum_n (\xi_n(t)\widehat{p}_n - \eta_n(t)\widehat{u}_n)\right\} |0>_{\text{ph}}, \qquad (11.10)$$

here $\hbar$ is Planck's constant, $|0>_{\text{ex}}$ and $|0>_{\text{ph}}$ are vacuum states of the exciton and phonon sub-systems and $\xi_n(t)$ is the diagonal matrix element of the displacement operator $\widehat{u}_n$ in the basis defined by Eq. (11.8), while $\eta_n(t)$ is diagonal matrix element of the momentum operator $\widehat{p}_n$ in the same basis, that is

$$\xi_n(t) = <\phi(t)|\widehat{u}_n|\phi(t)>, \qquad \eta_n(t) = <\phi(t)|\widehat{p}_n|\phi(t)> . \qquad (11.11)$$

The displacement $\widehat{u}_n$ and momentum $\widehat{p}_n$ operators satisfy the commutation relations

$$[\widehat{u}_n, \widehat{p}_m] = i\hbar \delta_{n,m}, \qquad (11.12)$$

where $\hbar$ is Planck's constant and $\delta_{n,m}$ is the Kronecker symbol,

$$\delta_{n,m} = \begin{cases} 1 & n = m, \\ 0 & n \neq m. \end{cases}$$

The state vector $|\phi(t)>$ satisfies the normalization condition

$$<\phi(t)|\phi(t)> = \sum_n |\psi_n(t)|^2 = N, \qquad (11.13)$$

with $|\psi_n(t)|^2$ being the probability to find exciton on the nth site and N is the total number of excitons.

Therefore, the dynamics of the exciton-photon system with Hamiltonian given by Eq. (11.1) can be described in terms of the functions $\psi_n(t)$, $\xi_n(t)$ and $\eta_n(t)$. In other words, Davydov's ansatz defined by Eqs. (11.8)-(11.11) allows us to go from the quantum Hamilton operator introduced by Eq. (11.1) to classical Hamiltonian function developed below. In the basis of the vectors $|\phi(t)>$, the Hamilton operators $\widehat{H}_{\text{ex}}$, $\widehat{H}_{\text{ph}}$ and $\widehat{H}_{\text{int}}$ become the functions $H_{\text{ex}}(\psi_n, \psi_n^*)$, $H_{\text{ph}}(\xi_n, \eta_n)$ and $H_{\text{int}}(\psi_n, \psi_n^*, \xi_n)$ of classical dynamic variables $\psi_n(t)$, $\psi_n^*(t)$, $\xi_n(t)$ and $\eta_n(t)$.

Thus, the function $H_{\text{ex}}(\psi_n, \psi_n^*)$ is introduced as

$$H_{\text{ex}}(\psi_n, \psi_n^*) = < \phi(t)|\widehat{H}_{\text{ex}}|\phi(t) > \tag{11.14}$$

$$= \varepsilon \sum_{n=-\infty}^{\infty} \psi_n^*(t)\psi_n(t) - \sum_{n,m=-\infty}^{\infty} \mathcal{J}_{n-m}\psi_n^*(t)\psi_m(t),$$

here $\mathcal{J}_{n-m}$ is given by Eq. (11.7).

The function $H_{\text{ph}}(\xi_n, \eta_n)$ is introduced as

$$H_{\text{ph}}(\xi_n, \eta_n) = < \phi(t)|\widehat{H}_{\text{ph}}|\phi(t) > \tag{11.15}$$

$$= \sum_{n=-\infty}^{\infty} \left(\frac{\eta_n^2}{2m} + \frac{w}{2}(\xi_{n+1} - \xi_n)^2 \right).$$

The function $H_{\text{int}}(\psi_n, \psi_n^*, \xi_n)$ is introduced as

$$H_{\text{int}}(\psi_n, \psi_n^*, \xi_n) = < \phi(t)|\widehat{H}_{\text{int}}|\phi(t) > \tag{11.16}$$

$$= \frac{\chi}{2} \sum_{n=-\infty}^{\infty} (\xi_{n+1} - \xi_n)\psi_n^*(t)\psi_n(t).$$

Combining together Eqs. (11.14)-(11.16) we obtain the Hamiltonian function $H(\psi_n, \psi_n^*, \xi_n, \eta_n)$ of the exciton-phonon system under consideration

$$H(\psi_n, \psi_n^*, \xi_n, \eta_n) = < \phi(t)|\widehat{H}|\phi(t) >$$

$$= H_{\text{ex}}(\psi_n, \psi_n^*) + H_{\text{ph}}(\xi_n, \eta_n) + H_{\text{int}}(\psi_n, \psi_n^*, \xi_n) \tag{11.17}$$

$$= \varepsilon \sum_{n=-\infty}^{\infty} \psi_n^*(t)\psi_n(t) - \sum_{n,m=-\infty}^{\infty} \mathcal{J}_{n-m}\psi_n^*(t)\psi_m(t)$$

$$+ \sum_{n=-\infty}^{\infty} \left(\frac{\eta_n^2}{2m} + \frac{w}{2}(\xi_{n+1} - \xi_n)^2 \right)$$

$$+ \frac{\chi}{2} \sum_{n=-\infty}^{\infty} (\xi_{n+1} - \xi_n)\psi_n^*(t)\psi_n(t).$$

Having the Hamiltonian function $H(\psi_n, \psi_n^*, \xi_n, \eta_n)$ we can develop the motion equations for dynamic variables $\psi_n(t)$, $\psi_n^*(t)$, $\xi_n(t)$ and $\eta_n(t)$.

11.2.2 *Motion equations*

Following [135] and identifying conjugate coordinates and momenta from the set of dynamic variables $\psi_n(t)$, $\psi_n^*(t)$, $\xi_n(t)$ and $\eta_n(t)$, we come up with the system of motion equations in terms of variational derivatives $\delta/\delta\psi_n^*(t)$, $\delta/\delta\psi_n(t)$, $\delta/\delta\eta_n(t)$ and $\delta/\delta\xi_n(t)$. For the variable $\psi_n(t)$ the motion equation is

$$i\hbar\frac{\partial\psi_n(t)}{\partial t} = \frac{\delta}{\delta\psi_n^*(t)}H(\psi_n,\psi_n^*,\xi_n,\eta_n). \tag{11.18}$$

For the complex conjugate variable $\psi_n^*(t)$ we have

$$i\hbar\frac{\partial\psi_n^*(t)}{\partial t} = -\frac{\delta}{\delta\psi_n(t)}H(\psi_n,\psi_n^*,\xi_n,\eta_n). \tag{11.19}$$

For the coordinate $\xi_n(t)$ the motion equation reads

$$\frac{\partial\xi_n(t)}{\partial t} = \frac{\delta}{\delta\eta_n(t)}H(\psi_n,\psi_n^*,\xi_n,\eta_n), \tag{11.20}$$

and for conjugate momenta $\eta_n(t)$ the motion equation is

$$\frac{\partial\eta_n(t)}{\partial t} = -\frac{\delta}{\delta\xi_n(t)}H(\psi_n,\psi_n^*,\xi_n,\eta_n), \tag{11.21}$$

here $H(\psi_n,\psi_n^*,\xi_n,\eta_n)$ is given by Eq. (11.16).

For our purposes we will need the system of dynamic equations for $\psi_n(t)$, $\xi_n(t)$ and $\eta_n(t)$. Calculating the variational derivatives yields the following system of three coupled equations

$$i\hbar\frac{\partial\psi_n(t)}{\partial t} = \Lambda\psi_n(t) - \sum_{\substack{m\\(n\neq m)}} \mathcal{J}_{n-m}\psi_m(t) + \frac{\chi}{2}(\xi_{n+1} - \xi_n)\psi_n(t), \tag{11.22}$$

$$\frac{\partial\xi_n(t)}{\partial t} = \frac{\eta_n}{\mathrm{m}}, \tag{11.23}$$

and

$$\frac{\partial\eta_n(t)}{\partial t} = w(\xi_{n+1}(t) - 2\xi_n(t) + \xi_{n-1}(t)) + \frac{\chi}{2}(|\psi_{n+1}(t)|^2 - |\psi_n(t)|^2|), \tag{11.24}$$

where the constant Λ is introduced by

$$\Lambda = \varepsilon + \frac{1}{2} \sum_{n=-\infty}^{\infty} \left(m \left(\frac{\partial \xi_n(t)}{\partial t} \right)^2 + w(\xi_{n+1} - \xi_n)^2 \right).$$

Further, substituting $\eta_n(t)$ from Eq. (11.23) into Eq. (11.24) yields

$$m \frac{\partial^2 \xi_n(t)}{\partial t^2} = w(\xi_{n+1} - 2\xi_n + \xi_{n-1}) + \frac{\chi}{2}(|\psi_{n+1}(t)|^2 - |\psi_n(t)|^2. \quad (11.25)$$

Our focus now is the system of two coupled discrete dynamic equations (11.22) and (11.25).

11.3 From lattice to continuum

To obtain the dynamical equations in continuum space from discrete equations (11.22) and (11.25) we need a map to go from discrete functions $\psi_n(t)$ and $\xi_n(t)$ to their continuum counterparts. Aiming to develop the map we define two functions $\varphi(k,t)$ and $\nu(k,t)$ in k space by means of the following equations

$$\varphi(k,t) = \sum_{n=-\infty}^{\infty} a e^{-ikna} \psi_n(t), \quad (11.26)$$

and

$$\nu(k,t) = \sum_{n=-\infty}^{\infty} a e^{-ikna} \xi_n(t). \quad (11.27)$$

With the help of identity

$$\sum_{n=-\infty}^{\infty} a e^{-ikna} \left(\frac{1}{2\pi} \int_{-\pi}^{\pi} dk' e^{ik'na} f(k',t) \right) = f(k,t)$$

we conclude that the function $\psi_n(t)$ is related to $\varphi(k,t)$ as follows

$$\psi_n(t) = \frac{1}{2\pi} \int_{-\pi}^{\pi} dk e^{ikna} \varphi(k,t), \quad (11.28)$$

and the function $\xi_n(t)$ is related to $\nu(k,t)$ as follows

$$\xi_n(t) = \frac{1}{2\pi} \int_{-\pi}^{\pi} dk e^{ikna} \nu(k,t), \quad (11.29)$$

here a is the lattice parameter, $a > 0$.

The lattice can be treated as a continuum media in the physical situation when wavelength $\lambda = 2\pi/k$ of dynamical processes in the exciton-phonon system exceeds the scale na, $\lambda \gg na$. In other words, this is the case when $k \ll (na)^{-1}$, and we can substitute discrete functions $\psi_n(t)$ and $\xi_n(t)$ with their continuum counterparts $\psi(x,t)$ and $\xi(x,t)$, that is

$$\psi_n(t) \xrightarrow[k \ll (na)^{-1}]{} \psi(x,t), \tag{11.30}$$

and

$$\xi_n(t) \xrightarrow[k \ll (na)^{-1}]{} \xi(x,t), \tag{11.31}$$

where x is continuous variable $x = na$.

In the long-wavelength approximation $k \ll (na)^{-1}$ the integration in Eqs. (11.28) and (11.29) can be expanded over the whole k space,

$$\frac{1}{2\pi} \int_{-\pi}^{\pi} dk e^{ikna} \dots \xrightarrow[k \ll (na)^{-1}]{} \frac{1}{2\pi} \int_{-\infty}^{\infty} dk e^{ikx} \dots, \tag{11.32}$$

while the summation in Eqs. (11.26) and (11.27) has to be substituted with integration over x space,

$$\sum_{n=-\infty}^{\infty} a e^{-ikna} \dots \xrightarrow[k \ll (na)^{-1}]{} \int_{-\infty}^{\infty} dx e^{-ikx} \dots. \tag{11.33}$$

It is easy to see that the integrations over k and x spaces support the identity

$$\int_{-\infty}^{\infty} dx e^{-ikx} \left(\frac{1}{2\pi} \int_{-\infty}^{\infty} dk' e^{ik'x} f(k',t) \right) = f(k,t).$$

The equations (11.30)-(11.33) present the mapping to go from lattice to continuum media. We apply this mapping to go from discrete to continuous nonlocal dynamical equations for the exciton-phonon system under consideration.

In the case of continuum media Eqs. (11.28) and (11.29) become

$$\psi(x,t) = \frac{1}{2\pi} \int_{-\infty}^{\infty} dk e^{ikx} \varphi(k,t), \tag{11.34}$$

and

$$\xi(x,t) = \frac{1}{2\pi} \int\limits_{-\infty}^{\infty} dk e^{ikx} \nu(k,t). \tag{11.35}$$

Functions $\varphi(k,t)$ and $\nu(k,t)$ are expressed by the following two equations

$$\varphi(k,t) = \int\limits_{-\infty}^{\infty} dx e^{-ikx} \psi(x,t) \tag{11.36}$$

and

$$\nu(k,t) = \int\limits_{-\infty}^{\infty} dx e^{-ikx} \xi(x,t). \tag{11.37}$$

Therefore, we conclude that in terms of the functions $\psi(x,t)$ and $\xi(x,t)$ Eqs. (11.22) and (11.25) become continuous equations of motion

$$i\hbar \frac{\partial \psi(x,t)}{\partial t} = \lambda \psi(x,t) \tag{11.38}$$

$$- \int\limits_{-\infty}^{\infty} dy \frac{\partial}{\partial x} K(x-y) \frac{\partial}{\partial y} \psi(y,t) + \chi \frac{\partial \xi(x,t)}{\partial x} \psi(x,t),$$

and

$$m \frac{\partial^2 \xi(x,t)}{\partial t^2} = w \frac{\partial^2 \xi(x,t)}{\partial x^2} + \chi \frac{\partial |\psi(x,t)|^2}{\partial x}. \tag{11.39}$$

Quantum equation (11.38) is nonlocal due to the presence of the integral term. The kernel $K(x)$ in the integral term of Eq. (11.38) has a form

$$K(x) = \frac{1}{\pi} \int\limits_{-\infty}^{\infty} dk e^{ikx} \frac{\mathcal{V}(k)}{k^2},$$

where the function $\mathcal{V}(k)$ is introduced as

$$\mathcal{V}(k) = \mathcal{J}(0) - \mathcal{J}(k), \tag{11.40}$$

and $\mathcal{J}(k)$ is defined by

$$\mathcal{J}(k) = \sum_{\substack{n=-\infty \\ (n \neq 0)}}^{\infty} e^{-ikna} \mathcal{J}_n, \tag{11.41}$$

with $\mathcal{J}_n$ given by Eq. (11.7) and, finally, $\lambda = \Lambda - J(0)$.

Thus, we come to a new system of coupled dynamic equations (11.38) and (11.39) to model 1D exciton phonon dynamics with long-range exciton-exciton interaction introduced by Eq. (11.7). The field $\psi(x,t)$ describes the exciton sub-system, while the field $\xi(x,t)$ describes the phonon sub-system. Equation (11.38) is the integro-differential equation while Eq. (11.39) is the differential one. The integral term in Eq. (11.38) originates from the long-range exciton-exciton interaction term - the second term in $H_{\text{ex}}(\psi_n, \psi_n^*)$ introduced by Eq. (11.14).

11.4 Fractional nonlinear quantum equations

To transform the system (11.38) and (11.39) into the system of coupled differential equations of motion we use the properties of function $\mathcal{V}(k)$ defined by Eq. (11.40) in the continuum limit $a \to 0$ (long wave limit $k \to 0$), which can be obtained from the asymptotics of the polylogarithm (see, Appendix D)

$$\mathcal{V}(k) \sim D_s |ak|^{s-1}, \qquad 2 \leq s < 3, \tag{11.42}$$

$$\mathcal{V}(k) \sim -\mathcal{J}(ak)^2 \ln ak, \qquad s = 3, \tag{11.43}$$

$$\mathcal{V}(k) \sim \frac{\mathcal{J}\zeta(s-2)}{2}(ak)^2, \qquad s > 3, \tag{11.44}$$

where $\Gamma(s)$ is the Gamma function, $\zeta(s)$ is the Riemann zeta function and coefficient D_s is defined by

$$D_s = \frac{\pi \mathcal{J}}{\Gamma(s) \sin(\pi(s-1)/2)}. \tag{11.45}$$

It is seen from Eq. (11.42) that the fractional power of k occurs for interactions with $2 \leq s \leq 3$ only. In the coordinate space fractional power of $|ak|^{s-1}$ gives us the fractional Riesz derivative of order $s-1$ [37], [133], [138], and we come to a fractional differential equation [134]

$$i\hbar\frac{\partial\psi(x,t)}{\partial t} = \lambda\psi(x,t)$$ (11.46)

$$-D_s a^{s-1}\partial_x^{s-1}\psi(x,t) + \chi\frac{\partial\xi(x,t)}{\partial x}\psi(x,t),$$

here ∂_x^{s-1} is the Riesz fractional derivative

$$\partial_x^{s-1}\psi(x,t) = -\frac{1}{2\pi}\int\limits_{-\infty}^{\infty} dk e^{ikx}|k|^{s-1}\varphi(k,t), \qquad 2 \le s < 3,$$ (11.47)

where $\psi(x,t)$ and $\varphi(k,t)$ are related to each other by the Fourier transforms defined by Eqs. (11.34) and (11.36).

Thus, we obtained the system of coupled equations, Eqs. (11.39) and (11.46) to study one-dimensional exciton-phonon dynamics with long-range interaction.

11.4.1 *Fractional nonlinear Schrödinger equation*

The system of coupled equations, Eqs. (11.39) and (11.46) can be further elaborated to get nonlinear fractional quantum and classical equations. Indeed, by assuming the existence of a stationary solution $\partial\xi(x,t)/\partial t = 0$ to Eq. (11.39) we obtain from Eq. (11.46) the following quantum fractional differential equation for wave function $\psi(x,t)$,

$$i\hbar\frac{\partial\psi(x,t)}{\partial t} = \lambda\psi(x,t)$$ (11.48)

$$-D_s a^{s-1}\partial_x^{s-1}\psi(x,t) - \frac{\chi^2}{w}|\psi(x,t)|^2\psi(x,t), \qquad 2 \le s < 3,$$

which can be rewritten in the form of fractional nonlinear Schrödinger equation [134],

$$i\hbar\frac{\partial\phi(x,t)}{\partial t} = -D_s a^{s-1}\partial_x^{s-1}\phi(x,t) - \frac{\chi^2}{w}|\phi(x,t)|^2\phi(x,t),$$ (11.49)

where $2 \leq s < 3$ and the wave function $\phi(x,t)$ is related to the wave function $\psi(x,t)$ by

$$\phi(x,t) = \exp\{i\lambda t/\hbar\}\psi(x,t). \tag{11.50}$$

It is easy to see that Eq. (11.49) supports normalization condition for wave function $\phi(x,t)$, and the normalization condition is

$$\int\limits_{-\infty}^{\infty} dx|\phi(x,t)|^2 = 1. \tag{11.51}$$

Thus, using the system of coupled equations, Eqs. (11.39) and (11.46), we discovered the fractional generalization of nonlinear Schrödinger equation (11.49) with a cubic focusing nonlinearity. The equation (11.49) has to be supplied with initial condition $\phi(x,t)|_{t=0} = \phi(x,0)$.

To express Eq. (11.49) in dimensionless form let's apply the scaling transforms

$$t' = \omega t, \qquad x' = \frac{x}{l}, \qquad \phi'(x',t') = \sqrt{l}\phi(x,t), \tag{11.52}$$

where we introduce dimensionless time t' and length x', while ω is characteristic frequency and l is characteristic space scale. The transform $\phi'(x',t') = \sqrt{l}\phi(x,t)$ supports the normalization condition Eq. (11.51).

Applying the scale transforms to Eq. (11.49) results in

$$i\hbar\omega \frac{\partial \phi'(x',t')}{\partial t'}$$

$$= -D_s a^{s-1} \frac{1}{l^{s-1}} \partial_{x'}^{s-1} \phi(x',t') - \frac{\chi^2}{wl}|\phi'(x',t')|^2 \phi'(x',t'), \tag{11.53}$$

$$2 \leq s < 3.$$

By choosing

$$l = \left(\frac{D_s}{\hbar\omega}\right)^{1/(s-1)} a, \tag{11.54}$$

we obtain the equation

$$i\frac{\partial \phi'(x',t')}{\partial t'} = -\partial_{x'}^{s-1}\phi'(x',t') - \varkappa|\phi'(x',t')|^2\phi'(x',t'),$$

where dimensionless parameter $\varkappa$ is introduced by

$$\varkappa = \frac{\chi^2(\hbar\omega)^{(2-s)/(s-1)}}{waD_s^{1/(s-1)}}. \tag{11.55}$$

Finally, by renaming $x' \to x$, $t' \to t$ and $\phi' \to \phi$ we come to dimensionless fractional nonlinear Schrödinger equation

$$i\frac{\partial \phi(x,t)}{\partial t} = -\partial_x^{s-1}\phi(x,t) - \varkappa|\phi(x,t)|^2\phi(x,t), \qquad 2 \leq s < 3, \tag{11.56}$$

with $\varkappa$ given by Eq. (11.55) and ∂_x^{s-1} being the Riesz fractional derivative defined by Eq. (11.47).

Due to Eq. (11.44) for $s > 3$, Eq. (11.49) turns into the nonlinear Schrödinger equation with a cubic focusing nonlinearity [134],

$$i\hbar\frac{\partial \phi(x,t)}{\partial t} = -\frac{\mathcal{J}\zeta(s-2)}{2}a^2\partial_x^2\phi(x,t) - \frac{\chi^2}{w}|\phi(x,t)|^2\phi(x,t), \tag{11.57}$$

where $s > 3$ and $\partial_x^2 = \partial^2/\partial x^2$.

11.4.2 Nonlinear Hilbert–Schrödinger equation

It follows from Eqs. (11.42) and (11.45) that in the case $s = 2$ the function $\mathcal{V}(k)$ in the long wave limit $k \to 0$ takes the form

$$\mathcal{V}(k) \sim \pi\mathcal{J}|ak|, \qquad s = 2. \tag{11.58}$$

In this case, assuming the existence of a stationary solution $\partial\xi(x,t)/\partial t = 0$ to Eq. (11.39) we find from Eq. (11.38) the following nonlinear quantum fractional differential equation for wave function $\psi(x,t)$,

$$i\hbar\frac{\partial \psi(x,t)}{\partial t} = \lambda\psi(x,t) \tag{11.59}$$

$$-\pi\mathcal{J}a\mathcal{H}\{\partial_x\psi(x,t)\} - \frac{\chi^2}{w}|\psi(x,t)|^2\psi(x,t), \qquad s = 2,$$

here $\mathcal{H}\{...\}$ is the Hilbert integral transform defined by

$$\mathcal{H}\{\psi(x,t)\} = \frac{1}{\pi}\text{PV} \int\limits_{-\infty}^{\infty} dy \frac{\psi(y,t)}{x-y}, \qquad (11.60)$$

where PV stands for the Cauchy principal value of the integral.

Introducing wave function $\phi(x,t)$ related to the wave function $\psi(x,t)$ by means of Eq. (11.50) brings us the nonlinear Hilbert–Schrödinger equation [133], [134]

$$i\hbar\frac{\partial\varphi(x,t)}{\partial t} = -\pi\mathcal{J}a\mathcal{H}\{\partial_x\varphi(x,t)\} - \frac{\chi^2}{w}|\varphi(x,t)|^2\varphi(x,t), \qquad (11.61)$$

with normalization condition given by Eq. (11.51).

11.4.3   *Fractional generalization of Zakharov system*

Introducing a new variable $\sigma(x,t) = \partial\xi(x,t)/\partial x$ turns Eqs. (11.39) and (11.48) into the following system of two coupled quantum equations [134] for the fields $\psi(x,t)$ and $\sigma(x,t)$,

$$i\hbar\frac{\partial\psi(x,t)}{\partial t} = \lambda\psi(x,t) \qquad (11.62)$$

$$-D_s a^{s-1}\partial_x^{s-1}\psi(x,t) + \chi\sigma(x,t)\psi(x,t),$$

and

$$\left(\frac{\partial^2}{\partial t^2} - \nu^2\frac{\partial^2}{\partial x^2}\right)\sigma(x,t) = \frac{\chi}{m}\frac{\partial^2}{\partial x^2}|\psi(x,t)|^2, \qquad (11.63)$$

where $\nu = \sqrt{w/m}$ is physical parameter with the unit of velocity, the parameter D_s is defined by Eq. (11.45), ∂_x^{s-1} is the Riesz fractional derivative introduced by Eq. (11.47) and parameter s is in the range $2 \leq s < 3$. Equation (11.62) is fractional differential equation, while Eq. (11.63) includes spatial and temporal derivatives of second order.

Considering wave function $\phi(x,t)$ related to the wave function $\psi(x,t)$ by means of Eq. (11.50), we rewrite Eqs. (11.62) and (11.63) in the form [134]

$$i\hbar\frac{\partial\phi(x,t)}{\partial t} = -D_s a^{s-1}\partial_x^{s-1}\phi(x,t) + \chi\sigma(x,t)\phi(x,t), \qquad 2 \leq s < 3, \quad (11.64)$$

and

$$\left(\frac{\partial^2}{\partial t^2} - \nu^2 \frac{\partial^2}{\partial x^2}\right)\sigma(x,t) = \frac{\chi}{\mathrm{m}}\frac{\partial^2}{\partial x^2}|\phi(x,t)|^2. \tag{11.65}$$

To express Eqs. (11.64) and (11.65) in dimensionless form, let's apply the scaling transforms

$$t' = \omega t, \qquad x' = \frac{x}{l}, \tag{11.66}$$

$$\phi'(x',t') = \sqrt{l}\phi(x,t), \qquad \sigma'(x',t') = l\sigma(x,t),$$

where we introduce dimensionless time t' and length x', while ω is characteristic frequency and l is characteristic space scale. The transform $\phi'(x',t') = \sqrt{l}\phi(x,t)$ supports the normalization condition Eq. (11.51).

Plugging those transforms into Eqs. (11.64) and (11.65) results in

$$i\hbar\omega\frac{\partial\phi'(x',t')}{\partial t'} = -D_s\left(\frac{a}{l}\right)^{s-1}\partial_{x'}^{s-1}\phi(x',t') + \chi\sigma'(x',t')\phi'(x',t'),$$

with $2 \le s < 3$, and

$$\left(\omega^2\frac{\partial^2}{\partial t'^2} - \frac{\nu^2}{l^2}\frac{\partial^2}{\partial x'^2}\right)\sigma'(x',t') = \frac{\chi}{\mathrm{m}l^2}\frac{\partial^2}{\partial x'^2}|\phi'(x',t')|^2.$$

Further, by choosing

$$l = \left(\frac{D_s}{\hbar\omega}\right)^{1/(s-1)}, \tag{11.67}$$

we have the system of coupled equations

$$i\frac{\partial\phi'(x',t')}{\partial t'} = -\partial_{x'}^{s-1}\phi'(x',t') + \gamma\sigma'(x',t')\phi'(x',t') \tag{11.68}$$

and

$$\left(\frac{\partial^2}{\partial t'^2} - \varkappa^2\frac{\partial^2}{\partial x'^2}\right)\sigma'(x',t') = \beta\frac{\partial^2}{\partial x'^2}|\phi'(x',t')|^2, \tag{11.69}$$

with dimensionless parameters γ, $\varkappa$ and β introduced by

$$\gamma = \frac{\chi}{\hbar\omega}, \qquad \varkappa = \frac{\nu}{\omega l}, \qquad \beta = \frac{\chi}{l^2\omega^2\mathrm{m}}. \tag{11.70}$$

Finally, by renaming $x' \to x$, $t' \to t$, $\phi' \to \phi$ and $\sigma' \to \sigma$ we come to the following system of coupled equations for the fields $\phi(x,t)$ and $\sigma(x,t)$

$$i\frac{\partial \phi(x,t)}{\partial t} = -\partial_x^{s-1}\phi(x,t) + \gamma\sigma(x,t)\phi(x,t), \qquad 2 \le s < 3, \qquad (11.71)$$

and

$$\left(\frac{\partial^2}{\partial t^2} - \varkappa^2\frac{\partial^2}{\partial x^2}\right)\sigma(x,t) = \beta\frac{\partial^2}{\partial x^2}|\phi(x,t)|^2, \qquad (11.72)$$

with γ, $\varkappa$ and β given by Eq. (11.70).

The system of coupled equations (11.71) and (11.72) is fractional generalization of the Zakharov system originally introduced in 1972 to study the Langmuir waves propagation in an ionized plasma [139].

In our approach, we came to the system of equations (11.71) and (11.72) by studying the quantum 1D exciton-phonon system assuming that the only exciton-exciton interaction has power-law long-range behavior.

11.4.4 *Fractional Ginzburg–Landau equation*

In the case of propagating waves one can search for the solution to the system of equations (11.64) and (11.65) in the form of travelling waves

$$\psi(x,t) = \psi(x - \mathrm{v}t) \qquad \text{and} \qquad \xi(x,t) = \xi(x - \mathrm{v}t), \qquad (11.73)$$

where v is wave velocity.

By introducing the notation $\zeta = (x - \mathrm{v}t)$ we can rewrite Eqs. (11.64) and (11.65) as

$$i\hbar\mathrm{v}\frac{\partial \phi(\zeta)}{\partial \zeta} = -D_s a^{s-1}\partial_\zeta^{s-1}\phi(\zeta) + \chi\sigma(\zeta)\phi(\zeta), \qquad 2 \le s < 3, \qquad (11.74)$$

and

$$(\mathrm{v}^2 - \nu^2)\frac{\partial^2}{\partial \zeta^2}\sigma(\zeta) = \frac{\chi}{\mathrm{m}}\frac{\partial^2}{\partial \zeta^2}|\phi(\zeta)|^2. \qquad (11.75)$$

It is easy to see that a solution to Eq. (11.75) is

$$\sigma(\zeta) = \frac{\chi}{\mathrm{m}(\mathrm{v}^2 - \nu^2)}|\phi(\zeta)|^2. \qquad (11.76)$$

Then Eq. (11.74) results in nonlinear fractional equation

$$i\hbar v \frac{\partial \phi(\zeta)}{\partial \zeta} = -D_s a^{s-1} \partial_\zeta^{s-1} \phi(\zeta) + \gamma |\phi(\zeta)|^2 \phi(\zeta), \qquad (11.77)$$

where nonlinearity parameter γ has been introduced as

$$\gamma = \frac{\chi^2}{m(v^2 - \nu^2)}.$$

We call Eq. (11.77) fractional Ginzburg–Landau equation. This is quantum nonlinear fractional differential equation in the framework of fractional quantum mechanics.

To express Eq. (11.77) in dimensionless form, let's apply the scaling transforms

$$\zeta' = \frac{\zeta}{b}, \qquad \phi'(\zeta') = \sqrt{b}\phi(\zeta), \qquad \sigma'(\zeta') = b\sigma(\zeta), \qquad (11.78)$$

where we introduce dimensionless length $\zeta'$and characteristic scale b. The transform $\phi'(\zeta') = \sqrt{b}\phi(\zeta)$ supports the normalization condition Eq. (11.51).

By performing the scaling transforms and then omitting the prime symbol, we obtain fractional nonlinear differential equation

$$i\frac{\partial \phi(\zeta)}{\partial \zeta} = -\epsilon_s \partial_\zeta^{s-1} \phi(\zeta) + \delta |\phi(\zeta)|^2 \phi(\zeta), \qquad 2 \le s < 3, \qquad (11.79)$$

with dimensionless coefficients ϵ_s and δ introduced by

$$\epsilon_s = \frac{D_s a^{s-1}}{b^{s-2}\hbar v}, \qquad \delta = \frac{\gamma b}{\hbar v} = \frac{\chi^2 b}{\hbar v m(v^2 - \nu^2)}. \qquad (11.80)$$

Thus, we came to the fractional Ginzburg–Landau equation (11.79) in dimensionless form.

11.5 Quantum lattice propagator

To study impact of long-range interaction on 1D quantum dynamics let us focus on the exciton sub-system only. Hence, follow [133], [134] we consider discrete linear problem associated with the exciton Hamiltonian $H_{\text{ex}}(\psi_n, \psi_n^*)$ given by Eq. (11.14).

Suppose that we know the solution $\psi_{n'}(t')$ to Eq. (11.14) at some time instant t' at the lattice site n'. Then the solution $\psi_n(t)$ at later time t, $(t > t')$ at the lattice site n will be

$$\psi_n(t) = \sum_{n'} G_{n,n'}(t|t')\psi_{n'}(t'), \tag{11.81}$$

where $G_{n,n'}(t|t')$ is quantum exciton 1D lattice propagator, that is the probability of exciton transition from site n' at the time moment t' to site n at the time moment t.

It follows from Eqs. (11.14) and (11.81) that $G_{n,n'}(t|t')$ is governed by the motion equation

$$i\hbar\frac{\partial G_{n,n'}(t|t')}{\partial t} = \varepsilon G_{n,n'}(t|t') - \sum_{\substack{m \\ (m\neq n)}} \mathcal{J}_{n-m}G_{m,n'}(t|t'), \qquad t \geq t', \tag{11.82}$$

with the initial condition

$$G_{n,n'}(t|t')|_{t=t'} = \delta_{n,n'}, \tag{11.83}$$

here $\delta_{n,n'}$ is the Kronecker symbol.

To avoid bulky notations let's choose $n' = 0$, $t' = 0$ and introduce $G_n(t)$ as

$$G_n(t) = G_{n,0}(t|0). \tag{11.84}$$

It yields

$$i\hbar\frac{\partial G_n(t)}{\partial t} = \varepsilon G_n(t) - \sum_{\substack{m \\ (m\neq n)}} \mathcal{J}_{n-m}G_n(t), \qquad t \geq t', \tag{11.85}$$

with the initial condition

$$G_n(t)|_{t=0} = \delta_{n,0}. \tag{11.86}$$

At this point we take into consideration the propagator $G(k,t)$ related to $G_n(t)$ as

$$G(k,t) = \sum_{n=-\infty}^{\infty} e^{-ikn} G_n(t), \qquad (11.87)$$

here we put the lattice parameter, $a = 1$ for simplify.

Hence, in terms of $G(k,t)$ the quantum lattice propagator $G_n(t)$ is given by[1]

$$G_n(t) = \frac{1}{2\pi} \int_{-\pi}^{\pi} dk e^{ikn} G(k,t). \qquad (11.88)$$

It follows from Eqs. (11.85), (11.87) and (11.41) that

$$i\hbar \frac{\partial G(k,t)}{\partial t} = (\omega + \mathcal{V}(k)) G(k,t), \qquad t \geq 0, \qquad (11.89)$$

where $\mathcal{V}(k)$ is defined by Eqs. (11.40) and (11.41), energy ω is given by

$$\omega = \varepsilon - \mathcal{J}(0) \qquad (11.90)$$

and

$$\mathcal{J}(0) = \sum_{\substack{n=-\infty \\ (n \neq 0)}}^{\infty} \mathcal{J}_n, \qquad (11.91)$$

with $\mathcal{J}_n$ given by Eq. (11.7). Then, it is easy to see from Eqs. (11.86) and (11.87) that the initial condition for $G(k,t)$ is

$$G(k,t=0) = 1. \qquad (11.92)$$

---

[1] The formula (1.17.17) [140]

$$\frac{1}{2\pi} \sum_{n=-\infty}^{\infty} e^{-ikn} \left(\int_{-\pi}^{\pi} dk' e^{ik'n} G(k',t) \right) = G(k,t)$$

has been used.

To solve Eq. (11.89) with the initial condition (11.92) let's introduce quantum propagator $g(k, t)$

$$g(k, t) = \exp(i\omega t/\hbar)G(k, t). \qquad (11.93)$$

Thus, $G(k, t)$ can be expressed in terms of $g(k, t)$ as

$$G(k, t) = \exp(-i\omega t/\hbar)g(k, t). \qquad (11.94)$$

It follows from Eq. (11.89) that the propagator $g(k, t)$ is governed by the equation

$$i\hbar\frac{\partial g(k, t)}{\partial t} = \mathcal{V}(k)g(k, t), \qquad t \geq 0, \qquad (11.95)$$

with the initial condition

$$g(k, t = 0) = 1. \qquad (11.96)$$

The solution to the problem defined by Eqs. (11.95) and (11.96) is

$$g(k, t) = \exp(-i\mathcal{V}(k)t/\hbar). \qquad (11.97)$$

By substituting Eqs. (11.94) and (11.97) into Eq. (11.88) we obtain

$$G_n(t) = \frac{1}{2\pi} \int\limits_{-\pi}^{\pi} dk \exp(ikn - i(\omega + \mathcal{V}(k))t/\hbar). \qquad (11.98)$$

Similarly to (11.94) we write

$$G_n(t) = \exp(-i\omega t/\hbar)g_n(t), \qquad (11.99)$$

where the quantum lattice propagator $g_n(t)$ has been introduced by

$$g_n(t) = \frac{1}{2\pi} \int\limits_{-\pi}^{\pi} dk \exp(ikn - i\mathcal{V}(k)t/\hbar). \qquad (11.100)$$

The generalization to quantum lattice propagator $g_{n,n'}(t|t')$ which describes transition of an exciton from site n' at the time moment t' to site n at the time moment t ($t > t'$), is obvious

$$g_{n,n'}(t|t') = \frac{1}{2\pi} \int\limits_{-\pi}^{\pi} dk \exp(ik(n - n') - i\mathcal{V}(k)(t - t')/\hbar). \qquad (11.101)$$

Therefore, the quantum lattice propagator $g_{n,n'}(t|t')$ can be considered as quantum transition amplitude. The propagator $g_{n,n'}(t|t')$ describes discrete in space and continuous in time 1D exciton transport.

Let us note that $g_{n,n'}(t|t')$ satisfies the following fundamental criteria:

1. Normalization condition

$$\sum_{n=-\infty}^{\infty} g_{n,n'}(t|t') = 1. \tag{11.102}$$

2. Rule for two events successive in time

$$g_{n,n'}(t_1|t_2) = \sum_{m=-\infty}^{\infty} g_{n,m}(t_1|t')g_{m,n'}(t'|t_2). \tag{11.103}$$

The last condition means that for exciton propagations occurring in succession in time the quantum transition amplitudes are multiplied.

Our intend now is to study behavior of $g_n(t)$ at large $|n|$, when the main contribution to the integral Eq. (11.100) comes from small k. Therefore, we can expand integral over k from $-\infty$ up to ∞,

$$g_n(t) = \frac{1}{2\pi} \int_{-\infty}^{\infty} dk \exp(ikn - i\gamma_s |k|^{\upsilon(s)} t/\hbar), \tag{11.104}$$

where

$$\upsilon(s) = \begin{cases} 2, & \text{for } s > 3, \\ s - 1, & \text{for } 2 < s < 3, \end{cases} \tag{11.105}$$

and

$$\gamma_s = \begin{cases} \frac{\mathcal{J}\zeta(s-2)}{2}, & \text{for } s > 3, \\ D_s, & \text{for } 2 < s < 3, \end{cases} \tag{11.106}$$

here D_s is given by Eq. (11.45).

Asymptotic behavior of the lattice quantum 1D propagator $g_n(t)$ at large $|n|$ depends on the value of the parameter s. Indeed, when $s > 3$ Eq. (11.104) goes to

$$g_n(t) = (\hbar/2\pi i \mathcal{J}\zeta(s-2)t)^{1/2} \exp\{-\frac{\hbar}{2i\mathcal{J}\zeta(s-2)t}|n|^2\}, \qquad s > 3, \tag{11.107}$$

if we take into account Eqs. (11.105) and (11.106).

When $2 < s < 3$ the integral in Eq. (11.104) is expressed in terms of the Fox's H-function [93] and [104],

$$g_n(t) = \frac{1}{|n|(s-1)} H_{2,2}^{1,1} \left[\left(\frac{\hbar}{iD_s t} \right)^{1/(s-1)} |n| \left| \begin{array}{c} (1, 1/(s-1)), (1, 1/2) \\ (1, 1), (1, 1/2) \end{array} \right. \right],$$

(11.108)

On the other hand, in the long-wave limit for $2 < s < 3$ the integral in Eq. (11.104) can be estimated as

$$g_n(t) \simeq \frac{1}{\pi} \Gamma(s) \sin(\frac{\pi(s-1)}{2}) \left(\frac{iD_s t}{\hbar} \right)^{s/(s-1)} \frac{1}{|n|^s}.$$

(11.109)

Thus, the asymptotics of the lattice quantum exciton propagator $g_n(t)$ exhibits the power-law behavior at large $|n|$ for $2 < s < 3$. Transition from Gaussian-like behavior Eq. (11.107) to the power-law decay Eq. (11.109) is due to long-range interaction (the second term on the right of Eqs. (11.14)). This transition can be interpreted as phase transition. In fact, when $s > 3$ the propagator $g_n(t)$ defined by Eq. (11.107) has correlation length[2] Δn

$$\Delta n \simeq \left(\frac{2\mathcal{J}\zeta(s-2)t}{\hbar} \right)^{1/2}, \qquad s > 3.$$

(11.110)

When $2 < s < 3$ the propagator $g_n(t)$ defined by Eq. (11.109) exhibits the power-law behavior with infinite correlation length, that is in the case $2 < s < 3$ the correlation length doesn't exist for 1D lattice system with long-range exciton-exciton interaction given by Eq. (11.7).

11.5.1 *Crossover in random walk on 1D lattice*

Exciton-exciton 1D lattice quantum dynamics initiates a 1D classic dynamics random walk model. Indeed, if we put $it \to \tau$,

$$w(k) = \mathcal{V}(k)/\hbar$$

(11.111)

---

[2]Let us remark, that the lattice scale a is equal 1, $a = 1$.

and rename

$$g_{n,n'}(t|t')|_{it \to \tau, it' \to \tau'} \to P_{n,n'}(\tau|\tau'),$$

then Eq. (11.101) becomes

$$P_{n,n'}(\tau|\tau') = \frac{1}{2\pi} \int\limits_{-\pi}^{\pi} dk \exp\{ik(n - n') - w(k)(\tau - \tau')\}, \qquad (11.112)$$

which is the definition of random walk transition probability $P_{n,n'}(\tau|\tau')$.

The transition probability answers the question: What is the probability that walker will be on site n at the time moment τ if at some previous time moment τ' ($\tau' < \tau$) he was on site n'. It is easy to see that $P_{n,n'}(\tau|\tau')$ is normalized

$$\sum_{n=-\infty}^{\infty} P_{n,n'}(\tau|\tau') = 1, \qquad (11.113)$$

and satisfies the Chapman-Kolmogorov equation

$$P_{n,n'}(\tau_1|\tau_2) = \sum_{m=-\infty}^{\infty} P_{n,m}(\tau_1|\tau')P_{m,n'}(\tau'|\tau_2), \qquad (11.114)$$

with the initial condition

$$P_{n,n'}(\tau|\tau')|_{\tau=\tau'} = \delta(n - n'). \qquad (11.115)$$

We came to a continuous in time τ and discrete in space (1D lattice) random walk model [133]. From Eq. (11.114) we conclude that the random walk model under consideration is Markov random process. As we will see, the obtained random walk model exhibits a cross-over from the Brownian random walk ($s > 3$) with finite correlation length to the symmetric α-stable (with $\alpha = s - 1$ and $2 < s < 3$) Lévy random process with an infinite correlation length.

Let's choose $n' = 0$, $\tau' = 0$ and introduce $P_n(\tau)$ as

$$P_n(\tau) = P_{n,n'}(\tau|\tau')|_{n'=0,\tau'=0} = \frac{1}{2\pi} \int\limits_{-\pi}^{\pi} dk \exp\{ikn - w(k)\tau\}. \qquad (11.116)$$

The probability $P_n(\tau)$ can be expressed as

$$P_n(\tau) = \frac{1}{2\pi} \int\limits_{-\pi}^{\pi} dk \exp(ikn) P(k,\tau), \qquad (11.117)$$

where $P(k,\tau)$ is given by

$$P(k,\tau) = \sum_{n=-\infty}^{\infty} e^{-ikn} P_n(\tau) = \exp\{-w(k)\tau\}, \qquad (11.118)$$

and $w(k)$ is defined by Eq. (11.111).

It follows straightforwardly from Eq. (11.118) that the probability $P(k,\tau)$ satisfies motion equation

$$\frac{\partial}{\partial \tau} P(k,\tau) = -w(k) P(k,\tau), \qquad (11.119)$$

with the initial condition

$$P(k, t = 0) = 1.$$

In the case of continuum space the integral over k Eq. (11.117) can be expanded from $-\infty$ up to ∞ and we obtain probability distribution function $P(x,\tau)$

$$P(x,\tau) = \frac{1}{2\pi} \int\limits_{-\infty}^{\infty} dk \exp(ikx) P(k,\tau), \qquad (11.120)$$

where $P(k,\tau)$ is the inverse Fourier transform

$$P(k,\tau) = \int\limits_{-\infty}^{\infty} dx \exp(-ikx) P(x,\tau). \qquad (11.121)$$

Thus, we have

$$P(x,\tau) = \frac{1}{2\pi} \int\limits_{-\infty}^{\infty} dk \exp\{ikx - w(k)\tau\}.$$

Further, using Eqs. (11.42) and (11.44) we write the probability distribution function $P(x, \tau)$ in the form

$$P(x, \tau) = \frac{1}{2\pi} \int\limits_{-\infty}^{\infty} dk \exp(ikx - \gamma_s |k|^{\upsilon(s)} \tau), \qquad (11.122)$$

which satisfies the initial condition

$$P(x, \tau = 0) = \delta(x), \qquad (11.123)$$

where $\upsilon(s)$ and γ_s are defined by Eq. (11.105) and Eq. (11.106) respectively. This probability distribution function bears dependency on index s and exhibits cross-over from Gaussian behavior at $s > 3$ to α-stable $(\alpha = s - 1)$ Lévy law at $2 < s < 3$.

It follows from Eq. (11.122) that

$$\frac{\partial}{\partial \tau} P(x, \tau) = -\frac{\gamma_s}{2\pi} \int\limits_{-\infty}^{\infty} dk |k|^{\upsilon(s)} \exp(ikx - \gamma_s |k|^{\upsilon(s)} \tau) \qquad (11.124)$$

$$= \gamma_s \frac{\partial^{\upsilon(s)}}{\partial x^{\upsilon(s)}} P(x, \tau),$$

where $\partial^{\upsilon(s)}/\partial x^{\upsilon(s)}$ stands for the *Riesz fractional derivative* introduced by

$$\frac{\partial^{\upsilon(s)}}{\partial x^{\upsilon(s)}} P(x, \tau) = -\frac{1}{2\pi} \int\limits_{-\infty}^{\infty} dk |k|^{\upsilon(s)} \exp(ikx) P(k, \tau). \qquad (11.125)$$

Here $P(x, \tau)$ and $P(k, \tau)$ are related to each other by the Fourier transforms defined by Eqs. (11.120) and (11.121). The initial condition to fractional differential equation (11.124) is given by Eq. (11.123).

Thus, it has been shown that the presented random walk model exhibits cross-over from normal to fractional mode, that is from Gaussian behavior at $s > 3$ to α-stable $(\alpha = s - 1)$ Lévy law at $2 < s < 3$.

Chapter 12

Time Fractional Quantum Mechanics

12.1 Introductory remarks

The crucial manifestation of fractional quantum mechanics is *fractional Schrödinger equation*. The fractional Schrödinger equation includes a spatial derivative of fractional order instead of the second order spatial derivative in the well-known Schrödinger equation. Thus, only the spatial derivative becomes fractional in the fractional Schrödinger equation, while the time derivative is the first order time derivative. Due to the presence of the first order time derivative in the fractional Schrödinger equation, fractional quantum mechanics supports all quantum mechanics fundamentals.

Inspired by the work of Laskin [67], [92], Naber invented *time fractional Schrödinger equation* [141]. The time fractional Schrödinger equation involves the time derivative of fractional order instead of the first-order time derivative, while the spatial derivative is the second-order spatial derivative as it is in the well-known Schrödinger equation. To obtain the time fractional Schrödinger equation, Naber mapped the time fractional diffusion equation into the time fractional Schrödinger equation, similarly to the map between the well-known diffusion equation and the standard Schrödinger equation. The mapping implemented by Naber can be considered as a "fractional" generalization of the Wick rotation [142]. To get the time fractional Schrödinger equation, Naber implemented the Wick rotation in complex t-plane by rising the imaginary unit i to the same fractional power as the fractional order of the time derivative in the time fractional diffusion equation. The time fractional derivative in the time fractional Schrödinger equation is the Caputo fractional derivative [143]. Naber has found the exact solutions to the time fractional Schrödinger equation for a free particle and a particle in a potential well [141].

Later on, Wang and Xu [144], and then Dong and Xu [145], combined both Laskin's fractional Schrödinger equation and Naber's time fractional Schrödinger equation and came up with *space-time fractional Schrödinger equation*. The space-time fractional Schrödinger equation includes both spatial and temporal fractional derivatives. Wang and Xu found exact solutions to the space-time fractional Schrödinger equation for a free particle and for an infinite square potential well. Dong and Xu found the exact solution to the space-time fractional Schrödinger equation for a quantum particle in δ-potential well.

Here we introduce time fractional quantum mechanics and develop its fundamentals. The wording "time fractional quantum mechanics" means that the time derivative in the fundamental quantum mechanical equations - Schrödinger equation and fractional Schrödinger equation, is substituted with a fractional time derivative. The time fractional derivative in our approach is the Caputo fractional derivative. To introduce and develop time fractional quantum mechanics we begin with our own version of the space-time fractional Schrödinger equation. Our space-time fractional Schrödinger equation involves two scale dimensional parameters, one of which can be considered as a time fractional generalization of the famous Planck's constant, while the other one can be interpreted as a time fractional generalization of the scale parameter emerging in fractional quantum mechanics [67], [96]. The fractional generalization of Planck's constant is a fundamental dimensional parameter of time fractional quantum mechanics, while the fractional generalization of Laskin's scale parameter [67], [96] plays a fundamental role in both time fractional quantum mechanics and time fractional classical mechanics.

In addition to the above mentioned dimensional parameters, time fractional quantum mechanics involves two dimensionless fractality parameters α, $1 < \alpha \leq 2$, and β, $0 < \beta \leq 1$. Parameter α is the order of the spatial fractional quantum Riesz derivative [67] and β is the order of the time fractional derivative. In other words, α is responsible for modelling *spatial fractality*, while parameter β, which is the order of Caputo fractional derivative, is responsible for modelling *temporal fractality*.

In the framework of time fractional quantum mechanics at particular choices of fractality parameters α and β, we rediscovered the following fundamental quantum equations:

1. Schrödinger equation (Schrödinger equation [6]), $\alpha = 2$ and $\beta = 1$;
2. Fractional Schrödinger equation (Laskin equation [67], [96]), $1 < \alpha \leq$

2 and $\beta = 1$;

3. Time fractional Schrödinger equation (Naber equation [141]), $\alpha = 2$ and $0 < \beta \leq 1$;

4. Space-time fractional Schrödinger equation (Wang and Xu [144] and Dong and Xu [145] equation), $1 < \alpha \leq 2$ and $0 < \beta \leq 1$.

12.1.1 *Shortcomings of time fractional quantum mechanics*

While fractional quantum mechanics supports all quantum mechanics fundamentals, time fractional quantum mechanics violates the following fundamental physical laws of Quantum Mechanics:

a. Quantum superposition law;

b. Unitarity of evolution operator;

c. Probability conservation law;

d. Existence of stationary energy levels of quantum system.

Fractional quantum dynamics is governed by a pseudo-Hamilton operator instead of the Hamilton operator in standard quantum mechanics. Eigenvalues of quantum pseudo-Hamilton operator are not the energy levels of a time fractional quantum system.

12.1.2 *Benefits of time fractional quantum mechanics*

What benefits does time fractional quantum mechanics bring into quantum theory and its applications?

Despite the above listed shortcomings, the developments in time fractional quantum mechanics can be considered a newly emerged and attractive application of fractional calculus to quantum theory. Time fractional quantum mechanics helps to understand the significance and importance of the fundamentals of quantum mechanics such as Hamilton operator, unitarity of evolution operator, existence of stationary energy levels of quantum mechanical system, quantum superposition law, conservation of quantum probability, etc.

Besides, time fractional quantum mechanics invokes new mathematical tools, which have never been used in quantum theory before.

From a stand point of quantum mechanical fundamentals, time fractional quantum mechanics seems not as a quantum physical theory but rather as an adequate, convenient mathematical framework, well adjusted to study dissipative quantum systems interacting with environment [43].

The time fractional quantum mechanics is an adequate, convenient mathematical framework, well adjusted to study dissipative quantum systems interacting with environment [43].

12.2   Time fractional Schrödinger equation

12.2.1   *Naber approach*

Time fractional Schrödinger equation is the Schrödinger equation, where first order time derivative is substituted with time derivative of fractional order. To introduce time fractional Schrödinger equation we begin with a brief review of Naber's original approach [141]. Following by Naber we present the standard 1D Schrödinger equation given by Eq. (3.17) using Planck units,

$$iT_P \frac{\partial \psi(x,t)}{\partial t} = -\frac{L_P^2 M_P}{2m} \frac{\partial^2}{\partial x^2} \psi(x,t) + \frac{V(x)}{E_P} \psi(x,t), \qquad (12.1)$$

where L_P, T_P, M_P, and E_P are Planck length, time, mass and energy defined by

$$L_P = \sqrt{\frac{G\hbar}{c^3}}, \qquad T_P = \sqrt{\frac{G\hbar}{c^5}}, \qquad M_P = \sqrt{\frac{\hbar c}{G}}, \qquad E_P = M_P c^2, \quad (12.2)$$

with G being the gravitational constant and c being the speed of light.

By introducing dimensionless parameter N_m and function $N_V(x)$

$$N_m = \frac{m}{M_P} \quad \text{and} \quad N_V(x) = \frac{V(x)}{E_P}, \qquad (12.3)$$

we rewrite Eq. (12.1) as

$$iT_P \frac{\partial \psi(x,t)}{\partial t} = -\frac{L_P^2}{2N_m} \frac{\partial^2}{\partial x^2} \psi(x,t) + N_V(x)\psi(x,t). \qquad (12.4)$$

The parameter N_m is the measure of particle mass in Planck mass units, while function $N_V(x)$ is the measure of potential energy $V(x)$ in Planck energy units.

To introduce time fractional Schrödinger equation we substitute time derivative $\partial/\partial t$ in Eq. (12.1) to fractional time derivative of order ν defined by (for details see Appendix B)

$$\partial_t^\nu f(t) = \frac{1}{\Gamma(1-\nu)} \int_0^t d\tau \frac{f'(\tau)}{(t-\tau)^\nu}, \qquad 0 < \nu < 1, \qquad (12.5)$$

and

$$\partial_t^\nu f(t) = \frac{d}{dt} f(t), \qquad \nu = 1, \qquad (12.6)$$

where $f'(\tau)$ is first order time derivative, $f'(\tau) = df(\tau)/d\tau$ and $\Gamma(1 - \nu)$ is the Gamma function.

Defined by Eqs. (12.5) and (12.6) fractional time derivative is called Caputo fractional derivative [143].

This substitution can be realized by two alternative ways,

$$(iT_P)^\nu \partial_t^\nu \psi(x,t) = -\frac{L_P^2}{2N_m} \frac{\partial^2}{\partial x^2} \psi(x,t) + N_V(x)\psi(x,t) \qquad (12.7)$$

or

$$i(T_P)^\nu \partial_t^\nu \psi(x,t) = -\frac{L_P^2}{2N_m} \frac{\partial^2}{\partial x^2} \psi(x,t) + N_V(x)\psi(x,t) \qquad (12.8)$$

as far as T_P must be raised to the same order as the order of time fractional derivative to save appropriate units. Note that the range for parameter ν in these equations is $0 < \nu \leq 1$.

The question of whether or not to rise imaginary unit i to the order of fractional time derivative was addressed by Naber [141]. Naber explained as well why one should raise the power of imaginary unit i to the order ν of fractional time derivative. From a stand point of a Wick rotation the imaginary unit has to be raised to the same power as the time variable. Another fundamental reason to raise imaginary unit to the same power as the time variable is the criteria of appropriateness (from stand point of quantum physics) of temporal behavior of the solution to time fractional Schrödinger equation. Naber wrote [141], "*When solving for the time component of Eq. (12.7) or Eq. (12.8) the Laplace transform is the preferred method. For Eq. (12.7), changing the order of the derivative moves the pole (from the inverse Laplace transform) up or down the negative imaginary axis. Hence, the temporal behavior of the solution will not change. For Eq. (12.8), changing the order of the derivative moves the pole to almost any desired location in the complex plane. Physically, this would mean that a small change in the order of the time derivative, in Eq. (12.8), could change the temporal behavior from sinusoidal to growth or to decay. Due to the simpler physical behavior of Eq. (12.7) and the role of 'i' in a Wick rotation, Eq. (12.7) is the best candidate for a time fractional Schrödinger equation.*"

Thus, Eq. (12.7) is 1D time fractional Schrödinger equation first obtained by Naber [141].

12.3 Space-time fractional Schrödinger equation

12.3.1 *Wang and Xu, and Dong and Xu approach*

Inspired by works [67], [92] and [141], Wang and Xu [144] and then Dong and Xu [145] combined both Laskin's fractional Schrödinger equation and Naber's time fractional Schrödinger equation and came up with space-time fractional Schrödinger equation. Dong and Xu [145] coined the term *space-time fractional Schrödinger equation* for Schrödinger equation with involvement of both space and time fractional derivatives.

To review Wang and Xu, and Dong and Xu approach we will follow the work by Dong and Xu [145]. With the help of Planck units defined by Eq. (12.2) the 1D fractional Schrödinger equation (3.20) can be rewritten in dimensionless form as [145]

$$iT_P \frac{\partial \psi(x,t)}{\partial t} = \frac{D_\alpha T_P^{2-2\alpha}(-\hbar^2 \Delta)^{\alpha/2}}{M_P^{1-\alpha} E_P^\alpha L_P^{2-2\alpha}} \psi(x,t) + \frac{V(x)}{E_P} \psi(x,t), \qquad (12.9)$$

$$1 < \alpha \le 2.$$

Substituting the time derivative $\partial/\partial t$ with the Caputo fractional time derivative ∂_t^β of order β, yields

$$(iT_P)^\beta \partial_t^\beta \psi(x,t) = \frac{D_\alpha T_P^{2-2\alpha}(-\hbar^2 \Delta)^{\alpha/2}}{M_P^{1-\alpha} E_P^\alpha L_P^{2-2\alpha}} \psi(x,t) + \frac{V(x)}{E_P} \psi(x,t), \qquad (12.10)$$

$$1 < \alpha \le 2, \qquad 0 < \beta \le 1,$$

here ∂_t^β stands for fractional time derivative introduced by Eqs. (12.5) and (12.6) with renaming $\nu \to \beta$.

Thus, Eq. (12.10) is first obtained by Wang and Xu [144] 1D space-time fractional Schrödinger equation expressed in the form presented by Dong and Xu [145].

12.3.2 Laskin approach

Considering a 1D spatial dimensional, Laskin launched the *space-time fractional Schrödinger equation* in the following form [146],

$$i^\beta \hbar_\beta \partial_t^\beta \psi(x,t) = D_{\alpha,\beta}(-\hbar_\beta^2 \Delta)^{\alpha/2}\psi(x,t) + V(x,t)\psi(x,t), \qquad (12.11)$$

$$1 < \alpha \le 2, \qquad 0 < \beta \le 1,$$

here $\psi(x,t)$ is the wave function, i is imaginary unit, $i = \sqrt{-1}$, $V(x,t)$ is potential energy, $\hbar_\beta$ and $D_{\alpha,\beta}$ are two scale coefficients, which we introduce into the framework of time fractional quantum mechanics, Δ is the 1D Laplace operator, $\Delta = \partial^2/\partial x^2$, and, finally, ∂_t^β is the left Caputo fractional derivative [143] of order β defined by

$$\partial_t^\beta f(t) = \frac{1}{\Gamma(1-\nu)} \int_0^t d\tau \frac{f'(\tau)}{(t-\tau)^\nu}, \qquad 0 < \beta < 1, \qquad (12.12)$$

and

$$\partial_t^\beta f(t) = \frac{\partial}{\partial t} f(t), \qquad \beta = 1, \qquad (12.13)$$

where $f'(\tau)$ is first order time derivative, $f'(\tau) = df(\tau)/d\tau$, $\Gamma(1-\beta)$ is the Gamma function.

The operator $(-\hbar_\beta^2 \Delta)^{\alpha/2}$ in Eq. (12.11) is a *time fractional quantum Riesz derivative* introduced first in [146]

$$(-\hbar_\beta^2 \Delta)^{\alpha/2}\psi(x,t) = \frac{1}{2\pi\hbar_\beta} \int dp_\beta \exp\{i\frac{p_\beta x}{\hbar_\beta}\}|p_\beta|^\alpha \varphi(p_\beta,t), \qquad (12.14)$$

where the wave functions in space and momentum representations, $\psi(x,t)$ and $\varphi(p,t)$, are related to each other by the Fourier transforms

$$\psi(x,t) = \frac{1}{2\pi\hbar_\beta} \int dp_\beta \exp\{i\frac{p_\beta x}{\hbar_\beta}\}\varphi(p_\beta,t) \qquad (12.15)$$

and

$$\varphi(p_\beta,t) = \int dr \exp\{-i\frac{p_\beta x}{\hbar_\beta}\}\psi(x,t). \qquad (12.16)$$

Quantum scale coefficient $\hbar_\beta$ has units of

$$[\hbar_\beta] = \text{erg} \cdot \text{sec}^\beta, \tag{12.17}$$

and scale coefficient $D_{\alpha,\beta}$ has units of

$$[D_{\alpha,\beta}] = \text{erg}^{1-\alpha} \cdot \text{cm}^\alpha \cdot \text{sec}^{-\alpha\beta}. \tag{12.18}$$

The introduction of the scale coefficient $D_{\alpha,\beta}$ was inspired by Bayin [147].

12.4   Pseudo-Hamilton operator in time fractional quantum mechanics

Aiming to obtain the operator form of space-time fractional Schrödinger equation Eq. (12.11), let us define time fractional quantum momentum operator $\widehat{p}_\beta$

$$\widehat{p}_\beta = -i\hbar_\beta \frac{\partial}{\partial x}, \tag{12.19}$$

and quantum operator of coordinate $\widehat{x}$,

$$\widehat{x} = x. \tag{12.20}$$

Hence, the commutation relation in the framework of time fractional quantum mechanics has the form

$$[\widehat{x}, \widehat{p}_\beta] = i\hbar_\beta, \tag{12.21}$$

where $[\widehat{x}, \widehat{p}_\beta]$ is the commutator of two quantum operators $\widehat{x}$ and $\widehat{p}_\beta$,

$$[\widehat{x}, \widehat{p}_\beta] = \widehat{x}\widehat{p}_\beta - \widehat{p}_\beta\widehat{x}.$$

Using quantum operators of momentum Eq. (12.19) and coordinate Eq. (12.20), we introduce a new time fractional quantum mechanical operator $\widehat{H}_{\alpha,\beta}(\widehat{p}_\beta, \widehat{x})$,

$$\widehat{H}_{\alpha,\beta}(\widehat{p}_\beta, \widehat{x}) = D_{\alpha,\beta}|\widehat{p}_\beta|^\alpha + V(\widehat{x}, t). \tag{12.22}$$

The operator $\widehat{H}_{\alpha,\beta}(\widehat{p}_\beta,\widehat{x})$ is not the Hamilton operator of the quantum mechanical system under consideration[1]. Following [145], we will call this operator *pseudo-Hamilton operator*.

Having the pseudo-Hamilton operator $\widehat{H}_{\alpha,\beta}$, let us rewrite Eq. (12.11) as

$$\hbar_\beta i^\beta \partial_t^\beta \psi(x,t) = \widehat{H}_{\alpha,\beta}(\widehat{p}_\beta,\widehat{x})\psi(x,t)$$

$$= (D_{\alpha,\beta}|\widehat{p}_\beta|^\alpha + V(\widehat{x},t))\,\psi(x,t), \qquad (12.23)$$

$$1 < \alpha \le 2, \qquad 0 < \beta \le 1,$$

which is operator form of the space-time fractional Schrödinger equation [146].

The pseudo-Hamilton operator $\widehat{H}_{\alpha,\beta}(\widehat{p}_\beta,\widehat{x})$ introduced by Eq. (3.18) is the Hermitian operator in space with scalar product

$$(\phi,\chi) = \int_{-\infty}^\infty dx\,\phi^*(x,t)\chi(x,t). \qquad (12.24)$$

In this space, operators $\widehat{p}_\beta$ and $\widehat{x}$ defined by Eqs. (12.19) and (12.20) are Hermitian operators. The proof can be found in any textbook on quantum mechanics, (see, for example, [94]).

To prove the Hermiticity of quantum mechanical operator $\widehat{H}_{\alpha,\beta}$ let us note that in accordance with the definition of the time fractional quantum Riesz derivative given by Eq. (3.5) there exists the integration-by-parts formula

$$(\phi,(-\hbar_\beta^2\Delta)^{\alpha/2}\chi) = ((-\hbar_\beta^2\Delta)^{\alpha/2}\phi,\chi). \qquad (12.25)$$

Therefore, using this integration-by-parts formula we prove straightforwardly Hermiticity of the term $D_{\alpha,\beta}|\widehat{p}_\beta|^\alpha = D_{\alpha,\beta}(-\hbar_\beta^2\Delta)^{\alpha/2}$. Next, potential energy operator $V(\widehat{x},t)$ in Eq. (3.18) is Hermitian operator by virtue of being a function of Hermitian operator $\widehat{x}$.

Thus, we complete the proof of Hermiticity of the pseudo-Hamilton operator $\widehat{H}_{\alpha,\beta}$ in the space with scalar product defined by Eq. (12.24).

---

[1] In time fractional quantum mechanics the eigenvalues of operator $\widehat{H}_{\alpha,\beta}(\widehat{p}_\beta,\widehat{x})$ are not energies of a quantum system. In classical time fractional mechanics the principle of least action with classical pseudo-Hamilton function $H_{\alpha,\beta}(p_\beta,x)$ does not result in Hamilton equations of motion.

12.5   3D generalization of space-time fractional Schrödinger equation

Considering the 3D spatial dimensional, we launch the space-time fractional Schrödinger equation of the following form [146],

$$i^\beta \hbar_\beta \partial_t^\beta \psi(\mathbf{r}, t) = D_{\alpha,\beta}(-\hbar_\beta^2 \Delta)^{\alpha/2} \psi(\mathbf{r}, t) + V(\mathbf{r}, t)\psi(\mathbf{r}, t), \qquad (12.26)$$

$$1 < \alpha \le 2, \qquad 0 < \beta \le 1,$$

where $\psi(\mathbf{r}, t)$ is the wave function, $\mathbf{r}$ is the 3D space vector, Δ is the Laplacian, $\Delta = (\partial/\partial \mathbf{r})^2$, all other notations are the same as for Eq. (3.1), and the 3D time fractional quantum Riesz derivative $(-\hbar_\beta^2 \Delta)^{\alpha/2}$ is defined by

$$(-\hbar_\beta^2 \Delta)^{\alpha/2}\psi(\mathbf{r}, t) = \frac{1}{(2\pi\hbar_\beta)^3} \int d^3r \exp\{i\frac{\mathbf{p}_\beta \mathbf{r}}{\hbar_\beta}\}|\mathbf{p}_\beta|^\alpha \varphi(\mathbf{p}_\beta, t), \qquad (12.27)$$

where the wave functions in space representation $\psi(\mathbf{r}, t)$ and momentum representation $\varphi(\mathbf{p}, t)$ are related to each other by the 3D Fourier transforms

$$\psi(\mathbf{r}, t) = \frac{1}{(2\pi\hbar_\beta)^3} \int d^3p \exp\{i\frac{\mathbf{p}_\beta \mathbf{r}}{\hbar_\beta}\}\varphi(\mathbf{p}_\beta, t) \qquad (12.28)$$

and

$$\varphi(\mathbf{p}_\beta, t) = \int d^3r \exp\{-i\frac{\mathbf{p}_\beta \mathbf{r}}{\hbar_\beta}\}\psi(\mathbf{r}, t). \qquad (12.29)$$

Further, the 3D generalization of the pseudo-Hamilton operator $\widehat{H}_{\alpha,\beta}$ in the framework of time fractional quantum mechanics is

$$\widehat{H}_{\alpha,\beta}(\widehat{\mathbf{p}}_\beta, \widehat{\mathbf{r}}) = D_{\alpha,\beta}|\widehat{\mathbf{p}}_\beta|^\alpha + V(\widehat{\mathbf{r}}, t), \qquad (12.30)$$

where $\widehat{\mathbf{r}}$ is the 3D quantum operator of coordinate

$$\widehat{\mathbf{r}} = \mathbf{r}, \qquad (12.31)$$

and $\widehat{\mathbf{p}}_\beta$ is the 3D time fractional quantum momentum operator introduced by

$$\widehat{\mathbf{p}}_\beta = -i\hbar_\beta \frac{\partial}{\partial \mathbf{r}}, \qquad (12.32)$$

with $\hbar_\beta$ being the scale coefficient appearing for the first time in Eqs. (3.1) and (12.26).

The basic canonical commutation relationships in the 3D case are

$$[\widehat{r}_k, \widehat{p}_{\beta j}] = i\hbar_\beta \delta_{kj}, \qquad [\widehat{r}_k, \widehat{r}_j] = 0, \qquad [\widehat{p}_{\beta k}, \widehat{p}_{\beta j}] = 0, \qquad (12.33)$$

where δ_{kj} is the Kronecker symbol,

$$\delta_{kj}, = \begin{cases} 1 & k = j, \\ 0 & k \neq j, \end{cases} \qquad (12.34)$$

and $k, j = 1, 2, 3$.

Having the 3D generalization of pseudo-Hamilton operator $\widehat{H}_{\alpha,\beta}(\widehat{\mathbf{p}}_\beta, \widehat{\mathbf{r}})$, let us present Eq. (12.26) in the form

$$\hbar_\beta i^\beta \partial_t^\beta \psi(\mathbf{r}, t) = \widehat{H}_{\alpha,\beta}(\widehat{\mathbf{p}}_\beta, \widehat{\mathbf{r}})\psi(\mathbf{r}, t) \qquad (12.35)$$

$$= \left(D_{\alpha,\beta}|\widehat{\mathbf{p}}_\beta|^\alpha + V(\widehat{\mathbf{r}}, t)\right)\psi(\mathbf{r}, t),$$

$$1 < \alpha \leq 2, \qquad 0 < \beta \leq 1,$$

which is the operator form of the 3D space-time fractional Schrödinger equation.

Operators (12.31) and (12.32) allow us to introduce time fractional angular momentum operator $\widehat{\mathbf{L}}_\beta$ as cross-product of the above two defined operators $\widehat{\mathbf{r}}$ and $\widehat{\mathbf{p}}_\beta$

$$\widehat{\mathbf{L}}_\beta = \widehat{\mathbf{r}} \times \widehat{\mathbf{p}}_\beta = -i\hbar_\beta \mathbf{r} \times \frac{\partial}{\partial \mathbf{r}}. \qquad (12.36)$$

This equation can be expressed in component form

$$\widehat{\mathbf{L}}_{\beta i} = \varepsilon_{ijk}\widehat{r}_j\widehat{p}_{\beta k}, \qquad i, k, j = 1, 2, 3, \qquad (12.37)$$

if we use the 3D Levi-Civita antisymmetric tensor ε_{ijk} which changes its sign under interchange of any pair of indices i, j, k.

It is obvious that the algebra of time fractional angular momentum operators $\widehat{\mathbf{L}}_{\beta i}$ and its commutation relationships with operators $\widehat{r}_j$ and $\widehat{p}_{\beta k}$ are the same as for the angular momentum operator of quantum mechanics [94].

12.5.1   *Hermiticity of pseudo-Hamilton operator*

To prove that defined by Eq. (12.26) $\widehat{H}_{\alpha,\beta}(\widehat{\mathbf{p}}_\beta, \widehat{\mathbf{r}})$ is Hermitian operator we introduce the space with scalar product defined by

$$(\phi(\mathbf{r},t), \chi(\mathbf{r},t)) = \int d^3r \phi^*(\mathbf{r},t)\chi(\mathbf{r},t), \qquad (12.38)$$

where $\phi^*(\mathbf{r},t)$ means complex conjugate function.

Then, the proof can be done straightforwardly by the 3D generalization of the considered presented in Sec. 12.4.

12.5.2   *The parity conservation law*

Here we study invariance of pseudo-Hamilton operator $\widehat{H}_{\alpha,\beta}(\widehat{\mathbf{p}}_\beta, \widehat{\mathbf{r}})$ under *inversion* transformation. Inversion, or to be precise, spatial inversion consists of the simultaneous change in the sign of all three spatial coordinates

$$\mathbf{r} \to -\mathbf{r}, \qquad x \to -x, \quad y \to -y, \quad z \to -z. \qquad (12.39)$$

It is easy to see that

$$(-\hbar_\beta^2 \Delta)^{\alpha/2} \exp\{i\frac{\mathbf{p}_\beta \mathbf{x}}{\hbar_\beta}\} = |\mathbf{p}_\beta|^\alpha \exp\{i\frac{\mathbf{p}_\beta \mathbf{x}}{\hbar_\beta}\}, \qquad (12.40)$$

which means that the function $\exp\{i\mathbf{p}_\beta \mathbf{x}/\hbar_\beta\}$ is the eigenfunction of the 3D time fractional quantum Riesz operator $(-\hbar_\beta^2 \Delta)^{\alpha/2}$ with eigenvalue $|\mathbf{p}_\beta|^\alpha$.

Thus, the operator $(-\hbar_\beta^2 \Delta)^{\alpha/2}$ is the symmetrized fractional derivative, that is

$$(-\hbar_\beta^2 \Delta_{\mathbf{r}})^{\alpha/2} ... = (-\hbar_\beta^2 \Delta_{-\mathbf{r}})^{\alpha/2} .... \qquad (12.41)$$

Assuming that the potential energy operator $V(\widehat{\mathbf{r}}, t)$ is invariant under spatial inversion $V(\widehat{\mathbf{r}}, t) = V(-\widehat{\mathbf{r}}, t)$, we conclude that pseudo-Hamilton operator $\widehat{H}_{\alpha,\beta}(\widehat{\mathbf{p}}_\beta, \widehat{\mathbf{r}})$ is invariant under inversion, or, in other words, it supports the parity conservation law. The inverse symmetry results in the fact that inversion operator $\widehat{P}_{\text{inv}}$ and the pseudo-Hamilton operator $\widehat{H}_{\alpha,\beta}(\widehat{\mathbf{p}}_\beta, \widehat{\mathbf{r}})$ commute

$$\widehat{P}_{\text{inv}}\widehat{H}_{\alpha,\beta} = \widehat{H}_{\alpha,\beta}\widehat{P}_{\text{inv}}. \qquad (12.42)$$

Hence, we can divide the wave functions of time fractional quantum mechanical states with defined eigenvalue of the operator $\widehat{P}_{\text{inv}}$ into two classes: (i) wave functions which are not changed upon the action of the inversion operator, $\widehat{P}_{\text{inv}}\psi_+(\mathbf{r}) = \psi_+(\mathbf{r})$, the corresponding states are called even states; (ii) wave functions which change sign under action of the inversion operator, $\widehat{P}_{\text{inv}}\psi_-(\mathbf{r}) = -\psi_-(\mathbf{r})$, the corresponding states are called odd states. Equation (12.42) represents the parity conservation law for time fractional quantum mechanics, that is, if the state of a time fractional quantum mechanical system has a given parity (i.e. if it is even, or odd), then this parity is conserved.

Thus, we conclude that time fractional quantum mechanics supports the parity conservation law.

12.5.3 *Space-time fractional Schrödinger equation in momentum representation*

To obtain the space-time fractional Schrödinger equation in momentum representation let us substitute the wave function $\psi(\mathbf{r}, t)$ from Eq. (12.28) into Eq. (12.26),

$$i^\beta \hbar_\beta \partial_t^\beta \frac{1}{(2\pi\hbar_\beta)^3} \int d^3 p'_\beta \exp\{i\frac{\mathbf{p}'_\beta \mathbf{r}}{\hbar_\beta}\}\varphi(\mathbf{p}'_\beta, t)$$

$$= \frac{D_{\alpha,\beta}}{(2\pi\hbar_\beta)^3} \int d^3 p'_\beta \exp\{i\frac{\mathbf{p}'_\beta \mathbf{r}}{\hbar_\beta}\}|\mathbf{p}'_\beta|^\alpha \varphi(\mathbf{p}'_\beta, t) \qquad (12.43)$$

$$+\frac{V(\mathbf{r},t)}{(2\pi\hbar_\beta)^3} \int d^3 p'_\beta \exp\{i\frac{\mathbf{p}'_\beta \mathbf{r}}{\hbar_\beta}\}\varphi(\mathbf{p}'_\beta, t).$$

Further, multiplying Eq. (12.43) by $\exp(-i\mathbf{p}_\beta \mathbf{r}/\hbar_\beta)$ and integrating over $d^3 r$ yields the equation for the wave function $\varphi(\mathbf{p}_\beta, t)$ in momentum representation

$$i^\beta \hbar_\beta \partial_t^\beta \varphi(\mathbf{p}_\beta, t) = D_{\alpha,\beta}|\mathbf{p}_\beta|^\alpha \varphi(\mathbf{p}_\beta, t) + \int d^3 p'_\beta U_{\mathbf{p}_\beta, \mathbf{p}'_\beta} \varphi(\mathbf{p}'_\beta, t), \qquad (12.44)$$

$$1 < \alpha \leq 2, \qquad 0 < \beta \leq 1,$$

where $U_{\mathbf{p}_\beta, \mathbf{p}'_\beta}$ is introduced as

$$U_{\mathbf{p}_\beta, \mathbf{p}'_\beta} = \frac{1}{(2\pi\hbar_\beta)^3} \int d^3 r \exp\{-i(\mathbf{p}_\beta - \mathbf{p}'_\beta)\mathbf{r}/\hbar_\beta\}V(\mathbf{r},t), \qquad (12.45)$$

and we used the following representation for the delta function $\delta(\mathbf{r})$

$$\delta(\mathbf{r}) = \frac{1}{(2\pi\hbar_\beta)^3} \int d^3p_\beta \exp(i\mathbf{p}_\beta\mathbf{r}/\hbar_\beta). \tag{12.46}$$

Equation (12.44) is the 3D space-time fractional Schrödinger equation in momentum representation.

Substituting the wave function $\psi(x,t)$ from Eq. (12.15) into Eq. (12.11), multiplying by $\exp(-ipx/\hbar)$ and integrating over dx bring us the 1D space-time fractional Schrödinger equation in momentum representation

$$i^\beta\hbar_\beta\partial_t^\beta\varphi(p_\beta,t) = D_{\alpha,\beta}|p_\beta|^\alpha\varphi(p_\beta,t) + \int dp'_\beta U_{p_\beta,p'_\beta}\varphi(p'_\beta,t), \tag{12.47}$$

$$1 < \alpha \le 2, \qquad 0 < \beta \le 1,$$

where U_{p_β,p'_β} is given by

$$U_{p_\beta,p'_\beta} = \frac{1}{2\pi\hbar_\beta} \int dx \exp\{-i(p_\beta-p'_\beta)x/\hbar_\beta\}V(x), \tag{12.48}$$

and we used the following representation for the delta function $\delta(x)$

$$\delta(x) = \frac{1}{2\pi\hbar_\beta} \int dp_\beta \exp(ip_\beta x/\hbar_\beta). \tag{12.49}$$

Equation (12.44) is the 1D space-time fractional Schrödinger equation in momentum representation for the wave function $\varphi(p_\beta,t)$.

12.6 Solution to space-time fractional Schrödinger equation

It is well known that if the Hamilton operator of a quantum mechanical system does not depend on time, then we can search for the solution to the Schrödinger equation in separable form. In the case when pseudo-Hamilton operator (12.30) doesn't depend on time we search for the solution to Eq. (12.11) assuming that the solution has form,

$$\psi(x,t) = \varphi(x)\chi(t), \tag{12.50}$$

where $\varphi(x)$ and $\chi(t)$ are spatial and temporal components of the wave function $\psi(x,t)$[2].

---

[2]We consider here 1D space-time fractional Schrödinger equation. The generalization to 3D case can be done straightforwardly.

It is assumed as well that initial wave function $\psi(x, t = 0) = \psi(x, 0)$ is normalized

$$\int_{-\infty}^{\infty} dx |\psi(x,0)|^2 = |\chi(0)|^2 \int_{-\infty}^{\infty} dx |\varphi(x)|^2 = 1. \qquad (12.51)$$

Substituting Eq. (12.50) into Eq. (12.11) we obtain two equations

$$\widehat{H}_{\alpha,\beta}(\widehat{p}_\beta, \widehat{x})\varphi(x) = \mathcal{E}\varphi(x) \qquad (12.52)$$

and

$$i^\beta \hbar_\beta \partial_t^\beta \chi(t) = \mathcal{E}\chi(t), \qquad \chi(t = 0) = \chi(0), \qquad (12.53)$$

where $\mathcal{E}$ is the eigenvalue of quantum mechanical pseudo-Hamilton operator $\widehat{H}_\beta(\widehat{p}_\beta, \widehat{x})$,

$$\widehat{H}_{\alpha,\beta}(\widehat{p}_\beta, \widehat{x}) = D_{\alpha,\beta} |\widehat{p}_\beta|^\alpha + V(\widehat{x}), \qquad 1 < \alpha \le 2, \qquad 0 < \beta \le 1. \qquad (12.54)$$

The solution to Eq. (12.52) depending on potential energy term $V(\widehat{x})$ can be obtained by the well-known methods of standard quantum mechanics [94]. To find the solution to time fractional equation (12.53) we can use the Laplace transform method[3]. In the Laplace transform domain Eq. (12.53) reads

$$i^\beta \hbar_\beta \left(s^\beta \widetilde{\chi}(s) - s^{\beta-1}\chi(0) \right) = \mathcal{E}\widetilde{\chi}(s), \qquad (12.57)$$

---

[3]The Laplace transform $\widetilde{\chi}(s)$ of a function $\chi(t)$ is defined as

$$\widetilde{\chi}(s) = \int_0^\infty dt e^{-st}\chi(t), \qquad (12.55)$$

where $\chi(t)$ is defined for $t \ge 0$ if the integral exists.
The inverse Laplace transform is defined by

$$\chi(t) = \frac{1}{2\pi i} \int_{\gamma i - \infty}^{\gamma + i\infty} ds e^{st}\widetilde{\chi}(s), \qquad (12.56)$$

where the integration is done along the vertical line $\text{Re}(s) = \gamma$ in the complex plane s such that γ is greater than the real part of all singularities in the complex plane of $\widetilde{\chi}(s)$. This requirement on integration path in the complex plane s ensures that the contour path is in the region of convergence. In practice, computing the complex integral can be done by using the Cauchy residue theorem.

where $\widetilde{\chi}(s)$ is defined by Eq. (12.55) and $\chi(0) = \chi(t = 0)$ is the initial condition on time dependent component $\chi(t)$ of the wave function $\psi(x,t)$ given by Eq. (12.50).

Further, from Eq. (12.57) we have

$$\widetilde{\chi}(s) = \frac{s^{-1}}{1 - \mathcal{E}s^{-\beta}/i^{\beta}\hbar_{\beta}}\chi(0), \qquad (12.58)$$

which can be presented as geometric series (the range of convergence is given by the criteria $|\mathcal{E}s^{-\beta}/i^{\beta}\hbar_{\beta}| < 1$, see [147]),

$$\widetilde{\chi}(s) = \chi(0)\sum_{m=0}^{\infty}(\frac{\mathcal{E}s^{-\beta}}{i^{\beta}\hbar_{\beta}})^{m}s^{-1} = \chi(0)\sum_{m=0}^{\infty}(\frac{\mathcal{E}}{i^{\beta}\hbar_{\beta}})^{m}s^{-m\beta-1}. \qquad (12.59)$$

Then the inverse Laplace transform yields

$$\chi(t) = \chi(0)\sum_{m=0}^{\infty}(\frac{\mathcal{E}}{i^{\beta}\hbar_{\beta}})^{m}\frac{t^{\beta m}}{\Gamma(\beta m + 1)} = \chi(0)E_{\beta}(\frac{(-it)^{\beta}\mathcal{E}}{\hbar_{\beta}}), \qquad (12.60)$$

here $E_{\beta}(z)$ is the Mittag-Leffler function [148], [149] defined by the series

$$E_{\beta}(z) = \sum_{m=0}^{\infty}\frac{z^{m}}{\Gamma(\beta m + 1)}, \qquad (12.61)$$

with $\Gamma(x)$ being the Gamma function given by Eq. (5.7).

In the limit case, when $\beta = 1$, the Mittag-Leffler function $E_{\beta}(z)$ becomes the exponential function

$$E_{\beta}(z)|_{\beta=1} = \sum_{m=0}^{\infty}\frac{z^{m}}{m!} = \exp(z), \qquad (12.62)$$

and $\chi(t)$ goes into

$$\chi(t)|_{\beta=1} = \chi(0)\exp(-i\frac{\mathcal{E}t}{\hbar}), \qquad (12.63)$$

here $\hbar$ is Planck's constant and $\mathcal{E}$ is an eigenvalue of quantum mechanical Hamilton operator $\widehat{H}_{\alpha}(\widehat{p},\widehat{x})$,

$$\widehat{H}_{\alpha}(\widehat{p},\widehat{x}) = \widehat{H}_{\alpha,\beta}(\widehat{p}_{\beta},\widehat{x})|_{\beta=1} = D_{\alpha}|\widehat{p}|^{\alpha} + V(\widehat{x}), \qquad 1 < \alpha \le 2, \quad (12.64)$$

where the scale coefficient $D_\alpha = D_{\alpha,\beta}|_{\beta=1}$ was introduced originally in [67]. It is easy to see that the time dependent component $\chi(t)$ (12.60) of the wave function $\psi(x,t)$ can be written as

$$\chi(t) = \chi(0) \left\{ Ec_\beta(\frac{\mathcal{E}(-t)^\beta}{\hbar_\beta}) + i^\beta Es_\beta(\frac{\mathcal{E}(-t)^\beta}{\hbar_\beta}) \right\}, \qquad (12.65)$$

if we use the following new expression for the Mittag-Leffler function $E_\beta(i^\beta z)$,

$$E_\beta(i^\beta z) = Ec_\beta(z) + i^\beta Es_\beta(z), \qquad 0 < \beta \leq 1, \qquad (12.66)$$

in terms of two functions $Ec_\beta(z)$ and $Es_\beta(z)$ introduced by Eqs. (E.5) and (E.6), see Appendix E.

It has been shown in Appendix E that Eq. (12.66) can be considered as a fractional generalization of the celebrated Euler's formula, which is recovered from Eq. (12.66) in the limit case $\beta = 1$.

Finally, we have the solution to the time fractional Schrödinger equation (12.11) given by

$$\psi(x,t) = \varphi(x) E_\beta(\frac{\mathcal{E}t^\beta}{i^\beta \hbar_\beta}) \qquad (12.67)$$

$$= \varphi(x)\chi(0) \left\{ Ec_\beta(\frac{\mathcal{E}(-t)^\beta}{\hbar_\beta}) + i^\beta Es_\beta(\frac{\mathcal{E}(-t)^\beta}{\hbar_\beta}) \right\}.$$

We see from Eq. (12.67) that the time fractional quantum mechanics does not support normalization condition for the wave function. If normalization condition (12.51) holds at the initial time moment $t = 0$, then at any time moment $t > 0$ it becomes time dependent

$$\int_{-\infty}^{\infty} dx |\psi(x,t)|^2 = \left| Ec_\beta(\frac{\mathcal{E}(-t^\beta)}{\hbar_\beta}) + i^\beta Es_\beta(\frac{\mathcal{E}(-t^\beta)}{\hbar_\beta}) \right|^2 \cdot |\chi(0)|^2 \int_{-\infty}^{\infty} dx |\varphi(x)|^2$$

$$= \mathcal{E} \left\{ \left| Ec_\beta(\frac{\mathcal{E}(-t^\beta)}{\hbar_\beta}) \right|^2 + \left| Es_\beta(\frac{\mathcal{E}(-t^\beta)}{\hbar_\beta}) \right|^2 \right. \qquad (12.68)$$

$$\left. + 2\operatorname{Re}\left(i^\beta Es_\beta(\frac{\mathcal{E}(-t^\beta)}{\hbar_\beta}) Ec_\beta^*(\frac{\mathcal{E}(-t^\beta)}{\hbar_\beta}) \right) \right\},$$

where $Ec^*_\beta(\frac{\mathcal{E}(-t^\beta)}{\hbar_\beta})$ stands for the complex conjugate of the function $Ec_\beta(\frac{\mathcal{E}(-t^\beta)}{\hbar_\beta})$, and we took into account the normalization condition given by Eq. (12.51).

Therefore, we come to the conclusion that in the framework of time fractional quantum mechanics total quantum mechanical probability $\int\limits_{-\infty}^{\infty} dx|\psi(x,t)|^2$ is time dependent. In other words, time fractional quantum mechanics does not support a fundamental property of the quantum mechanics - conservation of quantum mechanical probability.

12.7 Energy in the framework of time fractional quantum mechanics

Here, we introduce the concept of *energy of time fractional quantum system*. Having the pseudo-Hamilton operator $\widehat{H}_{\alpha,\beta}(\widehat{p}_\beta, \widehat{x})$ and the wave function $\psi_{\alpha,\beta}(x,t)$ given by Eq. (12.67) we define the energy of a quantum system in the framework of time fractional quantum mechanics as follows:

$$E_{\alpha,\beta} = \int dx \psi^*_{\alpha,\beta}(x,t) \widehat{H}_{\alpha,\beta}(\widehat{p}_\beta, \widehat{x}) \psi_{\alpha,\beta}(x,t), \qquad (12.69)$$

where $\psi^*_{\alpha,\beta}(x,t)$ stands for the complex conjugate of wave function $\psi_{\alpha,\beta}(x,t)$.

This definition is in line with the definition of stationary energy levels in the frameworks of standard quantum mechanics [94] and fractional quantum mechanics [67], [96]. Indeed, if the pseudo-Hamilton operator $\widehat{H}_{\alpha,\beta}(\widehat{p}_\beta, \widehat{x})$ does not depend on time, then at $\beta = 1$ we obtain time independent Hamilton operator $\widehat{H}_\alpha(\widehat{p}, \widehat{x}) = \widehat{H}_{\alpha,\beta}(\widehat{p}_\beta, \widehat{x})|_{\beta=1}$, while the wave function $\psi_{\alpha,\beta}(x,t)$ introduced by Eq. (12.50) becomes

$$\psi_{\alpha,\beta}(x,t)|_{\beta=1} = \psi_{\alpha,1}(x,t) = \varphi_{\alpha,1}(x)\chi(0)\exp(-i\frac{\mathcal{E}_{\alpha,1}t}{\hbar}), \qquad (12.70)$$

with the help of Eq. (12.63). Hence, we obtain from Eq. (12.69)

$$E_\alpha = E_{\alpha,\beta}|_{\beta=1} = \int dx \psi^*_{\alpha,1}(x,t)\widehat{H}_\alpha(\widehat{p}_\beta, \widehat{x})\psi_{\alpha,1}(x,t)|_{\beta=1} = \mathcal{E}_{\alpha,1}, \quad (12.71)$$

here $1 < \alpha \leq 2$, E_α is the energy of a physical quantum system, $\mathcal{E}_{\alpha,1}$ is the eigenvalue of quantum mechanical Hamilton operator $\widehat{H}_\alpha(\widehat{p}, \widehat{x})$ given by Eq. (12.64), and normalization condition (12.51) has been taken into account.

Hence, in the limit case $\beta = 1$ we recover the well-known statement of standard quantum mechanics [94] and fractional quantum mechanics [67], [96] that the energy spectrum of a quantum system is a set of eigenvalues of Hamilton operator.

To calculate $E_{\alpha,\beta}$ let us substitute $\psi_{\alpha,\beta}(x,t)$ given by Eq. (12.50) into Eq. (12.69)

$$E_{\alpha,\beta} = \mathcal{E}_{\alpha,\beta} |\chi(0)|^2 \tag{12.72}$$

$$\times \left| Ec_\beta(\frac{\mathcal{E}_{\alpha,\beta}(-t^\beta)}{\hbar_\beta}) + i^\beta Es_\beta(\frac{\mathcal{E}_{\alpha,\beta}(-t^\beta)}{\hbar_\beta}) \right|^2 \int\limits_{-\infty}^{\infty} dx |\varphi(x)|^2,$$

where we used definition (12.65) for the function $\chi_{\alpha,\beta}(t)$.

With the help of Eq. (12.51) the last equation can be rewritten as

$$E_{\alpha,\beta} = \mathcal{E}_{\alpha,\beta} \Bigg\{ \left| Ec_\beta(\frac{\mathcal{E}(-t^\beta)}{\hbar_\beta}) \right|^2 + \left| Es_\beta(\frac{\mathcal{E}(-t^\beta)}{\hbar_\beta}) \right|^2 \tag{12.73}$$

$$+ 2\,\mathrm{Re}\left(i^\beta Es_\beta(\frac{\mathcal{E}(-t^\beta)}{\hbar_\beta}) Ec_\beta^*(\frac{\mathcal{E}(-t^\beta)}{\hbar_\beta}) \right) \Bigg\}.$$

This equation defines the energy $E_{\alpha,\beta}$ of a time fractional quantum system with the pseudo-Hamilton operator $\widehat{H}_{\alpha,\beta}(\widehat{p}_\beta, \widehat{x})$ introduced by Eq. (12.22). The energy $E_{\alpha,\beta}$ is real due to the Hermiticity of the pseudo-Hamilton operator $\widehat{H}_{\alpha,\beta}(\widehat{p}_\beta, \widehat{x})$. We see that the energy $E_{\alpha,\beta}$ of a time fractional quantum system depends on time t, eigenvalue $\mathcal{E}_{\alpha,\beta}$ of pseudo-Hamilton operator, and fractality parameters α and β. Thus, we come to the conclusion that in the framework of time fractional quantum mechanics there are no stationary states and the eigenvalues of the pseudo-Hamilton operator are not the energy levels of time fractional quantum system.

Let us note that in the limit case $\beta = 1$ we have (see Eqs. (E.5) and (E.6) in Appendix E),

$$Ec_\beta(z)|_{\beta=1} = \cos(z), \qquad Es_\beta(z)|_{\beta=1} = \sin(z). \tag{12.74}$$

Hence, it follows from Eqs. (12.73) and (12.74) that

$$E_\alpha = E_{\alpha,\beta}|_{\beta=1} = \mathcal{E}_{\alpha,1}, \qquad 1 < \alpha \le 2,$$

which is in line with Eq. (12.71).

Chapter 13

Applications of Time Fractional Quantum Mechanics

13.1 A free particle wave function

For a free particle when $V(x,t) = 0$, the 1D space-time fractional Schrödinger equation (12.11) reads

$$i^\beta \hbar_\beta \partial_t^\beta \psi(x,t) = D_{\alpha,\beta}(-\hbar_\beta^2 \Delta)^{\alpha/2} \psi(x,t), \qquad (13.1)$$

$$1 < \alpha \le 2, \qquad 0 < \beta \le 1,$$

here $\psi(x,t)$ is the wave function, i is imaginary unit, $i = \sqrt{-1}$, $\hbar_\beta$ is scale coefficient, Δ is 1D Laplace operator, $\Delta = \partial^2/\partial x^2$, and, finally, ∂_t^β is the left Caputo fractional derivative of order β defined by Eqs. (12.12) and (12.13).

We are searching for solution to Eq. (13.1) with the initial condition $\psi_0(x)$,

$$\psi_0(x) = \psi(x, t = 0). \qquad (13.2)$$

By applying the Fourier transform to the wave function $\psi(x,t)$,

$$\psi(x,t) = \frac{1}{2\pi\hbar_\beta} \int\limits_{-\infty}^{\infty} dp_\beta \exp\{ip_\beta x/\hbar_\beta\} \varphi(p_\beta, t), \qquad (13.3)$$

with $\varphi(p_\beta, t)$ defined by

$$\varphi(p_\beta, t) = \int\limits_{-\infty}^{\infty} dx \exp\{-ip_\beta x/\hbar_\beta\} \psi(x,t), \qquad (13.4)$$

we obtain from Eq. (13.1)

$$i^{\beta}\hbar_{\beta}\partial_t^{\beta}\varphi(p_{\beta},t) = D_{\alpha,\beta}|p_{\beta}|^{\alpha}\varphi(p_{\beta},t), \qquad (13.5)$$

with the initial condition $\varphi_0(p_{\beta})$ given by

$$\varphi_0(p_{\beta}) = \varphi(p_{\beta}, t = 0) = \int\limits_{-\infty}^{\infty} dx \exp\{-ip_{\beta}x/\hbar_{\beta}\}\psi_0(x). \qquad (13.6)$$

The equation (13.5) is the 1D space-time fractional Schrödinger equation in momentum representation.

The solution to the problem introduced by Eqs. (13.5) and (13.6) is

$$\varphi(p_{\beta}, t) = E_{\beta}\left(D_{\alpha,\beta}|p_{\beta}|^{\alpha}\frac{t^{\beta}}{i^{\beta}\hbar_{\beta}}\right)\varphi_0(p_{\beta}), \qquad (13.7)$$

where E_{β} is the Mittag-Leffler function given by Eq. (12.61).

Hence, the solution to the 1D space-time fractional Schrödinger equation Eq. (13.1) with initial condition given by Eq. (13.2) can be presented as

$$\psi(x, t) = \frac{1}{2\pi\hbar_{\beta}} \int\limits_{-\infty}^{\infty} dx' \int\limits_{-\infty}^{\infty} dp_{\beta} \exp\left\{i\frac{p_{\beta}(x - x')}{\hbar_{\beta}}\right\} \qquad (13.8)$$

$$\times E_{\beta}\left(D_{\alpha,\beta}|p_{\beta}|^{\alpha}\frac{t^{\beta}}{i^{\beta}\hbar_{\beta}}\right)\psi_0(x').$$

13.2 Infinite potential well problem in time fractional quantum mechanics

Considering a particle in the symmetric infinite potential well defined by Eq. (10.25) we can use the results of Sec. 10.2 to come up with solution to space-time fractional Schrödinger equation (12.11). Since the potential field defined by Eq. (10.25) does not depend on time t, we are searching for solution $\psi(x, t)$ to Eq. (12.11) in separable form given by Eq. (12.50), $\psi(x, t) = \varphi(x)\chi(t)$, where $\varphi(x)$ is spatial and $\chi(t)$ is temporal components of the solution.

It is assumed as well, that initial wave function $\psi(x, t = 0) = \psi(x, 0)$ is normalized and the normalization condition is given by Eq. (12.51).

The spatial component $\varphi(x)$ of the solution $\psi(x,t)$ satisfies

$$-D_\alpha(\hbar\nabla)^\alpha\phi(x) = \mathcal{E}\phi(x), \qquad \phi(-a) = \phi(a) = 0, \qquad (13.9)$$

where $\mathcal{E}$ is the eigenvalue of fractional quantum mechanical operator $-D_\alpha(\hbar\nabla)^\alpha$.

The solution to the problem (13.9) was found in Sec. 10.2. The normalized spatial wave function $\phi_n(x)$ is

$$\phi_n(x) = \frac{1}{2\sqrt{a}}\{\exp(-i\frac{n\pi}{2a}x) - (-1)^n \exp(i\frac{n\pi}{2a}x)\}, \qquad (13.10)$$

and quantized parameter $\mathcal{E}_n$ with units of energy is given by

$$\mathcal{E}_n = D_\alpha(\frac{\pi\hbar}{2a})n^\alpha, \qquad (13.11)$$

where n is positive integer and $2a$ is width of the symmetric infinite potential well defined by Eq. (10.25).

In the framework of time fractional quantum mechanics the energy of a particle in the symmetric infinite potential well has the form

$$E_n = \mathcal{E}_n\left\{ \left| Ec_\beta(\frac{\mathcal{E}_n(-t^\beta)}{\hbar_\beta}) \right|^2 + \left| Es_\beta(\frac{\mathcal{E}_n(-t^\beta)}{\hbar_\beta}) \right|^2 \right. \qquad (13.12)$$

$$\left. + 2\,\text{Re}\left(i^\beta Es_\beta(\frac{\mathcal{E}_n(-t^\beta)}{\hbar_\beta})Ec_\beta^*(\frac{\mathcal{E}_n(-t^\beta)}{\hbar_\beta}) \right) \right\},$$

where functions Ec_β and Es_β are defined by Eqs. (E.5) and (E.6) (see, Appendix E).

Since the potential $V(x)$ defined by Eq. (10.25) is symmetric potential

$$V(x) = V(-x),$$

we can introduce even $\phi_m^{\text{even}}(x)$ and odd $\phi_m^{\text{odd}}(x)$ under reflection $x \to -x$ solutions to the problem (13.9). It has been found in Sec. 10.2 that $\phi_m^{\text{even}}(x)$ and $\phi_m^{\text{odd}}(x)$ are given by Eqs. (10.40) and (10.41) respectively.

The temporal component $\chi(t)$ of solution $\psi_n(x,t)$ is given by Eq. (12.60). Then the normalized solution to the problem (10.28) in the framework of time fractional quantum mechanics is

$$\psi_n(x,t) = \frac{1}{2\sqrt{a}}\{\exp(-i\frac{n\pi}{2a}x) - (-1)^n \exp(i\frac{n\pi}{2a}x)\} \qquad (13.13)$$

$$\times \chi(0) E_\beta \left(\frac{(-it)^\beta \mathcal{E}}{\hbar_\beta} \right),$$

here $E_\beta((-it)\mathcal{E}/\hbar_\beta)$ is the Mittag-Leffler function and $\chi(0)$ is the initial (at $t = 0$) value of the temporal component $\chi(0) = \chi(t)|_{t=0}$. We present $\psi_n(x, t)$ given by Eq. (13.13) as

$$\psi_n(x, t) = \varphi_n(x) \chi_\beta(t), \qquad (13.14)$$

where two functions $\varphi_n(x)$ and $\chi_\beta(t)$ have been introduced by

$$\varphi_n(x) = \frac{1}{2\sqrt{a}} \{ \exp(-i\frac{n\pi}{2a}x) - (-1)^n \exp(i\frac{n\pi}{2a}x) \} \qquad (13.15)$$

and

$$\chi_\beta(t) = \chi(0) E_\beta \left(\frac{(-it)^\beta \mathcal{E}}{\hbar_\beta} \right). \qquad (13.16)$$

It is assumed that at the initial time moment $t = 0$, $\psi_n(x, 0) = \psi_n(x, t)|_{t=0}$ is normalized, that is

$$\int_{-\infty}^{\infty} dx |\psi_n(x, 0)|^2 = |\chi(0)|^2 \int_{-\infty}^{\infty} dx |\varphi_n(x)|^2 = 1. \qquad (13.17)$$

Taking into account that $\varphi_n(x)$ is normalized spatial wave function (see, Sec. 10.2) we conclude that $|\chi(0)|^2 = 1$.

The wave function $\psi_n(x, t)$ given by Eq. (13.17) is solution to the space-time fractional Schrödinger equation for a particle in symmetric infinite potential well defined by Eq. (10.25).

It follows from Eq. (13.13) that even time independent solution to the space-time fractional Schrödinger equation for a particle in symmetric infinite potential well-defined by Eq. (10.25) is

$$\psi_m^{\text{even}}(x, t) = \frac{1}{\sqrt{a}} \cos \left(\frac{(2m + 1)\pi}{2a} x \right) \chi(0)$$

$$\times E_\beta \left\{ D_\alpha \left(\frac{(2m + 1)\pi\hbar}{2a} \right)^\alpha (-it)^\beta / \hbar_\beta \right\}, \qquad (13.18)$$

$$m = 0, 1, 2, ...,$$

and the odd time independent solution is

$$\psi_m^{\text{odd}}(x,t) = -\frac{i}{\sqrt{a}} \sin(\frac{m\pi}{a}x)\chi(0) \tag{13.19}$$

$$\times E_\beta \left\{ D_\alpha \left(\frac{2m\pi\hbar}{a} \right)^\alpha (-it)^\beta / \hbar_\beta \right\},$$

$$m = 1, 2, ...$$

In the infinite potential well the ground state is represented by the wave function $\psi_{\text{ground}}(x,t)$, which is $\psi_n(x,t)$ given by Eq. (13.13) at $n = 1$, $\psi_{\text{ground}}(x,t) = \psi_n(x,t)|_{n=1} = \psi_1(x,t)$, and it has the form

$$\psi_{\text{ground}}(x,t) = \psi_1(x,t) = \frac{1}{\sqrt{a}} \cos\{\frac{\pi x}{2a}\}\chi(0)E_\beta \left\{ D_\alpha \left(\frac{\pi\hbar}{2a} \right)^\alpha (-it)^\beta / \hbar_\beta \right\}. \tag{13.20}$$

Obviously, the ground state wave function $\psi_{\text{ground}}(x,t)$ can be presented as

$$\psi_{\text{ground}}(x,t) = \psi_m^{\text{even}}(x,t)|_{m=0}, \tag{13.21}$$

where $\psi_m^{\text{even}}(x,t)$ is defined by Eq. (13.18).

In the framework of time fractional quantum mechanics ground state energy E_{ground} is

$$E_{\text{ground}} = E_n|_{n=1} = \mathcal{E}_1 \left\{ \left| Ec_\beta(\frac{\mathcal{E}_1(-t^\beta)}{\hbar_\beta}) \right|^2 + \left| Es_\beta(\frac{\mathcal{E}_1(-t^\beta)}{\hbar_\beta}) \right|^2 \right. \tag{13.22}$$

$$\left. + 2\,\text{Re}\left(i^\beta Es_\beta(\frac{\mathcal{E}_1(-t^\beta)}{\hbar_\beta})Ec_\beta^*(\frac{\mathcal{E}_1(-t^\beta)}{\hbar_\beta}) \right) \right\},$$

where E_n is defined by Eq. (13.12) and $\mathcal{E}_1$ comes from Eq. (13.11) at $n = 1$,

$$\mathcal{E}_1 = \mathcal{E}_n|_{n=1} = D_\alpha(\frac{\pi\hbar}{2a}). \tag{13.23}$$

It is easy to see that Eqs. (13.14), (13.18) and (13.21) allow us to conclude that time independent wave function of ground state $\phi_{\text{ground}}(x)$ is given by

$$\phi_{\text{ground}}(x) \equiv \psi_0^{\text{even}}(x) = \frac{1}{\sqrt{a}} \cos\{\frac{\pi x}{2a}\}. \tag{13.24}$$

13.3   A free particle space-time fractional quantum mechanical kernel

The solution (13.8) to space-time fractional Schrödinger equation can be expressed in the form

$$\psi(x,t) = \int\limits_{-\infty}^{\infty} dx' K_{\alpha,\beta}^{(0)}(x - x', t)\psi_0(x'), \qquad (13.25)$$

if we introduce into consideration a free particle space-time fractional quantum mechanical kernel $K_{\alpha,\beta}^{(0)}(x,t)$ defined by

$$K_{\alpha,\beta}^{(0)}(x,t) = \frac{1}{2\pi\hbar_\beta} \int\limits_{-\infty}^{\infty} dp_\beta \exp\left\{i\frac{p_\beta x}{\hbar_\beta}\right\} E_\beta\left(D_{\alpha,\beta}|p_\beta|^\alpha \frac{t^\beta}{i^\beta\hbar_\beta}\right), \qquad (13.26)$$

$$1 < \alpha \le 2, \qquad 0 < \beta \le 1.$$

The space-time fractional quantum mechanical kernel $K_{\alpha,\beta}^{(0)}(x,t)$ satisfies

$$K_{\alpha,\beta}^{(0)}(x,t)|_{t=0} = K_{\alpha,\beta}^{(0)}(x,0) = \delta(x), \qquad (13.27)$$

$$1 < \alpha \le 2, \qquad 0 < \beta \le 1,$$

where $\delta(x)$ is delta function.

It follows immediately from Eq. (13.26) that the Fourier transform $K_{\alpha,\beta}^{(0)}(p_\beta, t)$ of space-time fractional quantum mechanical kernel is

$$K_{\alpha,\beta}^{(0)}(p_\beta, t) = \int\limits_{-\infty}^{\infty} dx \exp\left\{-i\frac{p_\beta x}{\hbar_\beta}\right\} K_{\alpha,\beta}^{(0)}(x,t)$$

$$= E_\beta\left(D_{\alpha,\beta}|p_\beta|^\alpha \frac{t^\beta}{i^\beta\hbar_\beta}\right), \qquad (13.28)$$

$$1 < \alpha \le 2, \qquad 0 < \beta \le 1.$$

This is the space-time fractional quantum mechanical kernel in momentum representation.

Applying the Laplace transform with respect to the time variable t we obtain the Fourier-Laplace transform $K_{\alpha,\beta}^{(0)}(p_\beta, u)$ of space-time fractional quantum mechanical kernel

$$K_{\alpha,\beta}^{(0)}(p_\beta, s) = \int\limits_0^\infty dt \exp\{-st\} K_{\alpha,\beta}^{(0)}(p_\beta, t) \qquad (13.29)$$

$$= \frac{s^{\beta-1}}{s^\beta - D_{\alpha,\beta}|p_\beta|^\alpha/i^\beta \hbar_\beta} K_{\alpha,\beta}^{(0)}(p_\beta, 0).$$

Taking into account Eq. (13.27) we have

$$K_{\alpha,\beta}^{(0)}(p_\beta, s) = \frac{s^{\beta-1}}{s^\beta - D_{\alpha,\beta}|p_\beta|^\alpha/i^\beta \hbar_\beta}. \qquad (13.30)$$

13.3.1 Renormalization properties of the space-time fractional quantum mechanical kernel

Aiming to study renormalization properties of the space-time fractional quantum mechanical kernel, let us re-write $K_{\alpha,\beta}^{(0)}(p_\beta, s)$ given by Eq. (13.30) in integral form

$$K_{\alpha,\beta}^{(0)}(p_\beta, s) = s^{\beta-1} \int\limits_0^\infty du \exp\{-u(s^\beta - D_{\alpha,\beta}|p_\beta|^\alpha/i^\beta \hbar_\beta)\} \qquad (13.31)$$

$$= \int\limits_0^\infty du N_{\alpha,\beta}(p_\beta, u) L_\beta(u, s),$$

where we introduced two functions $N_{\alpha,\beta}(p_\beta, u)$ and $L_\beta(u, s)$ defined by

$$N_{\alpha,\beta}(p_\beta, u) = \exp\{u D_{\alpha,\beta}|p_\beta|^\alpha/i^\beta \hbar_\beta\} \qquad (13.32)$$

and

$$L_\beta(u, s) = s^{\beta-1} \exp\{-us^\beta\}. \qquad (13.33)$$

Having new function $N_{\alpha,\beta}(p_\beta, u)$ we obtain

$$N_{\alpha,\beta}(x, u) = \frac{1}{2\pi\hbar_\beta} \int\limits_{-\infty}^\infty dp_\beta \exp\{i\frac{p_\beta x}{\hbar_\beta}\} N_{\alpha,\beta}(p_\beta, u) \qquad (13.34)$$

$$= \frac{1}{2\pi\hbar_\beta} \int\limits_{-\infty}^{\infty} dp_\beta \exp\{i\frac{p_\beta x}{\hbar_\beta}\} \exp\{u D_{\alpha,\beta} |p_\beta|^\alpha / i^\beta \hbar_\beta\}.$$

It is easy to see that the scaling

$$N_{\alpha,\beta}(x,u) = \frac{1}{u^{1/\alpha}} N_{\alpha,\beta}(\frac{x}{u^{1/\alpha}}, 1) \qquad (13.35)$$

holds for function $N_{\alpha,\beta}(x,u)$ given by Eq. (13.34).

Having new function $L_\beta(u,s)$ we obtain

$$L_\beta(u,t) = \frac{1}{2\pi i} \int\limits_{\sigma-i\infty}^{\sigma+i\infty} ds e^{st} L_\beta(u,s) \qquad (13.36)$$

$$= \frac{1}{2\pi i} \int\limits_{\sigma-i\infty}^{\sigma+i\infty} ds e^{st} s^{\beta-1} \exp\{-us^\beta\},$$

where σ is greater than the real part of all singularities of $L_\beta(u,s)$ in the complex plane.

It is easy to see that the scaling

$$L_\beta(u,t) = \frac{1}{t^\beta} L_\beta(\frac{u}{t^\beta}, 1), \qquad (13.37)$$

holds for function $L_\beta(u,t)$ given by Eq. (13.36).

In terms of the above introduced functions $N_{\alpha,\beta}(x,u)$ and $L_\beta(u,t)$ Eq. (13.26) reads

$$K_{\alpha,\beta}^{(0)}(x,t) = \int\limits_{0}^{\infty} du N_{\alpha,\beta}(x,u) L_\beta(u,t), \qquad (13.38)$$

where functions $N_{\alpha,\beta}(x,u)$ and $L_\beta(u,t)$ are defined by Eqs. (13.34) and (13.36).

The equivalent representation for $K_{\alpha,\beta}^{(0)}(x,t)$ is

$$K_{\alpha,\beta}^{(0)}(x,t) = \frac{1}{t^\beta} \int\limits_{0}^{\infty} du \frac{1}{u^{1/\alpha}} N_{\alpha,\beta}(\frac{x}{u^{1/\alpha}}, 1) L_\beta(\frac{u}{t^\beta}, 1), \qquad (13.39)$$

where $1 < \alpha \leq 2$ and $0 < \beta \leq 1$.

Hence, one can consider Eq. (13.39) as an alternative representation for space-time fractional quantum mechanical kernel $K_{\alpha,\beta}^{(0)}(x,t)$. The interesting feature of this representation is that it separates space and time variables for space-time fractional quantum mechanical kernel $K_{\alpha,\beta}^{(0)}(x,t)$. It follows from Eqs. (13.34) and (13.38) that

$$K_{\alpha,\beta}^{(0)}(x,t) = K_{\alpha,\beta}^{(0)}(-x,t). \tag{13.40}$$

With the help of Eqs. (13.35) and (13.37) we conclude that the scaling

$$K_{\alpha,\beta}^{(0)}(x,t) = \frac{1}{t^{\beta/\alpha}} K_{\alpha,\beta}^{(0)}\left(\frac{x}{t^{\beta/\alpha}}, 1\right), \tag{13.41}$$

holds for space-time fractional quantum mechanical kernel $K_{\alpha,\beta}^{(0)}(x,t)$ given by Eq. (13.26).

Defined by Eqs. (13.37) and (13.36) function $L_\beta(u/t^\beta, 1)$ can be rewritten as

$$L_\beta(z,1) = \frac{1}{2\pi i} \int\limits_{\sigma-i\infty}^{\sigma+i\infty} ds e^s s^{\beta-1} \exp\{-zs^\beta\}, \qquad 0 < \beta \leq 1, \tag{13.42}$$

where we introduced a new variable $z = u/t^\beta$.

13.3.2 Wright L-function in time fractional quantum mechanics

Our intent now is to show that the function $L_\beta(z,1)$ defined by Eq. (13.42) can be presented in terms of the Wright function (see, for example, [150]). We call the function $L_\beta(z,1)$ the Wright L-function. By deforming the integration path on the right-hand side of Eq. (13.42) into the Hankel contour H_a we have

$$L_\beta(z,1) = \frac{1}{2\pi i} \int\limits_{H_a} ds e^s s^{\beta-1} \exp\{-zs^\beta\}, \qquad 0 < \beta \leq 1. \tag{13.43}$$

A series expansion for $\exp\{-zs^\beta\}$ yields

$$L_\beta(z,1) = \frac{1}{2\pi i} \int\limits_{H_a} ds e^s s^{\beta-1} \sum_{k=0}^{\infty} \frac{(-1)^k z^k}{k!} s^{\beta k} \tag{13.44}$$

$$= \sum_{k=0}^{\infty} \frac{(-1)^k z^k}{k!} \left(\frac{1}{2\pi i} \int_{H_a} ds e^s s^{\beta k + \beta - 1} \right).$$

Using the well-known Hankel representation of the Gamma function (see, for example, [151]),

$$\frac{1}{\Gamma(-\sigma k + \nu)} = \left(\frac{1}{2\pi i} \int_{H_a} ds e^s s^{\sigma k - \nu} \right), \tag{13.45}$$

we obtain for $L_\beta(z, 1)$

$$L_\beta(z, 1) = \sum_{k=0}^{\infty} \frac{(-1)^k z^k}{k! \Gamma(-\beta k + (1 - \beta))}. \tag{13.46}$$

In the notations of [150], the Wright function $\phi(a, \beta; z)$ is expressed as

$$\phi(a, \beta; z) = \sum_{k=0}^{\infty} \frac{z^k}{k! \Gamma(ak + \beta)}, \qquad a > -1, \qquad \beta \in \mathbb{C}, \tag{13.47}$$

where $\mathbb{C}$ stands for the field of complex numbers.

In terms of Fox's H-function the function $\phi(a, \beta; z)$ has the following representation

$$\phi(a, \beta; z) = H_{0,2}^{1,0} \left[-z \left|_{(0,1), (1 - \beta, a)}^{-} \right. \right]. \tag{13.48}$$

Therefore, we see that the Wright L-function $L_\beta(z, 1)$ introduced by Eq. (13.43) has the following representation in terms of the Wright function $\phi(-\beta, 1 - \beta; -z)$

$$L_\beta(z, 1) = \phi(-\beta, 1 - \beta; -z). \tag{13.49}$$

In terms of Fox's $H_{0,2}^{1,0}$-function the Wright L-function $L_\beta(z, 1)$ is defined by

$$L_\beta(z, 1) = H_{0,2}^{1,0} \left[-z \left|_{(0,1), (\beta, -\beta)}^{-} \right. \right], \qquad 0 < \beta \leq 1. \tag{13.50}$$

Knowledge of particular cases for the Wright function let us obtain the particular cases for the function $L_\beta(z, 1)$.

1. When $\beta = 1/2$ we have

$$L_{1/2}(z, 1) = L_\beta(z, 1)|_{\beta=1/2} = \frac{1}{\sqrt{\pi}} \exp(-\frac{z^2}{4}). \tag{13.51}$$

2. When $\beta = 1$ we have

$$L_1(z, 1) = L_\beta(z, 1)|_{\beta=1} = \delta(z - 1), \tag{13.52}$$

where $\delta(z)$ is delta function. The last property is useful to recover fractional quantum mechanics and standard quantum mechanics formulas for a free particle quantum kernel from the general equation (13.38).

Thus, function $L_\beta(z, 1)$ involved into representation (13.39) of space-time fractional quantum kernel, is expressed by Eq. (13.49) as the Wright L-function. The Wright L-function introduces into the time fractional quantum mechanics a well-developed mathematical tool - Wright function. In other words, many well-known results related to the Wright function can be applied to study fundamental properties of the space-time fractional quantum kernel and can be used to develop a variety of new applications of time fractional quantum mechanics.

13.3.3   *Fox H-function representation for a free particle space-time fractional quantum mechanical kernel*

The space-time fractional quantum kernel given by Eq. (13.26) can be expressed in terms of the Fox H-function. The definition of the H-function and its fundamental properties can be found in [152]. We presented the definition of the Fox H-function and some of its properties in Appendix A.

In terms of the Fox H-function the Mittag-Leffler function $E_\beta(z)$ is presented by Eq. (B.11), see Appendix B. Then, $K_{\alpha,\beta}^{(0)}(x, t)$ given by Eq. (13.26) reads

$$K_{\alpha,\beta}^{(0)}(x, t) = \frac{1}{\pi \hbar_\beta} \int_0^\infty dp_\beta \cos(\frac{p_\beta x}{\hbar_\beta}) \tag{13.53}$$

$$\times H_{1,2}^{1,1} \left[-D_{\alpha,\beta} p_\beta^\alpha \frac{t^\beta}{i^\beta \hbar_\beta} \left| \begin{array}{c} (0, 1) \\ (0, 1), (0, \beta) \end{array} \right. \right].$$

With the help of the cosine transform of the H-function defined by of Eq. (A.20), see Appendix A, we obtain

$$K_{\alpha,\beta}^{(0)}(x,t) = \frac{1}{|x|} H_{3,3}^{2,1} \left[-\frac{i^\beta \hbar_\beta}{D_{\alpha,\beta} \hbar_\beta^\alpha t^\beta} |x|^\alpha \, \bigg| \, \begin{matrix} (1,1),(1,\beta),(1,\alpha/2) \\ (1,\alpha),(1,1),(1,\alpha/2) \end{matrix} \right], \quad (13.54)$$

$$1 < \alpha \leq 2, \qquad 0 < \beta \leq 1,$$

which is the expression for the space-time fractional quantum kernel $K_{\alpha,\beta}^{(0)}(x,t)$ in terms of $H_{3,3}^{2,1}$-function.

Alternatively, using Property 12.2.4 given by Eq. (A.13), see Appendix A, $K_{\alpha,\beta}^{(0)}(x,t)$ can be presented as

$$K_{\alpha,\beta}^{(0)}(x,t)$$

$$= \frac{1}{\alpha|x|} H_{3,3}^{2,1} \left[\frac{1}{\hbar_\beta} \left(-\frac{i^\beta \hbar_\beta}{D_{\alpha,\beta} t^\beta} \right)^{1/\alpha} |x| \, \bigg| \, \begin{matrix} (1,1/\alpha),(1,\beta/\alpha),(1,1/2) \\ (1,1),(1,1/\alpha),(1,1/2) \end{matrix} \right], \quad (13.55)$$

$$1 < \alpha \leq 2, \qquad 0 < \beta \leq 1.$$

Thus, Eqs. (13.54) and (13.55) introduce a new family of a free particle space-time fractional quantum mechanical kernels parametrized by two fractality parameters α and β. The kernel $K_{\alpha,\beta}^{(0)}(x,\tau)$ does not satisfy quantum superposition law defined by Eq. (24) in [92], due to the fractional time derivative of order β in Eq. (13.1).

13.4   Special cases of time fractional quantum mechanics

With particular choice of fractality parameters α and β the space-time fractional Schrödinger equation Eq. (12.11) covers the following three special cases: the Schrödinger equation ($\alpha = 2$ and $\beta = 1$), the fractional Schrödinger equation ($1 < \alpha \leq 2$ and $\beta = 1$) and the time fractional Schrödinger equation ($\alpha = 2$ and $0 < \beta \leq 1$).

13.4.1   The Schrödinger equation

When $\alpha = 2$ and $\beta = 1$, we have

$$D_{\alpha,\beta}|_{\alpha=2,\beta=1} = D_{2,1} = \frac{1}{2m}, \qquad p_\beta|_{\beta=1} = p, \quad (13.56)$$

and

$$\hbar_\beta|_{\beta=1} = \hbar, \tag{13.57}$$

where m is the mass of a quantum particle, p is momentum of a quantum particle and $\hbar$ is the well-known Planck's constant.

In this case space-time fractional Schrödinger equation Eq. (12.11) goes into the celebrated Schrödinger equation [6]

$$i\hbar\partial_t\psi(x,t) = -\frac{\hbar^2}{2m}\Delta\psi(x,t) + V(x,t)\psi(x,t). \tag{13.58}$$

For the quantum mechanical kernel $K_{2,1}^{(0)}(x,t) = K_{\alpha,\beta}^{(0)}(x,t)|_{\alpha=2,\beta=1}$, where $K_{\alpha,\beta}^{(0)}(x,t)$ is defined by Eq. (13.38) we obtain

$$K_{2,1}^{(0)}(x,t) = \int_0^\infty du N_{2,1}(x,u)L_1(u,t), \tag{13.59}$$

here

$$N_{2,1}(x,u) = \frac{1}{2\pi\hbar} \int_{-\infty}^\infty dp \exp\{i\frac{px}{\hbar}\}\exp\{uD_{2,1}p^2/i\hbar\} \tag{13.60}$$

$$= \frac{1}{2\pi\hbar} \int_{-\infty}^\infty dp \exp\{i\frac{px}{\hbar}\}\exp\{-iu\frac{p^2}{2m\hbar}\},$$

and

$$L_1(u,t) = \frac{1}{2\pi i} \int_{\sigma-i\infty}^{\sigma+i\infty} ds e^{st} \exp\{-us\} = \delta(t-u). \tag{13.61}$$

Substitution of Eqs. (13.60) and (13.61) into Eq. (13.59) yields

$$K_{2,1}^{(0)}(x,t) = \frac{1}{2\pi\hbar} \int_{-\infty}^\infty dp \exp\{i\frac{px}{\hbar}\}\exp\{-it\frac{p^2}{2m\hbar}\}. \tag{13.62}$$

The integral in Eq. (13.62) can be evaluated analytically and the result is

$$K_F^{(0)}(x,t) = K_{2,1}^{(0)}(x,t) = \sqrt{\frac{m}{2\pi i\hbar t}} \exp\{i\frac{mx^2}{2\hbar t}\}, \tag{13.63}$$

which is Feynman's free particle quantum kernel given by Eq. (7.35).

13.4.2 *Fractional Schrödinger equation*

In the case when $\beta = 1$, $D_{\alpha,\beta}$ goes into the scale coefficient D_α emerging in fractional quantum mechanics [67], [92],

$$D_{\alpha,\beta}|_{\beta=1} = D_{\alpha,1} = D_\alpha, \qquad 1 < \alpha \le 2, \tag{13.64}$$

with units of,

$$[D_\alpha] = \mathrm{erg}^{1-\alpha} \cdot \mathrm{cm}^\alpha \cdot \mathrm{sec}^{-\alpha}.$$

In this case the space-time fractional Schrödinger equation Eq. (12.11) goes into Laskin's fractional Schrödinger equation [96]

$$i\hbar\partial_t\psi(x,t) = D_\alpha(-\hbar^2\Delta)^{\alpha/2}\psi(x,t) + V(x,t)\psi(x,t), \tag{13.65}$$

$$1 < \alpha \le 2,$$

with $\hbar$ being fundamental Planck's constant.

For the quantum mechanical kernel $K_{\alpha,1}^{(0)}(x,t) = K_{\alpha,\beta}^{(0)}(x,t)|_{\beta=1}$, where $K_{\alpha,\beta}^{(0)}(x,t)$ is defined by Eq. (13.38) we obtain

$$K_{\alpha,1}^{(0)}(x,t) = \int\limits_0^\infty du N_{\alpha,1}(x,u)L_1(u,t), \tag{13.66}$$

here

$$N_{\alpha,1}(x,u) = \frac{1}{2\pi\hbar} \int\limits_{-\infty}^\infty dp\exp\{i\frac{px}{\hbar}\}\exp\{uD_\alpha|p|^\alpha/i\hbar\} \tag{13.67}$$

and

$$L_1(u,t) = \frac{1}{2\pi i} \int\limits_{\sigma-i\infty}^{\sigma+i\infty} ds e^{st}\exp\{-us\} = \delta(t-u). \tag{13.68}$$

Substitution of Eqs. (13.67) and (13.68) into Eq. (13.66) yields

$$K_{\alpha,1}^{(0)}(x,t) = \frac{1}{2\pi\hbar} \int\limits_{-\infty}^\infty dp\exp\{i\frac{px}{\hbar}\}\exp\left\{-it\frac{D_\alpha|p|^\alpha}{\hbar}\right\}. \tag{13.69}$$

It was shown in Sec. 7.2 that the integral in Eq. (13.69) can be expressed in terms of the $H_{2,2}^{1,1}$-function,

$$K_{\alpha,1}^{(0)}(x,t) = \frac{1}{\alpha|x|} H_{2,2}^{1,1}\left[\frac{1}{\hbar}\left(-\frac{i\hbar}{D_\alpha t} \right)^{1/\alpha} |x| \,\middle|\, \begin{matrix} (1,1/\alpha),(1,1/2) \\ (1,1),(1,1/2) \end{matrix} \right], \quad (13.70)$$

$$1 < \alpha \le 2.$$

The expression (13.70) for a free particle quantum kernel $K_{\alpha,1}^{(0)}(x,t)$ can be obtained directly from Eq. (13.55). Indeed, it is easy to see, that when $\beta = 1$ Eq. (13.55) reads

$$K_{\alpha,1}^{(0)}(x,t) = K_{\alpha,\beta}^{(0)}(x,t)|_{\beta=1}$$

$$= \frac{1}{\alpha|x|} H_{3,3}^{2,1}\left[\frac{1}{\hbar}\left(-\frac{i\hbar}{D_\alpha t} \right)^{1/\alpha} |x| \,\middle|\, \begin{matrix} (1,1/\alpha),(1,1/\alpha),(1,1/2) \\ (1,1),(1,1/\alpha),(1,1/2) \end{matrix} \right], \quad (13.71)$$

$$1 < \alpha \le 2.$$

By using Property 12.2.1 (see Appendix A) we rewrite the above expression as

$$K_{\alpha,1}^{(0)}(x,t) \quad (13.72)$$

$$= \frac{1}{\alpha|x|} H_{3,3}^{2,1}\left[\frac{1}{\hbar}\left(-\frac{i\hbar}{D_\alpha t} \right)^{1/\alpha} |x| \,\middle|\, \begin{matrix} (1,1/\alpha),(1,1/2),(1,1/\alpha) \\ (1,1/\alpha),(1,1),(1,1/2) \end{matrix} \right].$$

The next step is to use Property 12.2.2 (see Appendix A) to come to Eq. (13.69) for a free particle fractional quantum kernel $K_{\alpha,1}^{(0)}(x,t)$. It was shown in [93] that at $\alpha = 2$ Eq. (13.70) goes into Eq. (13.63).

13.4.3 *Time fractional Schrödinger equation*

In the case when $\alpha = 2$ we have $D_{\alpha,\beta}|_{\alpha=2} = D_{2,\beta}, 0 < \beta \le 1$, where $D_{2,\beta}$ is the scale coefficient with units of $[D_{2,\beta}] = \mathrm{g}^{-1} \cdot \sec^{2-2\beta}$. In this case the space-time fractional Schrödinger equation Eq. (12.11) goes into the time fractional Schrödinger equation of the form

$$i^\beta \hbar_\beta \partial_t^\beta \psi(x,t) = -D_{2,\beta}\hbar_\beta^2 \Delta\psi(x,t) + V(x,t)\psi(x,t), \quad 0 < \beta \le 1. \quad (13.73)$$

This equation can be considered as an alternative to Naber's time fractional Schrödinger equation [141].

For the quantum mechanical kernel $K_{2,\beta}^{(0)}(x,t) = K_{\alpha,\beta}^{(0)}(x,t)|_{\alpha=2}$, where $K_{\alpha,\beta}^{(0)}(x,t)$ is defined by Eq. (13.38) we obtain

$$K_{2,\beta}^{(0)}(x,t) = \int\limits_0^\infty du N_{2,\beta}(x,u) L_\beta(u,t), \qquad (13.74)$$

here

$$N_{2,\beta}(x,u) = \frac{1}{2\pi\hbar_\beta} \int\limits_{-\infty}^{\infty} dp_\beta \exp\{i\frac{p_\beta x}{\hbar}\} \exp\{u D_{2,\beta} p_\beta^2 / i^\beta \hbar_\beta\}, \qquad (13.75)$$

and $L_\beta(u,t)$ is given by Eq. (13.36).

To express a free particle time fractional quantum kernel $K_{2,\beta}^{(0)}(x,t)$ in terms of the H-function let's put $\alpha = 2$ in Eq. (13.55)

$$K_{2,\beta}^{(0)}(x,t) = K_{\alpha,\beta}^{(0)}(x,t)|_{\alpha=2}$$

$$= \frac{1}{2|x|} H_{3,3}^{2,1} \left[\frac{1}{\hbar_\beta} \left(-\frac{i^\beta \hbar_\beta}{D_{\alpha,\beta} t^\beta} \right)^{1/2} |x| \left| \begin{array}{c} (1,1/2),(1,\beta/2),(1,1/2) \\ (1,1),(1,1/2),(1,1/2) \end{array} \right. \right], \qquad (13.76)$$

$$0 < \beta \leq 1.$$

Using the Property 12.2.2 (see Appendix A) yields

$$K_{2,\beta}^{(0)}(x,t) = \frac{1}{2|x|} H_{2,2}^{2,0} \left[\frac{1}{\hbar_\beta} \left(-\frac{i^\beta \hbar_\beta}{D_{\alpha,\beta} t^\beta} \right)^{1/2} |x| \left| \begin{array}{c} (1,\beta/2),(1,1/2) \\ (1,1),(1,1/2) \end{array} \right. \right].$$

Next, by using the Property 12.2.1 and then Property 12.2.2 (see Appendix A) we find

$$K_{2,\beta}^{(0)}(x,t) = \frac{1}{2|x|} H_{1,1}^{1,0} \left[\frac{1}{\hbar_\beta} \left(-\frac{i^\beta \hbar_\beta}{D_{\alpha,\beta} t^\beta} \right)^{1/2} |x| \left| \begin{matrix} (1,\beta/2) \\ (1,1) \end{matrix} \right. \right]. \tag{13.77}$$

Thus, we find a new expression of a free particle time fractional quantum kernel $K_{2,\beta}^{(0)}(x,t)$ in terms of $H_{1,1}^{1,0}$-function. It can be shown that at $\beta = 1$ Eq. (13.77) goes into Eq. (13.63).

Chapter 14

Fractional Statistical Mechanics

But although, as a matter of history, statistical mechanics owes its origin to investigations in thermodynamics, it seems eminently worthy of an independent development, both on account of the elegance and simplicity of its principles, and because it yields new results and places old truths in a new light in departments quite outside of thermodynamics.

J. Willard Gibbs (1902)
Preface to *Elementary Principles in Statistical Mechanics*

14.1 Density matrix

14.1.1 *Phase space representation*

In order to develop quantum *fractional statistical mechanics* let us introduce fractional density matrix in terms of a path integral over Lévy flights. It can be done by means of the following fundamental relationship

$$\rho(x, \beta | x_0, 0) = K(x_b t_b | x_a t_a) \Big|_{x_a = x_0, t_a = 0}^{x_b = x, \, t_b = -i\hbar\beta} , \qquad (14.1)$$

where $K(x_b t_b | x_a t_a)$ is quantum kernel given by Eq. (6.5), $\hbar$ is Planck's constant and β is "inverse temperature" defined as

$$\beta = 1/k_B T, \qquad (14.2)$$

here k_B is Boltzmann's constant and T is the temperature of statistical mechanical system.

It is easy to see from Eqs. (14.1) and (6.5) that $\rho(x\beta|x_0 0)$ can be written as

$$\rho(x, \beta|x_0, 0) \tag{14.3}$$

$$= \int_{x_0}^{x} \mathrm{D}x(\tau) \int \mathrm{D}p(\tau) \exp\left\{ \frac{i}{\hbar} \int_{0}^{-i\hbar\beta} d\tau [p(\tau)\dot{x}(\tau) - H_\alpha(p(\tau), x(\tau), \tau)] \right\},$$

where integration over τ is performed along the negative imaginary time axis. Introducing a new integration variable

$$u = i\tau, \tag{14.4}$$

allows us to present the above equation for the density matrix as the following path integral

$$\rho(x, \beta|x_0, 0) \tag{14.5}$$

$$= \int_{x(0)=x_0}^{x(\hbar\beta)=x} \mathrm{D}x(u) \int \mathrm{D}p(u) \exp\left\{ -\frac{1}{\hbar} \int_{0}^{\hbar\beta} du [p(u)\dot{x}(u) - H_\alpha(p(u), x(u), u)] \right\},$$

where fractional Hamiltonian $H_\alpha(p, x)$ has the form (3.24) and $p(u), x(u)$ may be considered as a path (in phase space representation) evolving over "imaginary time" u and $\hbar$ is Planck's constant. A real variable u defined by Eq. (14.4) has units of time and it is called "imaginary time" since when the time τ is imaginary, u is real.

The exponential expression of Eq. (14.5) is very similar to the fractional canonical action given by (6.4). Since it governs the quantum-statistical path integral (14.5) it may be called the fractional quantum-statistical action or fractional Euclidean action $S_\alpha^{(e)}(p, x)$, indicated by the superscript (e) and subscript α,

$$S_\alpha^{(e)}(p, x) = \int_{0}^{\hbar\beta} du \{ p(u)\dot{x}(u) + H_\alpha(p(u), x(u)) \}. \tag{14.6}$$

Hence, the density matrix $\rho(x, \beta|x_0, 0)$ can be written as

$$\rho(x, \beta|x_0, 0) \tag{14.7}$$

$$= \int\limits_{x(0)=x_0}^{x(\hbar\beta)=x} \mathrm{D}x(u) \int \mathrm{D}p(u) \exp\left\{ -\frac{S_\alpha^{(e)}(p,x)}{\hbar} \right\}.$$

The path integral "measure" $\int\limits_{x(0)=x_0}^{x(\hbar\beta)=x} \mathrm{D}x(u) \int \mathrm{D}p(u)...$ in Eqs. (14.5) and (14.7) is defined by

$$\int\limits_{x(0)=x_0}^{x(\hbar\beta)=x} \mathrm{D}x(u) \int \mathrm{D}p(u)... \tag{14.8}$$

$$= \lim_{N\to\infty} \int\limits_{-\infty}^{\infty} dx_1...dx_{N-1} \frac{1}{(2\pi\hbar)^N} \int\limits_{-\infty}^{\infty} dp_1...dp_N ....$$

The parameter u in Eq. (14.5) is not the true time in any sense. It is just a parameter in an expression for the density matrix (see, for instance, [19]). Let us call u the "time", leaving the quotation marks to remind us that it is not real time (although u does have units of time). Likewise $x(u)$ will be called the "coordinate" and $p(u)$ the "momentum". Then Eq. (14.5) may be interpreted in the following way: Consider all the possible paths by which the system can travel between the initial $x(0) = x_0$ and final $x(\hbar\beta) = x$ configurations in "time" $\hbar\beta$. Then the density matrix $\rho(x\beta|x_0 0)$ introduced by Eq. (14.5) is the path integral over all possible paths, the contribution from a particular path is $\exp(S_\alpha^{(e)}(p,x)/\hbar)$, where $S_\alpha^{(e)}(p,x)$ is fractional Euclidean action (14.6).

It is easy to see that with the help of Eqs. (6.5) and (6.25) we can present the density matrix $\rho(x\beta|x_0 0)$ defined by Eq. (14.5) in the form

$$\rho(x,\beta|x_0,0) = \sum_{n=1}^{\infty} \phi_n(x)\phi_n^*(x_0)e^{-\beta E_n}, \tag{14.9}$$

where $\phi_n(x)$ are eigenfunctions and E_n are eigenvalues of time independent fractional Schrödinger equation with the fractional Hamilton operator given by Eq. (3.23).

14.1.2   *Coordinate representation*

In coordinate representation the path integral representation for the density matrix can be obtained from Eq. (6.18) by transition from t to $-i\hbar\beta$, where β is inverse temperature and then substituting $\tau = -iu$ with u being "imaginary time".

$$\rho(x,\beta|x_0,0) = \int\limits_{x(0)=x_0}^{x(\hbar\beta)=x} \mathcal{D}x(u)\exp\{-\frac{1}{\hbar}\int\limits_0^{\hbar\beta} du V(x(u),u)\}, \qquad (14.10)$$

where $V(x(u),u)$ is the potential energy as a functional of the Lévy flights path $x(u)$ and "imaginary time" u, and $\int\limits_{x(t_a)=x_a}^{x(t_b)=x_b} \mathcal{D}x(\tau)...$ is the path integral measure in coordinate space first introduced by Laskin [67],

$$\int\limits_{x(0)=x_0}^{x(\hbar\beta)=x} \mathcal{D}x(u)... \qquad (14.11)$$

$$= \lim_{N\to\infty} \int\limits_{-\infty}^{\infty} dx_1...dx_{N-1}\hbar^{-N}\left(D_\alpha\nu\right)^{-N/\alpha}$$

$$\times \prod_{j=1}^{N} L_\alpha\left\{\frac{1}{\hbar}\left(\frac{1}{D_\alpha\nu}\right)^{1/\alpha}|x_j - x_{j-1}|\right\}...,$$

here $\hbar$ denotes Planck's constant, $\nu = \beta/N$, $\beta = 1/k_B T$, and the Lévy probability distribution function L_α is given by Eq. (6.15).

The Lévy probability distribution function L_α is expressed in terms of Fox's $H_{2,2}^{1,1}$ function by Eq. (A.28). Hence, the path integral measure (14.11) can be alternatively presented as

$$\int\limits_{x(0)=x_0}^{x(\hbar\beta)=x} \mathcal{D}x(u)...$$

$$= \lim_{N\to\infty} \int\limits_{-\infty}^{\infty} dx_1...dx_{N-1}\prod_{j=1}^{N}\frac{1}{\alpha|x_j - x_{j-1}|} \qquad (14.12)$$

$$\times H_{2,2}^{1,1} \left[\frac{1}{\hbar} \left(\frac{1}{D_\alpha \nu} \right)^{1/\alpha} |x_j - x_{j-1}| \,\middle|\, \begin{matrix} (1,1/\alpha),(1,1/2) \\ (1,1),(1,1/2) \end{matrix} \right] \dots$$

The path integral measure defined by Eq. (14.11) or by Eq. (14.12), is generated by the Lévy flights stochastic process.

14.1.3 *Motion equation for the density matrix*

To obtain "motion equation"[1] for the density matrix $\rho(x,\beta|x_0) \equiv \rho(x,\beta|x_0,0)$ let us start from its path integral representation given by Eq. (14.10). By repeating the same steps which led us from the path integral (6.18) to fractional Schrödinger equation (6.53) we come to the following fractional differential equation [67], [92]

$$-\frac{\partial \rho(x,\beta|x_0)}{\partial \beta} = -D_\alpha (\hbar \nabla_x)^\alpha \rho(x,\beta|x_0) + V(x)\rho(x,\beta|x_0), \qquad (14.13)$$

$$\rho(x,0|x_0) = \delta(x - x_0),$$

which is "motion equation" for the density matrix $\rho(x,\beta|x_0)$. Note that quantum fractional Riesz operator $(\hbar \nabla_x)^\alpha$ acts on variable x.

Equation (14.13) can be rewritten as

$$-\frac{\partial \rho(x,\beta|x_0)}{\partial \beta} = H_\alpha \rho(x,\beta|x_0), \quad \rho(x,0|x_0) = \delta(x - x_0), \qquad (14.14)$$

where the fractional Hamiltonian H_α is defined by Eq. (3.19).

Having the density matrix $\rho(x,\beta|x_0)$, we introduce the density matrix in momentum representation $\rho^{(0)}(p,\beta|p_0)$ as

$$\rho(p,\beta|p_0) = \int_{-\infty}^{\infty} dx dx_0 \rho(x,\beta|x_0) \exp\left\{ -\frac{i}{\hbar}(px - p_0 x_0) \right\}. \qquad (14.15)$$

Then the inverse transform brings us

$$\rho(x,\beta|x_0) = \frac{1}{(2\pi\hbar)^2} \int_{-\infty}^{\infty} dp dp_0 \rho(p,\beta|p_0) \exp\left\{ \frac{i}{\hbar}(px - p_0 x_0) \right\}. \qquad (14.16)$$

---

[1] We use the term "motion equation" leaving the quotation marks in order to remind that the density matrix evolves with β.

Motion equation for the density matrix in momentum representation $\rho(p, \beta | p_0)$ has the form

$$-\frac{\partial \rho(p, \beta | p_0)}{\partial \beta} = -D_\alpha |p|^\alpha \rho(p, \beta | p_0) \qquad (14.17)$$

$$+ \int\limits_{-\infty}^{\infty} dp' V(p') \rho(p - p', \beta | p_0),$$

where the following notation

$$V(p) = \frac{1}{2\pi\hbar} \int\limits_{-\infty}^{\infty} dx V(x) \exp\{-i\frac{px}{\hbar}\},$$

has been introduced.

14.1.4   Fundamental properties of a free particle density matrix

When $V(x) = 0$, the solution to Eq. (14.13) gives a free particle density matrix $\rho^{(0)}(x - x_0, \beta)$

$$\rho^{(0)}(x - x_0, \beta) = \frac{1}{2\pi\hbar} \int\limits_{-\infty}^{\infty} dp \exp\left\{ i\frac{p(x - x_0)}{\hbar} - \beta D_\alpha |p|^\alpha \right\}. \qquad (14.18)$$

In terms of Fox's H-function $\rho^{(0)}(x - x_0, \beta)$ can be written as [67], [92]

$$\rho^{(0)}(x - x_0, \beta) = \frac{1}{\alpha |x - x_0|} H_{2,2}^{1,1}\left[\frac{|x - x_0|}{\hbar(D_\alpha \beta)^{1/\alpha}} \middle| \begin{array}{l} (1, 1/\alpha), (1, 1/2) \\ (1, 1), (1, 1/2) \end{array} \right], \qquad (14.19)$$

here $H_{2,2}^{1,1}$ is the Fox H-function (see [103], [104]), $\beta = 1/k_B T$, where k_B is the Boltzmann's constant, T is the temperature and D_α is scale coefficient with units of $[D_\alpha] = \text{erg}^{1-\alpha} \cdot \text{cm}^\alpha \cdot \text{sec}^{-\alpha}$.

Using H-function Property 12.2.5 given by Eq. (A.14) in Appendix A we can rewrite the solution given by Eq. (14.19) in alternative form

$$\rho^{(0)}(x - x_0, \beta) = \frac{1}{\alpha\hbar(D_\alpha \beta)^{1/\alpha}} \qquad (14.20)$$

$$\times H^{1,1}_{2,2} \left[\frac{|x - x_0|}{\hbar(D_\alpha \beta)^{1/\alpha}} \left| \begin{array}{c} (1 - 1/\alpha, 1/\alpha), (1/2, 1/2) \\ (0, 1), (1/2, 1/2) \end{array} \right. \right].$$

The momentum representation $\rho^{(0)}(p, \beta|p_0)$ of a free particle fractional density matrix $\rho^{(0)}(x, \beta|x_0)$ defined by

$$\rho^{(0)}(p, \beta|p_0) = \int\limits_{-\infty}^{\infty} dx dx_0 \rho^{(0)}(x, \beta|x_0) \exp\left\{ -\frac{i}{\hbar}(px - p_0 x_0) \right\}$$

has the form

$$\rho^{(0)}(p, \beta|p_0) = 2\pi\hbar\delta(p - p_0)\, e^{-\beta D_\alpha |p|^\alpha}. \tag{14.21}$$

The density matrix $\rho^{(0)}(x, \beta|x_0)$ is expressed in terms of $\rho^{(0)}(p, \beta|p_0)$ by the following way

$$\rho^{(0)}(x, \beta|x_0) = \frac{1}{(2\pi\hbar)^2} \int\limits_{-\infty}^{\infty} dp dp_0 \rho^{(0)}(p, \beta|p_0) \exp\left\{ \frac{i}{\hbar}(px - p_0 x_0) \right\}.$$

The density matrix $\rho^{(0)}(x - x_0, \beta)$ introduced by Eq. (14.18) has the following fundamental properties:

1. It satisfies the consistency equation

$$\rho^{(0)}(x_2 - x_1, \beta_1 + \beta_2) = \int\limits_{-\infty}^{\infty} dx' \rho^{(0)}(x_2 - x', \beta_2)\rho^{(0)}(x' - x_1, \beta_1). \tag{14.22}$$

2. It is non-negative

$$\rho^{(0)}(x - x_0, \beta) \geq 0. \tag{14.23}$$

3. It satisfies

$$\int\limits_{-\infty}^{\infty} dx \rho^{(0)}(x - x_0, \beta) = 1.$$

4. The symmetries hold

$$\rho^{(0)}(x - x_0, \beta) = \rho^{(0)}(-(x - x_0), \beta) \tag{14.24}$$

and

$$(\rho^{(0)}(x - x_0, \beta))^* = \rho^{(0)}(x - x_0, \beta), \qquad (14.25)$$

where $(\rho^{(0)}(x - x_0, \beta))^*$ stands for complex conjugate density matrix.

5. When $\beta = 0$, the density matrix $\rho^{(0)}(x - x_0, 0)$ is

$$\rho^{(0)}(x - x_0, 0) = \rho^{(0)}(x - x_0, \beta)|_{\beta=0} = \delta(x - x_0), \qquad (14.26)$$

where $\delta(x)$ is delta function.

6. When $x = x_0$, the density matrix $\rho^{(0)}(0, \beta)$ is

$$\rho^{(0)}(0, \beta) = \rho^{(0)}(x - x_0, \beta)|_{x=x_0} \qquad (14.27)$$

$$= \frac{1}{2\pi\hbar} \int\limits_{-\infty}^{\infty} dp \exp\left\{-\beta D_\alpha |p|^\alpha\right\} = \frac{1}{\alpha\pi\hbar} \left(\frac{1}{\beta D_\alpha}\right)^{1/\alpha} \Gamma(1/\alpha),$$

here $\Gamma(1/\alpha)$ is the Gamma function.

14.1.5   Scaling of density matrix: a free particle

To make general conclusions regarding space and inverse temperature dependencies of a free particle density matrix kernel $\rho^{(0)}(x, \beta)$, let's study its scaling.

For a free particle when $V(x) = 0$, Eq. (14.13) becomes

$$-\frac{\partial \rho^{(0)}(x, \beta|x_0)}{\partial \beta} = -D_\alpha(\hbar\nabla_x)^\alpha \rho^{(0)}(x, \beta|x_0), \qquad (14.28)$$

$$\rho^{(0)}(x, 0|x_0) = \delta(x - x_0).$$

Due to "initial" condition at $\beta = 0$ the solution will depend on $x - x_0$, that is $\rho^{(0)}(x, \beta|x_0) = \rho^{(0)}(x - x_0, \beta)$. To make general conclusions regarding solutions to the 1D fractional differential equation for density matrix for a free particle, let's study the scaling properties of $\rho^{(0)}(x - x_0, \beta; D_\alpha)$, where we keep the parameter D_α to remind that besides dependency on $x - x_0$, and β the density matrix depends on D_α as well. Scale transformations are written as

$$\beta = \lambda\beta', \qquad x = \lambda^\sigma x', \qquad x_0 = \lambda^\sigma x_0', \qquad D_\alpha = \lambda^\gamma D_\alpha', \qquad (14.29)$$

$$\rho^{(0)}(x - x_0, \beta; D_\alpha) = \lambda^\delta \rho^{(0)}(x' - x_0', \beta'; D_\alpha'),$$

where σ, γ, δ are exponents of the scale transformations which should leave a free particle solution to 1D fractional differential equation (14.28) invariant and save the condition $\int_{-\infty}^{\infty} dx \rho^{(0)}(x - x_0, \beta; D_\alpha) = 1$. It results in the relationships between scaling exponents,

$$\alpha\sigma - \gamma - 1 = 0, \qquad \delta + \sigma = 0, \qquad (14.30)$$

and reduces the number of exponents up to 2.

Hence, we obtain the two-parameters scale transformation group

$$\beta = \lambda\beta', \qquad x = \lambda^\sigma x', \qquad D_\alpha = \lambda^{\alpha\sigma - 1} D_\alpha',$$

$$\rho^{(0)}(\lambda^\sigma(x - x_0), \lambda\beta; \lambda^{\alpha\sigma - 1} D_\alpha) = \lambda^{-\delta} \rho(x - x_0, \beta; D), \qquad (14.31)$$

here σ and λ are two arbitrary group parameters.

To get the general scale invariant density matrix $\rho^{(0)}(x - x_0, \beta; D_\alpha)$ one can use the renormalization group framework. As far as the scale invariant solutions of Eq. (14.28) should satisfy the identity Eq. (14.31) for any arbitrary parameters σ and λ, the density matrix $\rho^{(0)}(x - x_0, \beta; D_\alpha)$ can depend on a combination of x and β to provide the independency on the group parameters of σ and λ. Therefore, due to Eq. (14.30) we obtain

$$\rho^{(0)}(x - x_0, \beta) = \frac{1}{|x - x_0|} \rho_1(|x - x_0|/\hbar(D_\alpha\beta)^{\frac{1}{\alpha}}) \qquad (14.32)$$

$$= \frac{1}{\hbar(D_\alpha\beta)^{\frac{1}{\alpha}}} \rho_2(|x - x_0|/\hbar(D_\alpha\beta)^{\frac{1}{\alpha}}),$$

where two arbitrary functions ρ_1 and ρ_2 are determined by the conditions, $\rho_1(.) = \rho^{(0)}(1, .)$ and $\rho_2(.) = \rho^{(0)}(., 1)$.

Thus, Eq. (14.32) brings us a general scale structure of density matrix for a free particle in the framework of factional statistical mechanics. Let us note that the form of the solution (14.19) is in line with the scale invariant solution expressed in terms of function ρ_1, while the solution (14.42) is in line with the scale invariant solution expressed in terms of function ρ_2.

14.1.6   Density matrix for a particle in a one-dimensional infinite well

As a physical application of the developed equation (14.13) let us consider a particle in symmetric one-dimensional infinite well described by the potential energy $V(x)$ which is zero for $-a \leq x \leq a$ and is infinite elsewhere, see Eq. (10.25).

To obtain density matrix in this case we use the solution given by Eq. (10.100) for a free particle quantum kernel developed in Sec. 11.5 and implement the substitution (14.1). Thus, we have

$$\rho_{\text{box}}(x, \beta | x_0) \qquad\qquad (14.33)$$

$$= \sum_{l=-\infty}^{\infty} \left\{ \rho^{(0)}(x + 4la, t | x_0 0) - (-1)^l \rho^{(0)}(x + 4la, t | -x_0 0) \right\},$$

where $\rho^{(0)}(x, \beta | x_0) \equiv \rho^{(0)}(x - x_0, \beta)$ is defined by Eq. (14.19).
In terms of Fox's $H_{2,2}^{1,1}$-function $\rho_{\text{box}}(x, \beta | x_0)$ is

$$\rho_{\text{box}}(x, \beta | x_0) = \frac{1}{\alpha} \left(\frac{\hbar}{(1/D_\alpha \beta)^{1/\alpha}} \right)^{-1}$$

$$\times \sum_{l=-\infty}^{\infty} \left\{ H_{2,2}^{1,1} \left[\frac{|x - x_0 + 4la|}{\hbar(\beta D_\alpha)^{1/\alpha}} \, \middle| \, \begin{array}{c} (1 - 1/\alpha, 1/\alpha), (1/2, 1/2) \\ (0, 1), (1/2, 1/2) \end{array} \right] \qquad (14.34)$$

$$- (-1)^l H_{2,2}^{1,1} \left[\frac{|x + x_0 + 4la|}{\hbar(\beta D_\alpha)^{1/\alpha}} \, \middle| \, \begin{array}{c} (1 - 1/\alpha, 1/\alpha), (1/2, 1/2) \\ (0, 1), (1/2, 1/2) \end{array} \right] \right\}.$$

Alternatively, $\rho_{\text{box}}(x, \beta | x_0)$ can be presented in terms of Fox's $H_{2,2}^{1,1}$-function as

$$\rho_{\text{box}}(x, \beta | x_0) = \frac{1}{\alpha} \frac{1}{|x - x_0|}$$

$$\times \sum_{l=-\infty}^{\infty} \left\{ H_{2,2}^{1,1} \left[\frac{|x - x_0 + 4la|}{\hbar(\beta D_\alpha)^{1/\alpha}} \, \middle| \, \begin{array}{c} (1, 1/\alpha), (1, 1/2) \\ (1, 1), (1, 1/2) \end{array} \right] \qquad (14.35)$$

$$-(-1)^l H_{2,2}^{1,1} \left[\frac{|x + x_0 + 4la|}{\hbar(\beta D_\alpha)^{1/\alpha}} \left| \begin{array}{c} (1, 1/\alpha), (1, 1/2) \\ (1, 1), (1, 1/2) \end{array} \right. \right] \right\},$$

where Property 12.2.5 of Fox's H-function has been used.

In the case when $\alpha = 2$ Eq. (14.35) gives us the solution for a density matrix of a free particle in the box $\rho_{\text{box}}(x, \beta|x_0)|_{\alpha=2}$ in the framework of standard quantum mechanics. Indeed, substituting $D_2 = 1/2m$, where m is particle mass, and using the identity

$$\frac{1}{\alpha\hbar(\beta D_\alpha)^{1/\alpha}} H_{2,2}^{1,1} \left[\frac{|x + 4la|}{\hbar(\beta D_\alpha)^{1/\alpha}} \left| \begin{array}{c} (1 - 1/\alpha, 1/\alpha), (1/2, 1/2) \\ (0, 1), (1/2, 1/2) \end{array} \right. \right] \right|_{\alpha=2} \quad (14.36)$$

$$= \sqrt{\frac{m}{2\pi\hbar^2\beta}} \exp\left\{ -\frac{m|x + 4la|^2}{2\hbar^2\beta} \right\},$$

we find

$$\rho_{\text{box}}(x, \beta|x_0)|_{\alpha=2} = \sqrt{\frac{m}{2\pi\hbar^2\beta}} \quad (14.37)$$

$$\times \sum_{l=-\infty}^{\infty} \left\{ \exp(-\frac{m|x - x_0 + 4la|^2}{2\hbar^2\beta}) - (-1)^l \exp(-\frac{m|x + x_0 + 4la|^2}{2\hbar^2\beta}) \right\}.$$

Thus, we obtain from Eq. (14.35) standard quantum mechanics solution for a free particle density matrix in a symmetric 1D box of length $2a$ confined by infinitely high walls at $x = -a$ and $x = a$.

14.2   Partition function

The partition function $Z(\beta)$ is defined as a trace of the density matrix $\rho(x, \beta|x_0)$ [67], [92], that is we have to take into account only those paths where the initial and the final configurations are the same $x = x_0$, and then integrate over dx,

$$Z(\beta) = \int dx \rho(x, \beta|x)$$

$$= \int dx \int_{x(0)=x(\hbar\beta)=x} \mathrm{D}x(\tau) \int \mathrm{D}p(\tau) \quad (14.38)$$

$$\times \exp\{-\frac{1}{\hbar}\int\limits_{0}^{\hbar\beta} du\left\{p(u)\dot{x}(u) + H_\alpha(p(u), x(u))\right\}$$

$$= \int\limits_{x(0)=x_0}^{x(\hbar\beta)=x} \mathrm{D}x(u) \int \mathrm{D}p(u) \exp\left\{-\frac{S_\alpha^{(e)}(p, x)}{\hbar}\right\},$$

where $H_\alpha(p(u), x(u))$ is classical mechanics Hamilton function $H_\alpha(p, x)$ defined by Eq. (3.24) with substitutions $p \to p(u)$ and $x \to x(u)$ and $S_\alpha^{(e)}(p, x)$ is given by Eq. (14.6).

Equation (14.38) may be interpreted in the following way: Consider all the possible paths by which the system can travel between the initial $x(0)$ and final $x(\hbar\beta)$ configurations in "time" $\hbar\beta$. The fractional density matrix ρ is a path integral over all possible paths, the contribution from a particular path being the "imaginary time" integral of the canonical action (14.6) (considered as the functional of the path $\{p(u), x(u)\}$ in the phase space) divided by $\hbar$. The partition function is derived by integrating over only those paths for which initial $x(0)$ and final $x(\beta)$ configurations are the same and after that we integrate over all possible initial (or final) configurations. The knowledge of the partition function allows us to build the thermodynamics of the system under consideration.

It is easy to see that with the help of Eq. (14.9) we can present the partition function $Z(\beta)$ in the form

$$Z(\beta) = \int dx \sum_{n=1}^{\infty} \phi_n(x)\phi_n^*(x)e^{-\beta E_n} = \sum_{n=1}^{\infty} e^{-\beta E_n}, \qquad (14.39)$$

because of normalization condition

$$\int dx \phi_n(x)\phi_n^*(x) = 1,$$

for an eigenfunction $\phi_n(x)$ corresponding to an eigenvalue E_n of time-independent fractional Schrödinger equation with the fractional Hamilton operator given by Eq. (3.23). Therefore, we have

$$\int dx \rho(x, \beta|x) = \sum_{n=1}^{\infty} e^{-\beta E_n}. \qquad (14.40)$$

For a free particle in 1D space we obtain from Eq. (14.18), that

$$\rho^{(0)}(x, \beta | x) = \rho^{(0)}(x - x_0, \beta)|_{x=x_0} \qquad (14.41)$$

$$= \frac{1}{2\pi\hbar} \int\limits_{-\infty}^{\infty} dp \exp\left\{-\beta D_\alpha |p|^\alpha\right\}$$

$$= \frac{1}{\pi\hbar} \int\limits_{0}^{\infty} dp \exp\left\{-\beta D_\alpha p^\alpha\right\} = \frac{1}{\alpha\pi\hbar} \frac{1}{(\beta D_\alpha)^{1/\alpha}} \Gamma(1/\alpha),$$

where $\Gamma(1/\alpha)$ is the Gamma function.

Therefore, where for the 1D system with space scale V the trace of Eq. (14.41) defines the partition function $Z(\beta)$ for a free particle

$$Z(\beta) = \int\limits_{V} dx \rho^{(0)}(x, \beta | x) = \frac{V}{\alpha\pi\hbar} \frac{1}{(\beta D_\alpha)^{1/\alpha}} \Gamma(1/\alpha), \qquad (14.42)$$

$$1 < \alpha \le 2.$$

In order to obtain a formula for the fractional partition function in the limit of fractional classical mechanics, let us study the case when $\hbar\beta$ is small. In this case the fractional density matrix $\rho(x, \beta | x_0)$ can be written as [67], [92]

$$\rho(x, \beta | x_0) = e^{-\beta V(x_0)} \frac{1}{2\pi\hbar} \int\limits_{-\infty}^{\infty} dp \exp\left\{i \frac{p(x - x_0)}{\hbar} - \beta D_\alpha |p|^\alpha\right\}. \qquad (14.43)$$

The partition function $Z(\beta)$ in the limit of classical statistical mechanics becomes

$$Z(\beta) = \int\limits_{-\infty}^{\infty} dx \rho(x, \beta | x) = \frac{\Gamma(1/\alpha)}{\alpha\pi\hbar(\beta D_\alpha)^{1/\alpha}} \int\limits_{-\infty}^{\infty} dx e^{-\beta V(x)}, \qquad (14.44)$$

$$1 < \alpha \le 2.$$

The partition function $Z(\beta)$ given by Eq. (14.44) is an approximation valid if the particles of the system cannot wander very far from their initial positions in the "time" $\hbar\beta$. The limit on the distance which the particles can wander before the approximation breaks down can be estimated from Eq. (14.43). We see that if the final point differs from the initial point by as much as

$$\Delta x \simeq \hbar(\beta D_\alpha)^{1/\alpha} = \hbar\left(\frac{D_\alpha}{kT}\right)^{1/\alpha},$$

the exponential function of Eq. (14.43) becomes greatly reduced. From this, we can infer that intermediate points only on paths which do not contribute greatly to the integral on the right-hand side of Eq. (14.43). Thus, we conclude that if the potential $V(x)$ does not alter very much as x moves over this distance, then the fractional classical statistical mechanics is valid.

14.3   Density matrix in 3D space

The above fractional statistical mechanics developments can be easily generalized to the 3D coordinate space. It can be done by means of the following fundamental relationship

$$\rho(\mathbf{r}, \beta|\mathbf{r}_0, 0) = K(\mathbf{r}_b t_b|\mathbf{r}_a t_a)\Big|_{\mathbf{r}_a=\mathbf{r}_0, t_a=0}^{\mathbf{r}_b=\mathbf{r}, t_b=-i\hbar\beta}, \qquad (14.45)$$

where $K(\mathbf{r}_b t_b|\mathbf{r}_a t_a)$ is quantum kernel given by Eq. (6.8), $\hbar$ is Planck's constant and β is "inverse temperature" defined by Eq. (14.2).

It is easy to see from Eqs. (14.45) and (6.8) that $\rho(\mathbf{r}, \beta|\mathbf{r}_0, 0)$ can be written as

$$\rho(\mathbf{r}, \beta|\mathbf{r}_0, 0) \qquad (14.46)$$

$$= \int_{\mathbf{r}(0)=\mathbf{r}_0}^{\mathbf{r}(\hbar\beta)=\mathbf{r}} \mathrm{D}\mathbf{r}(\tau) \int \mathrm{D}\mathbf{p}(\tau) \exp\left\{ \frac{i}{\hbar} \int_0^{-i\hbar\beta} du[\mathbf{p}(u)\dot{\mathbf{r}}(u) - H_\alpha(\mathbf{p}(u), \mathbf{r}(u), u)] \right\},$$

where integration over τ is performed along the negative imaginary time axis. Introducing a new integration variable $u = i\tau$, allows us to present Eq. (14.46) as the following path integral

$$\rho(\mathbf{r}, \beta|\mathbf{r}_0, 0) \qquad (14.47)$$

$$= \int_{\mathbf{r}(0)=\mathbf{r}_0}^{\mathbf{r}(\hbar\beta)=\mathbf{r}} \mathrm{D}\mathbf{r}(\tau) \int \mathrm{D}\mathbf{p}(\tau) \exp\left\{ -\frac{1}{\hbar} \int_0^{\hbar\beta} du[\mathbf{p}(u)\dot{\mathbf{r}}(u) - H_\alpha(\mathbf{p}(u), \mathbf{r}(u), u)] \right\},$$

where $\mathbf{r}$, $\mathbf{r}_0$ and $\mathbf{p}$ are the 3D vectors, fractional Hamiltonian $H_\alpha(\mathbf{p}, \mathbf{r})$ has form (3.4) and $\mathbf{p}(u), \mathbf{r}(u)$ may be considered as a path (in phase space representation) evolving over "imaginary time" u and $\hbar$ is Planck's constant. The exponential expression of Eq. (14.47) is very similar to the fractional classical mechanics action given by (6.11). Since it governs the quantum-statistical path integral (14.47) it may be called the fractional quantum-statistical action or fractional Euclidean action $S_\alpha^{(e)}(\mathbf{p}, \mathbf{r})$, indicated by the superscript (e) and subscript α,

$$S_\alpha^{(e)}(\mathbf{p}, \mathbf{r}) = \int_0^{\hbar\beta} du[\mathbf{p}(u)\dot{\mathbf{r}}(u) - H_\alpha(\mathbf{p}(u), \mathbf{r}(u), u)]. \qquad (14.48)$$

Therefore, we see that 3D density matrix $\rho(\mathbf{r}, \beta|\mathbf{r}_0, 0)$ reads

$$\rho(\mathbf{r}, \beta|\mathbf{r}_0, 0) \qquad (14.49)$$

$$= \int_{\mathbf{r}(0)=\mathbf{r}_0}^{\mathbf{r}(\hbar\beta)=\mathbf{r}} \mathrm{D}\mathbf{r}(\tau) \int \mathrm{D}\mathbf{p}(\tau) \exp\left\{ -\frac{S_\alpha^{(e)}(\mathbf{p}, \mathbf{r})}{\hbar} \right\}.$$

The path integral measure $\int_{\mathbf{r}(0)=\mathbf{r}_0}^{\mathbf{r}(\hbar\beta)=\mathbf{r}} \mathrm{D}\mathbf{r}(\tau) \int \mathrm{D}\mathbf{p}(\tau)...$ in Eqs. (14.47) and (14.49) is defined by

$$\int_{\mathbf{r}(0)=\mathbf{r}_0}^{\mathbf{r}(\hbar\beta)=\mathbf{r}} \mathrm{D}\mathbf{r}(\tau) \int \mathrm{D}\mathbf{p}(\tau)... \qquad (14.50)$$

$$= \lim_{N\to\infty} \int d\mathbf{r}_1...d\mathbf{r}_{N-1} \frac{1}{(2\pi\hbar)^{3N}} \int d\mathbf{p}_1...d\mathbf{p}_N....$$

One can interpret Eq. (14.49) in the following way: Consider all the possible paths by which the system can travel between the initial $\mathbf{r}(0) = \mathbf{r}_0$ and final $\mathbf{r}(\hbar\beta) = \mathbf{r}$ configurations in "time" $\hbar\beta$. We adopt the notations $d\mathbf{r}_i = d^3 r_i$, $(i = 1, 2, ..., N-1)$ and $d\mathbf{p}_j = d^3 p_j$, $(j = 1, 2, ..., N)$ while working with the path integral over Lévy flights in phase-space representation. Then the density matrix $\rho(\mathbf{r}, \beta|\mathbf{r}_0, 0)$ introduced by Eq. (14.47) is the path integral over all possible paths, the contribution from a particular path is given by $\exp(S_\alpha^{(e)}(\mathbf{p}, \mathbf{r})/\hbar)$, where $S_\alpha^{(e)}(\mathbf{p}, \mathbf{r})$ is fractional Euclidean action (14.48).

With the help of Eqs. (6.34) and (6.46) we can present the density matrix $\rho(\mathbf{r}, \beta | \mathbf{r}_0, 0)$ defined by Eq. (14.47) in the form

$$\rho(\mathbf{r}, \beta | \mathbf{r}_0, 0) = \sum_{n=1}^{\infty} \phi_n(\mathbf{r}) \phi_n^*(\mathbf{r}_0) e^{-\beta E_n},$$

where $\phi_n(\mathbf{r})$ are eigenfunctions and E_n are eigenvalues of time independent 3D fractional Schrödinger equation with the fractional Hamilton operator given by Eq. (3.3). Then, 3D generalization of Eq. (14.40) has form

$$\int d^3 r \rho(\mathbf{r}, \beta | \mathbf{r}, 0) = \sum_{n=1}^{\infty} e^{-\beta E_n}. \tag{14.51}$$

14.3.1 *Coordinate representation*

In coordinate representation the path integral representation for the density matrix can be obtained from Eq. (6.30) by transition from t to $-i\hbar\beta$, where β is inverse temperature given by Eq. (14.2) and then substituting $\tau = -iu$ with u being "imaginary time"

$$\rho(\mathbf{r}, \beta | \mathbf{r}_0, 0) \tag{14.52}$$

$$= \int_{\mathbf{r}(0)=0}^{\mathbf{r}(\hbar\beta)=\mathbf{r}} \mathcal{D}\mathbf{r}(u) \exp\{-\frac{1}{\hbar} \int_0^{\hbar\beta} du V(\mathbf{r}(u), u)\},$$

where $V(\mathbf{r}(u), u)$ is the potential energy as a functional of the Lévy flight path $\mathbf{r}(u)$ and "time" u, and $\int_{\mathbf{r}(0)=0}^{\mathbf{r}(\hbar\beta)=\mathbf{r}} \mathcal{D}\mathbf{r}(u)...$ is the path integral measure in 3D coordinate space,

$$\int_{\mathbf{r}(0)=0}^{\mathbf{r}(\hbar\beta)=\mathbf{r}} \mathcal{D}\mathbf{r}(u)... = \lim_{N \to \infty} \int d\mathbf{r}_1...d\mathbf{r}_{N-1} \hbar^{-3N} (D_\alpha \nu)^{-3N/\alpha} \tag{14.53}$$

$$\times \prod_{i=1}^{N} L_\alpha \left\{ \frac{|\mathbf{r}_i - \mathbf{r}_{i-1}|}{\hbar(D_\alpha \nu)^{1/\alpha}} \right\} ...,$$

here $\hbar$ denotes the Planck's constant, all possible paths go between the initial $\mathbf{r}(0) = \mathbf{r}_0$ and final $\mathbf{r}(\hbar\beta) = \mathbf{r}$ configurations in "time" $\hbar\beta$, $\nu = \beta/N$, and the Lévy probability distribution function L_α is defined by Eq. (6.28).

We adopt the notations $d\mathbf{r}_i = d^3 r_i$, $(i = 1, 2, ..., N - 1)$ while working with the path integral over Lévy flights in 3D space.

The Lévy probability distribution function $L_\alpha \left\{ (1/\hbar)(1/D_\alpha \nu)^{1/\alpha} |\mathbf{r}| \right\}$ is expressed in terms of Fox's $H_{3,3}^{1,2}$ function by Eq. (6.32). Hence, the path integral measure (14.53) can be alternatively presented as

$$\int_{\mathbf{r}(0)=0}^{\mathbf{r}(\hbar\beta)=\mathbf{r}} \mathcal{D}\mathbf{r}(u)...$$

$$= \lim_{N \to \infty} \int d\mathbf{r}_1 ... d\mathbf{r}_{N-1} \prod_{i=1}^{N} (-\frac{1}{2\pi\alpha |\mathbf{r}_i - \mathbf{r}_{i-1}|^3}) \tag{14.54}$$

$$\times H_{3,3}^{1,2} \left[\frac{|\mathbf{r}_i - \mathbf{r}_{i-1}|}{\hbar (D_\alpha \nu)^{1/\alpha}} \, \middle| \, \begin{array}{c} (1,1), (1,1/\alpha), (1,1/2) \\ (1,1), (1,1/2), (2,1) \end{array} \right] ....$$

The path integral measures defined by Eq. (14.53) and Eq. (14.54), are generated by the Lévy flights stochastic process.

It is obvious that a free particle density matrix $\rho^{(0)}(\mathbf{r}, \beta|\mathbf{r}_0, 0)$ for the 3D case has a form

$$\rho^{(0)}(\mathbf{r}, \beta|\mathbf{r}_0, 0) = \frac{1}{(2\pi\hbar)^3} \int d^3 p \, \exp \left\{ i \frac{\mathbf{p}(\mathbf{r} - \mathbf{r}_0)}{\hbar} - \beta D_\alpha |\mathbf{p}|^\alpha \right\}, \tag{14.55}$$

where $\mathbf{r}$, $\mathbf{r}_0$ and $\mathbf{p}$ are the 3D vectors.

To present the density matrix $\rho_L(\mathbf{r}, \beta|\mathbf{r}_0)$ in terms of the Fox H-function we rewrite Eq. (14.55) as

$$\rho^{(0)}(\mathbf{r}, \beta|\mathbf{r}_0, 0) = \frac{1}{2\pi^2\hbar^2 |\mathbf{r} - \mathbf{r}_0|} \int_0^\infty dp \, p \sin(\frac{p|\mathbf{r} - \mathbf{r}_0|}{\hbar}) \exp \left\{ -\beta D_\alpha |\mathbf{p}|^\alpha \right\}.$$

With the help of identity $\rho^{(0)}(\mathbf{r}, \beta|\mathbf{r}_0) = -\frac{1}{2\pi} \frac{\partial}{\partial x} \rho^{(0)}(x, \beta|0)|_{x=|\mathbf{r}-\mathbf{r}_0|}$, where $\rho^{(0)}(x, \beta|0)$ is the 1D density matrix given by Eq. (14.18), we find the equation for a free particle fractional density matrix in the 3D space,

$$\rho^{(0)}(\mathbf{r}, \beta|\mathbf{r}_0, 0) \tag{14.56}$$

$$= -\frac{1}{2\pi\alpha} \frac{1}{|\mathbf{r} - \mathbf{r}_0|^3} H_{3,3}^{1,2} \left[\frac{|\mathbf{r} - \mathbf{r}_0|}{\hbar (D_\alpha \beta)^{1/\alpha}} \, \middle| \, \begin{array}{c} (1,1), (1,1/\alpha), (1,1/2) \\ (1,1), (1,1/2), (2,1) \end{array} \right].$$

It is easy to see that with the help of Eqs. (6.5) and (6.25) we can present the density matrix $\rho^{(0)}(\mathbf{r}, \beta|\mathbf{r}_0)$ defined by Eq. (14.55) in the form

$$\rho^{(0)}(\mathbf{r}, \beta|\mathbf{r}_0, 0) = \sum_{n=1}^{\infty} \phi_n(\mathbf{r})\phi_n^*(\mathbf{r}_0)e^{-\beta E_n}, \qquad (14.57)$$

where $\phi_n(\mathbf{r})$ are eigenfunctions and E_n are eigenvalues of time independent 3D fractional Schrödinger equation with the fractional Hamilton operator given by Eq. (3.23).

The density matrix $\rho(\mathbf{r}, \beta|\mathbf{r}_0)$ obeys the fractional differential equation

$$-\frac{\partial\rho(\mathbf{r}, \beta|\mathbf{r}_0, 0)}{\partial\beta} = D_\alpha(-\hbar^2\Delta)^{\alpha/2}\rho(\mathbf{r}, \beta|\mathbf{r}_0, 0) + V(\mathbf{r})\rho(\mathbf{r}, \beta|\mathbf{r}_0, 0) \quad (14.58)$$

or

$$-\frac{\partial\rho(\mathbf{r}, \beta|\mathbf{r}_0, 0)}{\partial\beta} = H_\alpha\rho(\mathbf{r}, \beta|\mathbf{r}_0, 0), \qquad \rho(\mathbf{r}, \beta = 0|\mathbf{r}_0, 0) = \delta(\mathbf{r} - \mathbf{r}_0),$$
$$(14.59)$$

where the 3D fractional Hamiltonian H_α is defined by Eq. (3.23). The initial condition for Eqs. (14.58) and (14.59) is given by

$$\rho(\mathbf{r}, \beta = 0|\mathbf{r}_0) = \delta(\mathbf{r} - \mathbf{r}_0).$$

14.3.2 *A free particle partition function in 3D*

Generalization of Eq. (14.42) to the 3D space is given by [153]

$$Z(\beta) = \int_V d^3r\, \rho^{(0)}(\mathbf{r}, \beta|\mathbf{r}, 0) = \frac{V}{2\alpha\pi^2\hbar^3}\frac{\Gamma(3/\alpha)}{(\beta D_\alpha)^{3/\alpha}}, \qquad (14.60)$$

where V is space volume and $\Gamma(3/\alpha)$ is the Gamma function.

The last result can be transformed into $Z_1 = V/\lambda_\alpha^3$, if we introduce a *generalized thermal de Broglie wavelength* λ_α [153]

$$\lambda_\alpha = \left(2\alpha\pi^2\hbar^3\frac{(\beta D_\alpha)^{\frac{3}{\alpha}}}{\Gamma(3/\alpha)}\right)^{1/3}. \qquad (14.61)$$

When $\alpha = 2$, we retrieve the expression for the well-known thermal de Broglie wavelength $\lambda_2 = \lambda_\alpha|_{\alpha=2} = \sqrt{2\pi\hbar^2/mk_BT}$.

Thus, Eqs. (14.5), (14.10), (14.13), (14.38), (14.47), (14.52) are fundamental equations of fractional statistical mechanics.

14.3.3 Feynman's density matrix

When $\alpha = 2$ and $D_2 = 1/2m$, m is a particle mass, Eq. (14.18) gives[2] the well-known density matrix for the 1D free particle (see Eq. (10-46) of [12] or Eq. (2-61) of [19])

$$\rho_F^{(0)}(x, \beta | x_0) = \left(\frac{m}{2\pi\hbar^2\beta}\right)^{1/2} \exp\left\{-\frac{m}{2\hbar^2\beta}(x - x_0)^2\right\}. \qquad (14.62)$$

It follows from Eq. (14.21) that in the momentum representation $\rho_F^{(0)}(p, \beta | p_0)$ has form

$$\rho_F^{(0)}(p, \beta | p') = 2\pi\hbar\delta(p - p')\, e^{-\beta p^2/2m}. \qquad (14.63)$$

Feynman's density matrix in 3D case is

$$\rho_F^{(0)}(\mathbf{r}, \beta | \mathbf{r}_0, 0) = \left(\frac{m}{2\pi\hbar^2\beta}\right)^{3/2} \exp\left\{-\frac{m}{2\hbar^2\beta}(\mathbf{r} - \mathbf{r}_0)^2\right\}. \qquad (14.64)$$

In the momentum representation $\rho_F^{(0)}(\mathbf{r}, \beta | \mathbf{r}_0)$ has form

$$\rho_F^{(0)}(\mathbf{p}, \beta | \mathbf{p}_0) = (2\pi\hbar)^3\delta(\mathbf{p} - \mathbf{p}_0)\, e^{-\beta \mathbf{p}^2/2m}. \qquad (14.65)$$

14.4 Fractional thermodynamics

14.4.1 Ideal gas

The canonical partition function $Z_N(\beta)$ of an ideal gas composed of N particles occupying a volume V at a temperature T is given by

$$Z_N(\beta) = \frac{1}{N!}Z(\beta)^N, \qquad (14.66)$$

where $Z(\beta)$ is the individual partition function of a free particle given by Eq. (14.60) and $\beta = 1/(k_B T)$ with k_B being Boltzmann constant.

The Helmholtz energy $F(\beta)$ is defined by [154]

$$F(\beta) = -\frac{1}{\beta}\log Z_N(\beta). \qquad (14.67)$$

---

[2] Details of calculations can be found in Sec. 7.2.

Substituting Eq. (14.66) into Eq. (14.67) yields

$$F(\beta) = -\frac{N}{\beta} \log Z(\beta) + \frac{1}{\beta} \log N!. \tag{14.68}$$

Since N is large, we can use the formula

$$\log N! \simeq N \log(N/e), \tag{14.69}$$

here e is the base of the natural logarithm and it is approximately equal to 2.71828.

Hence, we have

$$F(\beta) = -\frac{N}{\beta} \log[\frac{e}{N} Z(\beta)] = -\frac{N}{\beta} \log[\frac{e}{N} \frac{V}{2\alpha\pi^2\hbar^3} \frac{\Gamma(3/\alpha)}{(\beta D_\alpha)^{3/\alpha}}]. \tag{14.70}$$

The knowledge of Helmholtz energy $F(\beta)$ allows us to compute the thermodynamical quantities of the system. Indeed, since the pressure P of the gas is calculated as

$$P = -\frac{\partial F(\beta)}{\partial V}, \tag{14.71}$$

we come to the equation of state

$$P = \frac{N}{\beta V} \quad \text{or} \quad PV = k_B N T, \tag{14.72}$$

where T is the temperature and k_B is the Boltzmann constant. As far as the dependency of Helmholtz energy $F(\beta)$ on volume of the system is not impacted by dispersion law

$$E_\alpha(\mathbf{p}) = D_\alpha |\mathbf{p}|^\alpha, \quad 1 < \alpha \le 2,$$

we see that equation of state (14.72) has the same form as the equation of state for ideal gas with the well-known dispersion law, when $\alpha = 2$,

$$E_2(\mathbf{p}) = E_\alpha(\mathbf{p})|_{\alpha=2} = \frac{\mathbf{p}^2}{2m}.$$

The internal energy U of gas is

$$U = -\frac{\partial \log Z_N(\beta)}{\partial \beta} = -\frac{\partial \beta F(\beta)}{\partial \beta} = \frac{3}{\alpha} \frac{N}{\beta} \quad \text{or} \quad U = \frac{3}{\alpha} N k_B T, \tag{14.73}$$

and the specific heat C_v at constant volume is [153]

$$C_v = (\frac{\partial U}{\partial T})_v = \frac{3}{\alpha} N k_B. \tag{14.74}$$

14.4.2 Ideal Fermi and Bose gases

It is well-known that at low temperatures, $T \to 0$, $\beta \to \infty$, at a given gas density, the distribution of number of identical particles over momentum has form (see Eq. (55.3) in [154])

$$dN_{\mathbf{p}} = \frac{g\mathrm{V}}{(2\pi\hbar)^3} \frac{d^3p}{e^{\beta(\varepsilon-\mu)} \pm 1}, \qquad (14.75)$$

where the '+' sign stands for Fermi statistics, while the '−' sign stands for Bose statistics, $g = 2S + 1$ (with S being spin of a particle), V is volume of system, μ is chemical potential, and, finally, ε stands for particle energy, which has the form

$$\varepsilon = D_\alpha |\mathbf{p}|^\alpha. \qquad (14.76)$$

The wave function of a system of N identical particles can be either antisymmetric or symmetric with respect to interchanges of any two particles. The antisymmetric wave function occurs for particles with half-integer spin, while symmetric case is for those with integer spin. For a system of particles described by antisymmetric wave function *Pauli's principle says:* two or more identical particles with half-integer spin cannot occupy the same quantum state within a quantum system simultaneously. The statistics based on this principle is called *Fermi statistics*. For a system of particles described by symmetric wave function, the statistics is called *Bose statistics.*

It follows immediately from Eqs. (14.75) and (14.76) that distribution of number of particles over energy ε is

$$dN_\varepsilon = \frac{g\mathrm{V}}{2\alpha\pi^2\hbar^3 D_\alpha^{3/\alpha}} \frac{\varepsilon^{\frac{3-\alpha}{\alpha}} d\varepsilon}{e^{\beta(\varepsilon-\mu)} \pm 1}. \qquad (14.77)$$

When $\alpha = 2$, Eq. (14.77) goes into Eq. (55.4) from [154].

Performing integration over ε we obtain the total number of particles in the gas

$$N = \frac{g\mathrm{V}}{2\alpha\pi^2\hbar^3 D_\alpha^{3/\alpha}} \int_0^\infty \frac{\varepsilon^{\frac{3-\alpha}{\alpha}} d\varepsilon}{e^{\beta(\varepsilon-\mu)} \pm 1}. \qquad (14.78)$$

Introducing a new integration variable $z = \beta\varepsilon$ let us rewrite Eq. (14.78) in the form

$$\frac{N}{V} = \frac{g\mathrm{V}}{2\alpha\pi^2\hbar^3 D_\alpha^{3/\alpha}\beta^{3/\alpha}} \int\limits_0^\infty \frac{z^{\frac{3-\alpha}{\alpha}}dz}{e^{z-\beta\mu}\pm 1}. \tag{14.79}$$

This formula defines chemical potential μ as function of β and gas density N/V.

The internal energy U of gas is defined as

$$U = \int\limits_0^\infty \varepsilon dN_\varepsilon, \tag{14.80}$$

where dN_ε is given by Eq. (14.77). Hence, we have

$$U = \frac{g\mathrm{V}}{2\alpha\pi^2\hbar^3 D_\alpha^{3/\alpha}} \int\limits_0^\infty \frac{\varepsilon^{\frac{3}{\alpha}}d\varepsilon}{e^{\beta(\varepsilon-\mu)}\pm 1}. \tag{14.81}$$

The thermodynamic potential Ω,

$$\Omega = -P\mathrm{V}, \tag{14.82}$$

with P being gas pressure and V being gas volume, has the following representation [154]

$$\Omega = \pm\frac{g\mathrm{V}}{(2\pi\hbar)^3\beta} \int d^3p \log(1 \mp e^{\beta(\mu-\varepsilon)}). \tag{14.83}$$

With the help of Eq. (14.76) we can present the thermodynamic potential Ω as

$$\Omega = \mp\frac{g\mathrm{V}}{2\alpha\pi^2\hbar^3\beta D_\alpha^{3/\alpha}} \int\limits_0^\infty d\varepsilon\,\varepsilon^{\frac{3-\alpha}{\alpha}} \log(1 \pm e^{\beta(\mu-\varepsilon)}).$$

Integration by parts yields

$$\Omega = -\frac{\alpha}{3}\frac{g\mathrm{V}}{2\alpha\pi^2\hbar^3\beta D_\alpha^{3/\alpha}} \int\limits_0^\infty \frac{\varepsilon^{\frac{3}{\alpha}}d\varepsilon}{e^{\beta(\varepsilon-\mu)}\pm 1}. \tag{14.84}$$

Then with the help of Eq. (14.81) we can express the thermodynamic potential Ω in terms of the internal energy U

$$\Omega = -\frac{\alpha}{3}U. \tag{14.85}$$

Taking into account Eq. (14.82) we find

$$PV = \frac{\alpha}{3}U. \tag{14.86}$$

It follows from Eqs. (14.82) and (14.84) that

$$P = \frac{g}{6\pi^2\hbar^3\beta D_\alpha^{3/\alpha}} \int\limits_0^\infty \frac{\varepsilon^{\frac{3}{\alpha}}d\varepsilon}{e^{\beta(\varepsilon-\mu)} \pm 1}. \tag{14.87}$$

Introducing a new integration variable $z = \beta\varepsilon$ let us rewrite Eq. (14.87) in the form

$$P = \frac{g}{6\pi^2\hbar^3\beta(\beta D_\alpha)^{3/\alpha}} \int\limits_0^\infty \frac{z^{\frac{3}{\alpha}}dz}{e^{z-\beta\mu} \pm 1}. \tag{14.88}$$

Equations (14.79) and (14.88) define the equation of state of the gas (the relation between P, V and T) in parametric form, with parameter μ. Let us study Eqs. (14.79) and (14.88) in the case when $e^{\beta\mu} \ll 1$, which is the limit case of Boltzmann gas. In other words, is the limit case $e^{\beta\mu} \ll 1$ Boltzmann statistics holds (see, paragraph 55 in [154]). For $e^{\beta\mu} \ll 1$ we expand the integrand in Eq. (14.88) in a power series of $e^{\beta\mu-z}$. For the first two terms of expansion we have

$$\int\limits_0^\infty \frac{z^{\frac{3}{\alpha}}dz}{e^{z-\beta\mu} \pm 1} \simeq \int\limits_0^\infty dz z^{\frac{3}{\alpha}} e^{\beta\mu-z}(1 \mp e^{\beta\mu-z}) \tag{14.89}$$

$$= \frac{3}{\alpha}\Gamma(3/\alpha)e^{\beta\mu}(1 \mp \frac{1}{2^{(3+\alpha)/\alpha}}e^{\beta\mu}),$$

where $\Gamma(3/\alpha)$ is the Gamma function.

By substituting Eq. (14.89) into Eq. (14.88) and taking into account Eq. (14.82) we find

$$\Omega = -PV = -\frac{gV}{2\alpha\pi^2\hbar^3\beta(\beta D_\alpha)^{3/\alpha}}\Gamma(3/\alpha)e^{\beta\mu}(1 \mp \frac{1}{2^{(3+\alpha)/\alpha}}e^{\beta\mu}),$$

which can be rewritten as

$$\Omega = \Omega_0 \pm \frac{g\text{V}}{2^{(3+2\alpha)/\alpha}\alpha\pi^2\hbar^3\beta(\beta D_\alpha)^{3/\alpha}}\Gamma(3/\alpha)e^{2\beta\mu}, \qquad (14.90)$$

where Ω_0 introduced by

$$\Omega_0 = -\frac{g\text{V}}{2\alpha\pi^2\hbar^3\beta(\beta D_\alpha)^{3/\alpha}}\Gamma(3/\alpha)e^{\beta\mu} \qquad (14.91)$$

can be considered as Boltzmann approximation ($e^{\beta\mu} \ll 1$) for thermodynamic potential Ω in the framework of fractional thermodynamics.

For Boltzmann ideal gas we have

$$\Omega_0 = -\frac{N}{\beta}. \qquad (14.92)$$

Then from Eqs. (14.91) and (14.92) we come to the expression for chemical potential μ as function of gas temperature T and density N/V.

$$\mu = \frac{1}{\beta}\log[\frac{N}{\text{V}}\frac{2\alpha\pi^2\hbar^3(\beta D_\alpha)^{3/\alpha}}{g\Gamma(3/\alpha)}]. \qquad (14.93)$$

Thus, the criteria of applicability of Boltzmann approximation in the framework of fractional thermodynamics is

$$\frac{N}{\text{V}}\hbar^3(\beta D_\alpha)^{3/\alpha} \ll 1. \qquad (14.94)$$

This criteria says that Boltzmann approximation is valid if gas density at given temperature is sufficiently low. In other words, gas has to be rarefied enough to be considered in Boltzmann approximation.

When $\alpha = 2$, Eq. (14.94) becomes the well-known expression for chemical potential (at $g = 1$ see, for example, Eq. (45.6) in [154]).

By expressing the second term in the right-hand side of Eq. (14.90) in terms of N and V, we come to the formula for a free energy

$$F = F_0 \pm \frac{N^2}{g\text{V}}\frac{\alpha\pi^2\hbar^3(\beta D_\alpha/2)^{3/\alpha}}{\beta\Gamma(3/\alpha)}, \qquad (14.95)$$

where F_0 is given by

$$F_0 = -\frac{N}{\beta}\log[\frac{e}{N}\frac{\text{V}}{2\alpha\pi^2\hbar^3}\frac{\Gamma(3/\alpha)}{(\beta D_\alpha)^{3/\alpha}}],$$

see Eq. (14.70).

Then, using Eq. (14.71), we obtain quantum correction to the equation of state of ideal Boltzmann gas in the framework of fractional thermodynamics

$$PV = \frac{N}{\beta}\left(1 \pm \frac{N}{gV}\frac{\alpha\pi^2\hbar^3(\beta D_\alpha/2)^{3/\alpha}}{\Gamma(3/\alpha)}\right) \qquad (14.96)$$

or

$$PV = k_B NT\left(1 \pm \frac{N}{gV}\frac{\alpha\pi^2\hbar^3(D_\alpha/2k_B T)^{3/\alpha}}{\Gamma(3/\alpha)}\right). \qquad (14.97)$$

This equation is applicable when the criteria (14.94) holds.

We see from Eq. (14.97) that the deviations of equation of state for quantum gases from the equation of state for classical Boltzmann gas depend on statistics of quantum gas. Indeed, in the case of Fermi statistics ('+' sign) the pressure of gas is higher comparing to the pressure for classical Boltzmann gas, while for the case of Bose statistics ('−' sign), the pressure is lower comparing to the pressure for classical Boltzmann gas. We can say that in the case of Fermi statistics the quantum exchange interaction leads to an additional effective "repulsion" between particles of Fermi gas resulting in pressure increase. In the case of Bose statistics we observe the pressure decrease comparing to Boltzmann gas. Hence, in the case of Bose statistics the quantum exchange interaction leads to an additional effective "attraction" between particles of Bose gas resulting in pressure decrease.

Chapter 15

Fractional Classical Mechanics

Fractional classical mechanics was introduced by Laskin [134] as a classical counterpart of fractional quantum mechanics. Here we present the Lagrange, Hamilton and Hamilton–Jacobi frameworks for fractional classical mechanics. Scaling analysis of fractional classical motion equations has been implemented based on the mechanical similarity. We discover and discuss fractional Kepler's third law which is a generalization of the well-known Kepler's third law.

Fractional classical oscillator model has been introduced and motion equations for the fractional classical oscillator have been integrated. We found an equation for the period of oscillations of fractional classical oscillator. The map between the energy dependence of the period of classical oscillations and the non-equidistant distribution of the energy levels for quantum fractional oscillator has been established.

In the case when $\alpha = 2$, all new developments are turned into the well-known results of the classical mechanics.

15.1 Introductory remarks

Now we introduce and explore fractional classical mechanics as a classical counterpart of the fractional quantum mechanics [67], [92], [96]. To begin with, let us consider the equation for canonical classical mechanics action $S_\alpha(p, q)$ defined by Eq. (6.4) (see Eq. (23) in [92]),

$$S_\alpha(p, q) = \int\limits_{t_a}^{t_b} d\tau (p(\tau)\dot{q}(\tau) - H_\alpha(p(\tau), q(\tau))), \tag{15.1}$$

where $H_\alpha(p(\tau), q(\tau))$ arrives from the classical mechanics Hamiltonian $H_\alpha(p, q)$ [67], [92], [96]

$$H_\alpha(p, q) = D_\alpha |p|^\alpha + V(q), \qquad 1 < \alpha \le 2, \qquad (15.2)$$

with substitutions $p \to p(\tau)$, $q \to q(\tau)$.

Since the Hamiltonian $H_\alpha(p, q)$ does not explicitly depend on time, it represents a conserved quantity which is in fact the total energy of fractional classical mechanics system.

Here we consider 1D fractional classical mechanics. Then the 3D generalization is straightforward and it is based on the Hamiltonian (3.3)

$$H_\alpha(\mathbf{p}, \mathbf{q}) = D_\alpha |\mathbf{p}|^\alpha + V(\mathbf{q}), \qquad 1 < \alpha \le 2, \qquad (15.3)$$

where $\mathbf{p}$ and $\mathbf{q}$ are the 3D vectors.

15.2 Fundamentals of fractional classical mechanics

15.2.1 *Lagrange outline*

The Lagrangian of fractional classical mechanics $L_\alpha(\dot{q}, q)$ is defined as usual

$$L_\alpha(\dot{q}, q) = p\dot{q} - H_\alpha(p, q), \qquad (15.4)$$

where the momentum p is

$$p = \frac{\partial L_\alpha(\dot{q}, q)}{\partial \dot{q}}, \qquad (15.5)$$

and $H_\alpha(p, q)$ is given by Eq. (15.2).

Hence, we obtain the Lagrangian of the fractional classical mechanics [134]

$$L_\alpha(\dot{q}, q) = \left(\frac{1}{\alpha D_\alpha}\right)^{\frac{1}{\alpha - 1}} \frac{\alpha - 1}{\alpha} |\dot{q}|^{\frac{\alpha}{\alpha - 1}} - V(q), \qquad 1 < \alpha \le 2. \qquad (15.6)$$

For a free particle, $V(q) = 0$, the Lagrangian of the fractional classical mechanics is

$$L_\alpha^{(0)}(\dot{q}, q) = \left(\frac{1}{\alpha D_\alpha}\right)^{\frac{1}{\alpha - 1}} \frac{\alpha - 1}{\alpha} |\dot{q}|^{\frac{\alpha}{\alpha - 1}}, \qquad 1 < \alpha \le 2. \qquad (15.7)$$

Further, the Euler-Lagrange equation of motion has the standard form

$$\frac{d}{dt}\frac{\partial L_\alpha(\dot{q}, q)}{\partial \dot{q}} - \frac{\partial L_\alpha(\dot{q}, q)}{\partial q} = 0. \tag{15.8}$$

By substituting Eq. (15.6) into Eq. (15.8) we find the motion equation in Lagrangian form

$$\left(\frac{1}{\alpha D_\alpha}\right)^{\frac{1}{\alpha-1}}\frac{1}{\alpha-1}\ddot{q}|\dot{q}|^{\frac{2-\alpha}{\alpha-1}} + \frac{\partial V(q)}{\partial q} = 0. \tag{15.9}$$

This equation has to be accompanied by the initial conditions. We impose the following initial conditions. At $t = 0$, the initial displacement is denoted by q_0 and the corresponding velocity is denoted by $\dot{q}_0$, that is we can write,

$$q(t = 0) = q_0 \quad \text{and} \quad \dot{q}(t = 0) = \dot{q}_0. \tag{15.10}$$

As one can see, Eq. (12.19) has a nonlinear kinematic term.

The new equation (12.19) is fractional generalization of the well-known equation of motion of classical mechanics in the Lagrange form.

In the special case when $\alpha = 2$, Eq. (12.19) goes into

$$m\ddot{q} + \frac{\partial V(q)}{\partial q} = 0, \tag{15.11}$$

where m is a particle mass $(D_2 = 1/2m)$ and the initial conditions are given by Eq. (15.10).

15.2.2   *Hamilton outline*

To obtain the Hamilton equations of motion for the fractional classical mechanics we apply variational principle,

$$\delta S_\alpha(p, q) = 0, \tag{15.12}$$

where the action $S_\alpha(p, q)$ is given by Eq. (3.1).

Considering the momentum p and coordinate q as independent variables we can write

$$\delta S_\alpha(p, q) = \int_{t_a}^{t_b} d\tau (\delta p\dot{q} + p\delta\dot{q} - \frac{\partial H_\alpha(p, q)}{\partial p}\delta p - \frac{\partial H_\alpha(p, q)}{\partial q}\delta q) = 0. \tag{15.13}$$

Upon integration by parts of the second term $p\delta\dot{q}$, the variation δS_α becomes

$$\delta S_\alpha(p, q) = \int\limits_{t_a}^{t_b} d\tau \delta p(\dot{q} - \frac{\partial H_\alpha(p, q)}{\partial p})$$

$$+p\delta q|_{t_a}^{t_b} - \int\limits_{t_a}^{t_b} d\tau \delta q(\dot{p} + \frac{\partial H_\alpha(p, q)}{\partial q}) = 0.$$

Since $\delta q(t_a) = \delta q(t_b) = 0$ at the end points of the trajectory, the term $p\delta q$ is 0. Between the end points δp and δq can take on any arbitrary value. Hence, the variation δS_α can be 0 if the following conditions are satisfied

$$\dot{q} = \frac{\partial H_\alpha(p, q)}{\partial p}, \qquad \dot{p} = -\frac{\partial H_\alpha(p, q)}{\partial q}. \tag{15.14}$$

This is, of course, the canonical Hamilton equations of motion. Thus, for the time-independent Hamiltonian given by Eq. (15.2) we obtain the equations

$$\dot{q} = \alpha D_\alpha |p|^{\alpha-1} \operatorname{sgn} p, \qquad \dot{p} = -\frac{V(q)}{\partial q}, \qquad 1 < \alpha \le 2, \tag{15.15}$$

where $\operatorname{sgn} p$ is the sign function which for nonzero values of p can be defined by the formula

$$\operatorname{sgn} p = \frac{p}{|p|}$$

and $|p|$ is the absolute value of p.

Equations (15.15) are the Hamilton equations for fractional classical mechanics system.

15.2.3   *Poisson bracket outline*

It is well-known that Hamiltonian classical mechanics could be reformulated in terms of the Poisson brackets. For the two arbitrary functions $u(p, q)$ and $v(p, q)$ of variables p and q, the Poisson bracket is defined as [122]

$$\{u(p, q), v(p, q)\} = \frac{\partial u}{\partial p}\frac{\partial v}{\partial q} - \frac{\partial u}{\partial q}\frac{\partial v}{\partial p}. \tag{15.16}$$

The Hamilton's equations of motion have an equivalent expression in terms of the Poisson bracket. Indeed, suppose that $f(p, q, t)$ is a function of momentum p, coordinate q and time t. Then we have

$$\frac{df}{dt} = \frac{\partial f}{\partial t} + \left(\frac{\partial f}{\partial p} \dot{p} + \frac{\partial f}{\partial q} \dot{q} \right). \tag{15.17}$$

Substituting $\dot{q}$ and $\dot{p}$ given by Eq. (12.23) yields

$$\frac{df}{dt} = \frac{\partial f}{\partial t} + \{H_\alpha, f\}, \tag{15.18}$$

where $\{H_\alpha, f\}$ is the Poisson bracket defined by

$$\{H_\alpha, f\} = \frac{\partial H_\alpha}{\partial p} \frac{\partial f}{\partial q} - \frac{\partial H_\alpha}{\partial q} \frac{\partial f}{\partial p}, \tag{15.19}$$

with H_α given by Eq. (15.2).

15.2.4 Hamilton–Jacobi outline

Having the Lagrangian (15.4), we can present classical mechanics action given by Eq. (15.1) as a function of coordinate q,

$$S_\alpha(q) = \int_{t_a}^{t_b} d\tau L_\alpha(\dot{q}, q) \tag{15.20}$$

$$= \int_{t_a}^{t_b} d\tau \left\{ \left(\frac{1}{\alpha D_\alpha} \right)^{\frac{1}{\alpha-1}} \frac{\alpha - 1}{\alpha} |\dot{q}|^{\frac{\alpha}{\alpha-1}} - V(q) \right\},$$

where $1 < \alpha \le 2$.

Let's now treat the action $S_\alpha(q, t_b)$ as a function of coordinate q and the upper limit of integration, t_b. Further, we suppose that motion goes along the actual path, that is we consider action as a functional of actual trajectory $q(\tau)$ in coordinate space. To evaluate the variance $\delta S_\alpha(q, t_b)$ we have to compare the values of $S_\alpha(q, t_b)$ for trajectories having a common start point at $q(t_a)$, but passing through different end points at t_b. Thus, for the variation of the action by the variation of the end point $\delta q(t_b)$ of the trajectory we have

$$\delta S_\alpha(q, t_b) = \delta \int_{t_a}^{t_b} d\tau L_\alpha(\dot{q}, q) \qquad (15.21)$$

$$= \frac{\partial L_\alpha}{\partial \dot{q}} \delta q |_{t_a}^{t_b} + \int_{t_a}^{t_b} d\tau \delta q \left(\frac{\partial L_\alpha}{\partial q} - \frac{d}{dt} \frac{\partial L_\alpha}{\partial \dot{q}} \right).$$

Because of Eq. (15.8) and $\delta q(t_a) = 0$ we obtain

$$\delta S_\alpha(q, t_b) = p \delta q(t_b)$$

or

$$p = \frac{\delta S_\alpha(q, t_b)}{\delta q(t_b)}. \qquad (15.22)$$

For simplicity, letting $\delta q(t_b) = \delta q$ and $t_b = t$, we can write

$$\frac{dS_\alpha(q, t)}{dt} = \frac{\partial S_\alpha(q, t)}{\partial t} + \frac{\partial S_\alpha(q, t)}{\partial q} \dot{q} = \frac{\partial S_\alpha(q, t)}{\partial t} + p\dot{q}. \qquad (15.23)$$

According to the definition of the action its total time derivative along the trajectory is $dS_\alpha/dt = L_\alpha$, and Eq. (15.23) can be presented as

$$\frac{\partial S_\alpha(q, t)}{\partial t} = L_\alpha - p\dot{q} = -H_\alpha(p, q), \qquad (15.24)$$

where Eq. (15.4) has been taken into account and p is given by Eq. (15.22). Therefore, we come to the equation

$$\frac{\partial S_\alpha(q, t)}{\partial t} + H_\alpha \left(\frac{\partial S_\alpha(q, t)}{\partial q}, q \right) = 0. \qquad (15.25)$$

With $H_\alpha(p, q)$ given by Eq. (15.2) it takes the form

$$\frac{\partial S_\alpha(q, t)}{\partial t} + D_\alpha \left| \frac{\partial S_\alpha(q, t)}{\partial q} \right|^\alpha + V(q) = 0, \qquad 1 < \alpha \le 2, \qquad (15.26)$$

where S_α is called the Hamilton's principal function.

For time-independent Hamiltonian the variables q and t in this equation can be separated. That is, we can search for solution of Eq. (15.26) in the form

$$S_\alpha(q, t, E) = S_\alpha^{(0)}(q, E) - Et, \qquad (15.27)$$

where $S_\alpha^{(0)}(q, E)$ is time-independent Hamilton's principal function and E, is the constant of integration, which has been identified with the total energy. Substituting Eq. (15.27) into Eq. (15.26) yields

$$D_\alpha |\frac{\partial S_\alpha^{(0)}(q, E)}{\partial q}|^\alpha + V(q) = E, \qquad 1 < \alpha \le 2. \qquad (15.28)$$

Equations (15.26) and (15.28) are the Hamilton–Jacobi equations of fractional classical mechanics.

As an example, let us consider a free particle, $V(q) = 0$. The Hamilton-Jacobi equation Eq. (15.26) for a free particle is

$$\frac{\partial S_\alpha(q, t)}{\partial t} + D_\alpha |\frac{\partial S_\alpha(q, t)}{\partial q}|^\alpha = 0, \qquad 1 < \alpha \le 2. \qquad (15.29)$$

Therefore, from Eqs. (15.27) and (15.28) we obtain a solution to the Hamilton–Jacobi equation Eq. (15.29),

$$S_\alpha(q, t, E) = (\frac{E}{D_\alpha})^{1/\alpha} q - Et. \qquad (15.30)$$

Follow the Hamilton-Jacobi fundamentals (see, for instance, [122]) we differentiate Eq. (15.30) over the energy E and put the derivative equal to a new constant δ,

$$\delta = \frac{\partial S_\alpha(q, t, E)}{\partial E} = \frac{1}{\alpha D_\alpha} (\frac{E}{D_\alpha})^{\frac{1}{\alpha} - 1} q - t, \qquad (15.31)$$

which yields

$$q = \alpha D_\alpha (\frac{E}{D_\alpha})^{1 - \frac{1}{\alpha}} (t + \delta) \qquad (15.32)$$

and

$$p = \frac{\partial S_\alpha(q, t, E)}{\partial q} = (\frac{E}{D_\alpha})^{1/\alpha}, \qquad (15.33)$$

where $E > 0$ and $1 < \alpha \le 2$.

The equations (15.32) and (15.33) define the trajectory of a free particle in fractional classical mechanics framework.

In limit case when $\alpha = 2$ and $D_2 = 1/2m$, Eqs. (15.9), (15.15), (15.26) and (15.28) are transformed into the well-known Hamilton, Lagrange and Hamilton–Jacobi equations of classical mechanics for a particle with mass m moving in the potential field $V(q)$.

15.3   Mechanical similarity

First of all, we intend to study the general properties of 1D fractional classical motion without integrating motion equations (15.9), (15.15), and (15.28).

When the potential energy $V(q)$ is a homogeneous function of coordinate q, it is possible to find some general similarity relationships.

Let us carry out a transformation in which the coordinates are changed by a factor ρ and the time by a factor τ: $q \to q' = \rho q$, $t \to t' = \tau t$. Then all velocities $\dot{q} = dq/dt$ are changed by a factor ρ/τ, and the kinetic energy by a factor $(\rho/\tau)^{\alpha/(\alpha-1)}$. If $V(q)$ is a homogeneous function of degree β then it satisfies,

$$V(\rho q) = \rho^{\beta} V(q). \tag{15.34}$$

It is obvious that if ρ and τ are such that $(\rho/\tau)^{\alpha/(\alpha-1)} = \rho^{\beta}$, i.e. $\tau = \rho^{1-\beta+\frac{\beta}{\alpha}}$, then the transformation leaves the motion equation unaltered. This invariance is called mechanical similarity.

A change of all the coordinates of the particles by the same scale factor corresponds to replacement of the classical mechanical trajectories of the particles by other trajectories, geometrically similar but different in size. Thus, we conclude that, if the potential energy of the system is a homogeneous function of degree β in Cartesian coordinates, the fractional equations of motion permit a series of geometrically similar trajectories, and the times of the motion between corresponding points are in the ratio

$$\frac{t'}{t} = \left(\frac{l'}{l}\right)^{1-\beta+\frac{\beta}{\alpha}}, \tag{15.35}$$

where $\frac{l'}{l}$ is the ratio of linear scales of the two paths. Not only the times but also any other mechanical quantities are in a ratio which is a power of $\frac{l'}{l}$. For example, the velocities and energies follow the scaling laws

$$\frac{\dot{q}'}{\dot{q}} = \left(\frac{l'}{l}\right)^{\beta-\frac{\beta}{\alpha}}, \qquad \frac{E'}{E} = \left(\frac{l'}{l}\right)^{\beta}, \tag{15.36}$$

and

$$\frac{t'}{t} = \left(\frac{E'}{E}\right)^{\frac{1}{\alpha}+\frac{1}{\beta}-1}. \tag{15.37}$$

The following are some examples of the foregoing.

It follows from Eq. (15.37) that at $\frac{1}{\alpha} + \frac{1}{\beta} = 1$ the period of oscillations does not depend on energy of oscillator. The condition $\frac{1}{\alpha} + \frac{1}{\beta} = 1$ with $1 < \alpha \leq 2$, $1 < \beta \leq 2$ brings $\alpha = 2$ and $\beta = 2$ only. It means, for instance, that considering the fractional classical oscillator model with Hamiltonian (15.41), we can conclude that the standard classical harmonic oscillator only,

$$H_2(p, q) = p^2/2m + g^2 q^2, \tag{15.38}$$

has the period of oscillations which does not depend on energy of oscillator.

In the uniform field of force, the potential energy is a linear function of the coordinates, i.e. $\beta = 1$. From Eq. (15.37) we have $\frac{t'}{t} = \left(\frac{l'}{l}\right)^{1/\alpha}$. Therefore, the time of motion in a uniform field ($\beta = 1$) is as the α-root of the initial altitude.

If the potential energy is inversely proportional to the distance apart, i.e. it is a homogeneous function of degree $\beta = -1$, then Eq. (15.35) becomes

$$\frac{t'}{t} = \left(\frac{l'}{l}\right)^{2 - \frac{1}{\alpha}}. \tag{15.39}$$

For instance, regarding the problem of orbital motion in the 3D fractional classical mechanics, we can state that the orbital period to the power α is proportional to the power $2\alpha - 1$ of its orbit scale. This statement is in fact a generalization of the well-known Kepler's third law. We call Eq. (15.39) as *fractional Kepler's third law*. In special case $\alpha = 2$, Eq. (15.39) turns into the well-known Kepler's third law [122].

In general, for negative β, $\beta = -\gamma$, in the space of the fractional Hamiltonians $H_{\alpha, -\gamma}$ at

$$1 + \gamma - \frac{\gamma}{\alpha} = \frac{3}{2}, \tag{15.40}$$

there exists the subset of fractional dynamic systems, orbital motion of which follows the well-known Kepler's third law, that is, the square of the orbital period is proportional to the cube of the orbit size.

15.4 Integration of the motion equations. Fractional classical oscillator

15.4.1 *Lagrange approach*

Quantum fractional oscillator model has been introduced in [67]. Fractional classical oscillator can be considered as a classical counterpart of the quantum fractional oscillator model.

We introduce fractional classical 1D oscillator as a mechanical system with the Hamiltonian

$$H_{\alpha,\beta}(p,q) = D_\alpha |p|^\alpha + g^2 |q|^\beta, \tag{15.41}$$

where g is a constant with physical dimensional $[g] = \text{erg}^{1/2} \cdot \text{cm}^{-\beta/2}$ and α and β are parameters, $1 < \alpha \leq 2$, $1 < \beta \leq 2$.

It follows from Eq. (15.6) that the Lagrangian of the fractional classical 1D oscillator is

$$L(\dot{q}, q) = \left(\frac{1}{\alpha D_\alpha} \right)^{\frac{1}{\alpha-1}} \frac{\alpha-1}{\alpha} |\dot{q}|^{\frac{\alpha}{\alpha-1}} - g^2 |q|^\beta, \tag{15.42}$$

$$1 < \alpha \leq 2, \quad 1 < \beta \leq 2.$$

Hence, the motion equation of fractional classical 1D oscillator is

$$\left(\frac{1}{\alpha D_\alpha} \right)^{\frac{1}{\alpha-1}} \frac{1}{\alpha-1} \ddot{q} |\dot{q}|^{\frac{2-\alpha}{\alpha-1}} + \beta g^2 |q|^{\beta-1} \text{sgn}\, q = 0. \tag{15.43}$$

This is a new nonlinear classical mechanics equation of motion. When $\alpha = 2$ and $\beta = 2$ it turns into the well-known linear equation of motion for classical mechanics oscillator which has the form,

$$m\ddot{q} + 2g^2 q = 0, \tag{15.44}$$

where m is a particle mass $(D_2 = 1/2m)$.

To integrate Eq. (15.43) we start from the law of energy conservation. For the Lagrangian Eq. (15.42) we have

$$D_\alpha \left(\frac{1}{\alpha D_\alpha} \right)^{\frac{\alpha}{\alpha-1}} |\dot{q}|^{\frac{\alpha}{\alpha-1}} + g^2 |q|^\beta = E, \tag{15.45}$$

where E is the total energy. Thus, we have

$$|\dot{q}| = \alpha D_\alpha^{1/\alpha}(E - g^2|q|^\beta)^{\frac{\alpha-1}{\alpha}},$$ (15.46)

which is a first-order differential equation, and it can be integrated.

Since the kinetic energy is positive, the total energy E always exceeds the potential energy, that is $E > g^2|q|^\beta$. The points where the potential energy equals the total energy $g^2|q|^\beta = E$ are turning points of classical trajectory. The 1D motion bounded by two turning points is oscillatory, the particle moves repeatedly between those two points.

Substituting $q = (E/g^2)^{1/\beta}y$ into Eq. (15.46) yields

$$|\dot{y}| = \alpha D_\alpha^{1/\alpha} g^{2/\beta} E^{1-(\frac{1}{\alpha}+\frac{1}{\beta})}(1 - |y|^\beta)^{1-\frac{1}{\alpha}}.$$ (15.47)

Hence, the period $T(\alpha, \beta)$ of oscillations is

$$T(\alpha, \beta) = 4\frac{E^{(\frac{1}{\alpha}+\frac{1}{\beta})-1}}{\alpha D_\alpha^{1/\alpha} g^{2/\beta}} \int\limits_0^1 \frac{dy}{(1 - y^\beta)^{1-\frac{1}{\alpha}}}.$$

By substituting $z = y^\beta$ we rewrite the last equation in the form

$$T(\alpha, \beta) = 4\frac{E^{(\frac{1}{\alpha}+\frac{1}{\beta})-1}}{\alpha\beta D_\alpha^{1/\alpha} g^{2/\beta}} \int\limits_0^1 dz z^{\frac{1}{\beta}-1}(1 - z)^{\frac{1}{\alpha}-1}.$$ (15.48)

With the help of the B-function definition [156]

$$B(\frac{1}{\beta}, \frac{1}{\alpha}) = \int\limits_0^1 dz z^{\frac{1}{\beta}-1}(1 - z)^{\frac{1}{\alpha}-1},$$

we finally find for the period of oscillations of fractional classical 1D oscillator [155]

$$T(\alpha, \beta) = 4\frac{E^{(\frac{1}{\alpha}+\frac{1}{\beta})-1}}{\alpha\beta D_\alpha^{1/\alpha} g^{2/\beta}} B(\frac{1}{\beta}, \frac{1}{\alpha}).$$ (15.49)

This equation shows that the period depends on the energy of fractional classical 1D oscillator. The dependency on energy is in agreement with the scaling law given by Eq. (15.37).

It follows from Eq. (15.49) that period $T(\alpha, \beta)$ doesn't depend on the energy of oscillator when $(1/\alpha + 1/\beta) = 1$. If $1 < \alpha \leq 2$ and $1 < \beta \leq 2$

then condition $(1/\alpha + 1/\beta) = 1$ gives us that $\alpha = 2$ and $\beta = 2$. Hence, we come to the standard classical mechanics harmonic oscillator with the Hamiltonian given by Eq. (15.38) and the energy independent oscillation period, $T(2,2) = \pi\sqrt{2m}/g$.

Table 2 shows that the energy dependency of the period of fractional classical oscillator is a classical counterpart of the fractional quantum mechanics statement on non-equidistant energy levels of quantum fractional oscillator . For classical mechanics, independence on energy of the period of classical oscillator is a classical counterpart of the quantum mechanics statement on equidistant energy levels of quantum oscillator.

	Period of oscillations T	Quantum energy levels E_n
$1<\alpha<2$ $1<\beta<2$	Fractional classical mechanics $T = 4 \dfrac{E^{(\frac{1}{\alpha} + \frac{1}{\beta}) - 1}}{\alpha\beta D_\alpha^{1/\alpha} g^{1/\beta}} B(\frac{1}{\beta}, \frac{1}{\alpha})$	Fractional quantum mechanics $E_n = \left(\dfrac{\pi\hbar\beta D_\alpha^{1/\alpha} g^{2/\beta}}{2B(\frac{1}{\beta}, \frac{1}{\alpha}+1)} \right)^{\frac{\alpha\beta}{\alpha+\beta}} (n+\frac{1}{2})^{\frac{\alpha\beta}{\alpha+\beta}}$
$\alpha=2$ $\beta=2$	Classical mechanics $T = \pi\sqrt{2m}/g$	Quantum mechanics $E_n = \hbar\sqrt{2/mg}(n+\frac{1}{2})$

Table 2. *Period of classical oscillations and quantum energy levels: fractional mechanics* $(1 < \alpha < 2, 1 < \beta < 2)$ *vs standard mechanics* $(\alpha = 2, \beta = 2)$[1].

15.4.2 Hamilton approach

For the fractional 1D oscillator the Hamilton equations of motion in accordance with Eqs. (15.14) and (15.41) are

$$\dot{q} = \frac{\partial H_{\alpha,\beta}}{\partial p} = \alpha D_\alpha |p|^{\alpha-1} \text{sgn}\, p, \qquad (15.50)$$

---

[1]The period of oscillations T and frequency of oscillations ω are related by $T = 2\pi/\omega$. Hence, the classical mechanics frequency is $\omega = \sqrt{2/mg}$, then the quantum mechanics energy levels are $E_n = \hbar\omega(n + \frac{1}{2})$.

The energy levels equation for the fractional quantum oscillator has been taken from [96].

$$\dot{p} = -\frac{\partial H_{\alpha,\beta}}{\partial q} = -\beta g^2 |q|^{\beta-1} \mathrm{sgn}\, q, \tag{15.51}$$

where $1 < \alpha \leq 2$, $1 < \beta \leq 2$.
Further, Hamilton equation (15.50) leads to

$$|p| = \left(\frac{1}{\alpha D_\alpha}\right)^{\frac{1}{\alpha-1}} |\dot{q}|^{\frac{1}{\alpha-1}}. \tag{15.52}$$

Substituting Eq. (15.52) into Eq. (15.41) yields exactly Eq. (15.45). Therefore, downstream integration can be done by the same way as it was done above in the framework of the Lagrange approach.

15.4.3 Hamilton–Jacobi approach

For fractional classical oscillator with the Hamilton function defined by Eq. (15.41) a complete integral of the Hamilton–Jacobi equation (15.27) is

$$S(q,t,E) = \int dq (\frac{1}{D_\alpha}(E - g^2|q|^\beta))^{1/\alpha} - Et, \tag{15.53}$$

where the integration constant E can be identified with the total energy of the fractional classical oscillator.

It is known that in standard classical mechanics [122], $\alpha = 2$, the integral in Eq. (15.53) is considered as an indefinite integral. At this point, to move forward, we introduce the ansatz to treat the integral in Eq. (15.53) as

$$S(q,t,E) = \int_0^q dq (\frac{1}{D_\alpha}(E - g^2|q|^\beta))^{1/\alpha} - Et. \tag{15.54}$$

Following the Hamilton–Jacobi fundamentals (see, for instance, [122]) we differentiate Eq. (15.53) over the energy E and put the derivative equal to a new constant δ,

$$\frac{\partial S(q,t,E)}{\partial E} = \frac{1}{\alpha D_\alpha^{1/\alpha}} \int_0^q dq' (E - g^2|q'|^\beta)^{(1-\alpha)/\alpha} - t = \delta. \tag{15.55}$$

Substituting integration variable q' with new variable z, $q' = (E/g^2)^{1/\beta} z^{1/\beta}$ yields

$$t + \delta = \frac{E^{(\frac{1}{\alpha}+\frac{1}{\beta})-1}}{\alpha\beta D_\alpha^{1/\alpha} g^{2/\beta}} \int\limits_0^{q^\beta g^2/E} dz z^{\frac{1}{\beta}-1}(1-z)^{\frac{1}{\alpha}-1}. \tag{15.56}$$

The integral in the above equation can be expressed as the incomplete Beta function[2] and we obtain

$$t + \delta = \frac{E^{(\frac{1}{\alpha}+\frac{1}{\beta})-1}}{\alpha\beta D_\alpha^{1/\alpha} g^{2/\beta}} B_{q^\beta g^2/E}(\frac{1}{\beta}, \frac{1}{\alpha}). \tag{15.58}$$

This equation is the solution of the fractional 1D oscillator as function $t(q)$ in terms of the incomplete Beta function.

The incomplete Beta function $B_x(\mu, \nu)$ has the hypergeometric representation

$$B_x(\mu, \nu) = \frac{x^\mu}{\mu} F(\mu, 1 - \nu; \mu + 1; x), \tag{15.59}$$

where $F(\mu, 1 - \nu; \mu + 1; x)$ is the hypergeometric function [158].
Now we can write Eq. (15.58) in the form

$$t + \delta = \frac{E^{\frac{1}{\alpha}-1}}{\alpha D_\alpha^{1/\alpha}} q F(\frac{1}{\beta}, 1 - \frac{1}{\alpha}; \frac{1}{\beta} + 1; q^\beta g^2/E). \tag{15.60}$$

Thus, in terms of the hypergeometric function we found solution of the fractional 1D oscillator as function $t(q)$.

Let us show that the ansatz given by Eq. (15.54) and based on it solution given by Eq. (15.60) allow us to reproduce the well-known solution to standard classical harmonic oscillator, $\alpha = 2$ and $\beta = 2$. Indeed, at $\alpha = 2$ and $\beta = 2$ we have

$$\omega(t + \delta) = x F(\frac{1}{2}, \frac{1}{2}; \frac{3}{2}; x^2), \tag{15.61}$$

---

[2]The incomplete Beta function is defined as [157]

$$B_x(\mu, \nu) = \int\limits_0^x dy y^{\mu-1}(1-y)^{\nu-1}. \tag{15.57}$$

where a new variable x has been introduced as,

$$x = g\sqrt{1/E}q, \tag{15.62}$$

and $\omega = \sqrt{2/m}g$ is the frequency of classical 1D oscillator.
Since

$$xF(\frac{1}{2}, \frac{1}{2}; \frac{3}{2}; x^2) = \arcsin x,$$

it follows from Eq. (15.61) that

$$x(t) = \sin\omega(t + \delta).$$

By restoring the original dynamic variable $q(t)$ from Eq. (15.62) we obtain,

$$q(t) = \sqrt{\frac{2E}{m\omega^2}} \sin\omega(t + \delta), \tag{15.63}$$

which is the solution of the Hamilton-Jacobi equation for standard 1D harmonic oscillator with the Hamiltonian given by Eq. (15.38).

Thus, we proved that the ansatz given by Eq. (15.54) gives us the well-known solution to standard classical mechanics harmonic oscillator.

Chapter 16

Fractional Dynamics in Polar Coordinate System

Following [159] we develop an alternative approach to study fractional classical mechanical oscillator described by the Hamilton function given by Eq.
(15.41). To simplify the notations we rewrite Eq. (15.41) in the unified
form to cover 1D and 2D cases,

$$H = \frac{1}{\alpha} p^\alpha + \frac{1}{\beta} r^\beta, \qquad p = |\mathbf{p}|, \quad \mathbf{p} \in \mathbf{P}^n, \qquad (16.1)$$

$$r = |\mathbf{r}|, \quad \mathbf{r} \in \mathbf{R}^n, \qquad n = 1, 2.$$

The idea behind the approach is to transform an integrable Hamiltonian
system with two degrees of freedom on the plane into a dynamic system
that is defined on the sphere and inherits the integrals of motion of the
original system.

The same class of dynamical systems may be treated as an extension of
"pseudobilliard" dynamical systems [159], which are closely related to the
case of $\alpha = 1$. Each system from this class corresponds to a point in the
plane of fractional parameters (α, β). Dynamical systems described by Eq.
(16.1) have evident first integrals of energy and angular momentum.

First, consider systems (16.1) with $\alpha > 0$, $\beta > 0$ in the one-dimensional
case. In these systems, there exist oscillations in the level $H = E$ with
period

$$T(\alpha, \beta; E) \propto E^{\gamma-1}, \qquad \gamma \equiv 1/\alpha + 1/\beta. \qquad (16.2)$$

In the plane (α, β), the period T becomes independent on energy E if
the condition holds

$$\gamma = \frac{1}{\alpha} + \frac{1}{\beta} = 1, \qquad (16.3)$$

which corresponds to isochronous dynamic systems. In the two-dimensional case, the period of finite oscillations in the systems corresponding to the same condition is also independent on energy (for $\alpha > 0, \beta > 0, E > 0$ and for $\alpha + \beta > 0$, $\alpha\beta < 0, E < 0$). Furthermore, for $\alpha > 0$ and $\beta < 0$, the points (α, β), where $\gamma = -\frac{1}{2}$ represent fractional dynamic systems, for which the period (of oscillations, in the one-dimensional case, or that of revolution along circular orbits, in the two-dimensional case) depends on energy exactly as in the classical mechanics Kepler problem: $T \propto E^{-3/2}$.

The traditional method for classification of orbits in $\mathbf{R}^2$ involves inversion of integrals that extend the elliptic integrals to the case of irrational values of the fractional parameters (α, β). For example, when $\alpha > 0, \beta > 0$, the dependence of the absolute value r of the radius–vector of a particle on the polar angle φ is determined by inverting the integral

$$\varphi = \mathcal{M} \int_{\rho_{\min}}^{\rho} \frac{d\left(\frac{1}{\rho}\right)}{\sqrt{(1 - \rho^\beta)^{\frac{2}{\alpha}} - \mathcal{M}^2}}, \qquad \rho_{\min} \leq \rho \leq \rho_{\max}.$$

Here , $\rho_{\min}$ and $\rho_{\max}$ are turning points of the trajectory, which are specified by the roots of the equation $\rho^\beta = 1 - \frac{M^\alpha}{\rho^\alpha}$; $\rho = \frac{r}{r_0}$, $r_0 = (\beta E)^\beta$, $\mathcal{M} = \frac{\beta}{\alpha^{1/\alpha}\beta^{1/\beta}}ME^{-\gamma}$, where E and M are constant values of the first integrals of motion.

We use an alternative approach to analyze orbits of dynamic systems introduced by Hamiltonian function given by Eq. (16.1). The key point of our approach is the passage to a dynamic system whose configurational manifold is a two-dimensional unit sphere [160].

Let us specify each of the vectors $\mathbf{p}$ and $\mathbf{r}$ by their coordinates in a respective polar coordinate system: $\mathbf{r} = (r, \varphi)$; $\mathbf{p} = (p, \psi)$. We change from the polar angles ψ and φ to the angular variables $\Psi = \psi + \varphi$ and $\Phi = \psi - \varphi$. The Hamiltonian function (16.1) leads to non-canonical equations of motion of the form

$$\dot{r} = p^{\alpha - 1} \cos \Phi, \quad \dot{\Phi} = \frac{1}{rp}(r^\beta - p^\alpha) \sin \Phi,$$

$$\dot{p} = -r^{\beta - 1} \cos \Phi, \dot{\Psi} = \frac{1}{rp}(r^\beta + p^\alpha) \sin \Phi. \tag{16.4}$$

They have two evident integrals, the energy $\mathcal{E} = \frac{1}{\alpha}p^\alpha + \frac{1}{\beta}r^\beta$ and the moment $M = rp \sin \Phi$.

Excluding p and r in (16.4) yields the following equations for the angular variables (Ψ, Φ):

$$\ddot{\Phi} = \cot(\Phi)\left\{2\dot{\Phi}^2 + \frac{1}{4}(\alpha + \beta)(\dot{\Psi}^2 - \dot{\Phi}^2)\right\},$$

$$\ddot{\Psi} = \cot(\Phi)\left\{2\dot{\Phi}\dot{\Psi} - \frac{1}{4}(\alpha - \beta)(\dot{\Psi}^2 - \dot{\Phi}^2)\right\}. \tag{16.5}$$

This system of equations describes the motion of a point with spherical coordinates (Ψ, Φ) on the unit sphere, and is the main system in our approach. Note that the parameters α and β enter only into the coefficients in these equations, they do not enter into exponents of derivatives. Moreover, only the derivatives of Ψ appear in system (16.5) and not this function itself.

When evolution of the angular variables (Ψ, Φ) is determined, the absolute values of the vectors $(\mathbf{p}, \mathbf{r})$ are restored from the relations

$$p^\alpha = \frac{1}{2rp} \frac{\dot{\Psi} - \dot{\Phi}}{\sin \Phi} > 0, \quad r^\beta = \frac{1}{2rp} \frac{\dot{\Psi} + \dot{\Phi}}{\sin \Phi} > 0.$$

The first integrals of system (16.5)

$$I_1 = (\sin \Phi)^{-2}\{(\alpha + \beta)\dot{\Psi} + (\alpha - \beta)\dot{\Phi}\},$$

$$I_2 = (\sin \Phi)^{1-2\gamma} \left[\dot{\Psi} + \dot{\Phi}\right]^{\frac{1}{\beta}} \cdot \left[\dot{\Psi} - \dot{\Phi}\right]^{\frac{1}{\alpha}} \tag{16.6}$$

can be expressed in terms of the integrals E and $\mathbf{M}$ of the original Hamiltonian system: $I_1 = \alpha\beta E/\mathbf{M}$, $I_2 = 2^\gamma \mathbf{M}^{1-\gamma}$.

Let us consider, in more detail, the possible motions of a point on the unit sphere that are yielded by system (16.5).

First, it can be shown that, for any orbit, the angular velocity $\dot{\Psi}$ does not change sign, and Φ varies within the range $0 \leq \Phi_- \leq \Phi \leq \Phi_+ \leq \pi$, where $\Phi_\pm$ are solutions to the system

$$\sin \Phi_\pm = \frac{\mathbf{M}}{r_0 p_0}, \quad p_0^\alpha = r_0^\beta = \frac{E}{\gamma}.$$

As $\mathbf{M} \to (E/\gamma)^\gamma$, this strip shrinks into the equatorial circle $\Phi = \frac{\pi}{2}$. The motion of the image point along the equator of the sphere corresponds to the circular orbit (one with $r = r_0$ and $p = p_0$; such circular orbits exist in systems with any nonzero α and β). The motion along the equator is uniform: the azimuthal angle Ψ varies linearly with time with velocity $\theta_0 \equiv \dot{\Psi} = (E/\gamma)^{1-\gamma}$. Note that in the systems corresponding to the curve $\gamma = 1$ the frequency of revolution along the circular orbit is independent of E, and the first integrals $(E, \mathbf{M})$ are functionally related.

Let us seek different closed orbits in a system with the Hamiltonian function (16.1). In terms of the dynamic system (16.5), closed orbits correspond to closed trajectories on the unit sphere. For such trajectories, the frequency of oscillation of Φ must be commensurable with the frequency of revolution (i.e., with the mean velocity of growth) of Ψ.

By linearizing system (16.5) in the neighborhood of the circular orbit $(\Phi = \frac{\pi}{2}, \Psi = \Psi_0 + \theta_0 t)$, we obtain the solution

$$\Phi = \frac{\pi}{2} + \varepsilon \sin \tau, \quad \Psi = \Psi_0 + \left[\frac{2}{\sqrt{\alpha + \beta}} + \varepsilon \frac{\theta_1}{\omega_0}\right] \tau - \varepsilon \frac{1}{\theta_0} \frac{\alpha - \beta}{\alpha + \beta} \sin \tau,$$

where $\varepsilon \ll 1$, $\tau = \omega_0 t$, $\omega_0 = \frac{1}{2}\sqrt{\alpha + \beta}\,\theta_0$ is the frequency of small oscillations around the equator of the sphere, and θ_1 is the shift of the revolution frequency. Thus, in the first-order approximation in ε, Φ is a 2π-periodic function of the variable τ. The value of Ψ changes by $4\pi/\sqrt{\alpha + \beta} + \mathcal{O}(\varepsilon)$ over the period of Φ. Let us define the rotation number d for a bounded trajectory as the angular distance $\varphi_2 - \varphi_1$ between two subsequent apocenters (points where $r = r_{\max}$) divided by 2π. Since, by virtue of Eqs. (16.4), $\Phi = \pi/2$ at apocenters, the rotation number is

$$d = \frac{1}{4\pi}\left(\Psi(\tau = 0) - \Psi(\tau = 2\pi)\right). \tag{16.7}$$

The rotation number for a circular orbit is not defined. It is natural to define it as the limit of rotation numbers of orbits that are close to the circular one: $d_0 = \lim_{\varepsilon \to 0} d(\varepsilon) = 1/\sqrt{\alpha + \beta}$. Closed orbits can branch off a circular one only for such values of (α, β) for which d_0 is rational.

While studying the system (16.5), it has been shown in [159] that, when a line $\alpha + \beta = r^2$ ($r \in Q$) in the plane (α, β) is crossed, a closed orbit with rotation number $1/r$ branches off the circular orbit. Note that the circular orbit is unstable for $\alpha + \beta < 0$.

As we move away from the circular orbit, the shape of trajectories oscillating on the sphere in the strip between two symmetrical "tropics" becomes more and more saw-toothed. In the limit, (as M $\to$ 0) they degenerate into the circuit of the contour consisting of the trajectories on which Φ and Ψ are linearly dependent. In the plane (Φ, Ψ), these trajectories are segments of straight lines; each segment is passed by the image point in an infinite time. To find such trajectories explicitly, it is convenient to write out a differential equation for Φ treated as a function of Ψ. By virtue of (16.5),

$$\Phi'' = \frac{1}{4}(\cot \Phi)\left[1 - (\Phi')^2\right]\left\{(\alpha + \beta) + (\alpha - \beta)\Phi'\right\}, \tag{16.8}$$

where prime denotes differentiation with respect to Ψ. This equation has the following integral:

$$\sin \Phi \, \frac{(1 - \Phi')^{\frac{1}{\alpha}}(1 + \Phi')^{\frac{1}{\beta}}}{\left[1 + \frac{\alpha - \beta}{\alpha + \beta}\Phi'\right]^\gamma} = \text{const.}$$

It is easy to see from (16.8) that, for each pair (α, β), there are three straight-line trajectories, with the angular coefficients $k = +1$, $k = -1$, and $k = \mu$, where $\mu(\alpha, \beta) = -\frac{\alpha + \beta}{\alpha - \beta}$. A limit contour consists of segments with two alternating values of k out of those three; they are "selected" by the signs of α and β. The trajectory with $k = \mu$ exists if $\alpha < 0$ or $\beta < 0$

and corresponds to motion along an open curve in the original system. In particular, when $\alpha > 0$, $\beta < 0$, this orbit is a generalization of a parabolic orbit in the Kepler problem; here, $r \to \infty$ as $t \to \pm\infty$. In the "conjugate" case of $\alpha < 0$ and $\beta > 0$, the corresponding orbit has the shape of a loop emanating from and entering into the origin; here, $r \to 0$ as $t \to \pm\infty$. Trajectories with $k = \pm 1$, which exist in system (16.5), do not have a physical pre-image in the original problem (for one, $r = \infty$, $p = 0$; for the other, $r = 0$, $p = \infty$). However, introducing corresponding "motions" in the original system is useful so that to define a formal closed contour that is a limit contour for closed orbits (for example, in the Kepler problem, the parabolic orbit supplemented by the "orbit" $r = \infty$, $p = 0$ forms a limit contour for the set of elliptic orbits).

An analogue of the rotation number can be defined for such limit contours by using definition (16.7). Specifically, for $\alpha > 0$, $\beta < 0$, we have $d_\infty = (2\alpha\gamma)^{-1}$; and for $\alpha < 0$, $\beta > 0$, we find that $d_\infty = (2\beta\gamma)^{-1}$. These values are limit ones for the rotation numbers of bounded trajectories that degenerate into a circuit of the contour.

The dynamic system (16.5) admits a formal solution in the form of asymptotic expansions of the angular variables (Φ, Ψ) with respect to the parameter ε, which is the measure of deviation from the circular orbit. These expansions take the form

$$\Phi = \tfrac{\pi}{2} + \varepsilon \, F(\varepsilon; \alpha, \beta; \{\sin(m\tau)\}),$$

$$\Psi = \Psi_0 + C(\varepsilon; \alpha, \beta)\,\tau + \varepsilon \, G(\varepsilon; \alpha, \beta; \{\sin(m\tau)\}),$$

(16.9)

where F and G are polynomials of harmonics $\sin(m\tau)$ and the mean frequency C is specified by the expression

$$C(\varepsilon; \alpha, \beta) = \frac{2}{\sqrt{\alpha + \beta}} + \sum_{n=1} \varepsilon^{2n} C_{2n}(\alpha, \beta).$$

(16.10)

The rotation number $d = d(\varepsilon; \alpha, \beta)$ equals $C/2$. If all the coefficients C_{2n} vanish for some (α, β), then all bounded orbits in this system have equal rotation numbers. In particular, if this number is rational, then all these orbits are closed and have identical topology.

It was shown in [159] that $C_2 = 0$ at the points of the plane (α, β) that lie on the ellipsis specified by the formula

$$(\alpha + \beta)^2 - 3(\alpha + \beta) = \alpha\beta.$$

(16.11)

Surprisingly, it has occurred that C_4 identically vanishes on this ellipsis too, since it can be factorized as $C_4(\alpha, \beta) = C_2(\alpha, \beta) \cdot R(\alpha, \beta)$. Therefore, for all the points on the ellipsis defined by (16.11), the rotation number anomalously weakly depends on the deviation from the circular orbit: $d^{(\text{ell})} = d_0 + \mathcal{O}(\varepsilon^6)$.

The results of numerical simulations presented in [159] show that, in systems corresponding to the points of the ellipsis (16.11), the relative deviation of the rotation number d from the rotation number d_0 assigned to the circular orbit is less than 10^{-5} for about a third of all possible initial condition scenarios corresponding to finite orbits. However, asymptotic expansion coefficient C_6 is nonzero for almost all systems on ellipsis (16.11) and equals

$$C_6^{\text{ell}} = \frac{1}{2880} \frac{(\alpha + \beta - 4)(\alpha + \beta - 1)^2}{\sqrt{\alpha + \beta}}. \tag{16.12}$$

It vanishes at three points of the ellipsis in (α, β) plane: $(2, 2)$ (it corresponds to the harmonic oscillator), $(2, -1)$ (it corresponds to the Kepler problem), and $(-1, 2)$ (it corresponds to the conjugate Kepler problem). On the other hand, for these three points, the rotation number d_0 of the circular orbit coincides with the rotation number d_∞ assigned to the limit contour. This is in agreement with the fact that, in each of the three systems, all the bounded orbits are closed and have equal rotation numbers. As is well known, bounded orbits have the shape of an ellipsis in the cases of the harmonic oscillator $(d = 1/2)$ and the Kepler problem $(d = 1)$. In the case of the conjugate Kepler problem, however, all the bounded orbits are circles with $d = 1$ (their centers are not at the origin, but the origin lies inside them). This implies that, in the Kepler case, the end of the vector of momentum moves along a circle (in the p-plane) shifted off the origin.

16.1 Generalized 2D fractional classical mechanics

It is well known [122], that in the class of Hamiltonian systems with two degrees of freedom the Kepler problem defined by the Hamilton function of the form

$$H = \frac{\mathbf{p}^2}{2} + U(|\mathbf{r}|), \qquad U(|\mathbf{r}|) = \frac{\sigma}{|\mathbf{r}|}, \qquad \sigma < 0, \qquad \mathbf{r}, \mathbf{p} \in \mathbb{R}^2, \tag{16.13}$$

is one of the two classical mechanical systems, together with the isotropic harmonic oscillator, where the orbits of all finite motions are closed. This is due to the hidden symmetry of the Kepler problem caused by the existence of the third integral of motion in addition to the well-known two integrals of motion - energy and momentum. Any structural perturbations in the Hamiltonians defined by Eq. (16.13) result in a precession of orbits and disappearance of the hidden symmetry. This behavior of the Hamiltonian systems with two degrees of freedom changes significantly for fractional Hamiltonian system like classical mechanical fractional oscillator introduced by Eq. (15.41). To explore this interesting feature we introduce generalized fractional classical mechanical 2D systems, the Hamilton function of which is [160]

$$H = p^{\alpha_1} r^{\beta_1} + \sigma p^{\alpha_2} r^{\beta_2} \qquad r = |\mathbf{r}|, \qquad p = |\mathbf{p}|, \qquad \mathbf{r}, \mathbf{p} \in \mathbb{R}^2, \qquad (16.14)$$

where α_1, β_1, α_2, β_2, and σ are real parameters. This class of fractional dynamic models includes, in particular, the Hamiltonian of a particle in a central field (16.13) with the potential $U \sim r^{\beta_2}$, when $\alpha_1 = 2$, $\beta_1 = \alpha_2 = 0$.

Our intent is to show that in the four-dimensional space of structural parameters $(\alpha_1, \beta_1, \alpha_2, \beta_2)$, there exists a one-dimensional manifold (containing the case of the planar Kepler problem) along which the closedness of the orbits of all finite motions and the third Kepler law are preserved. Similarly, there exists a one-dimensional manifold (containing the case of the two-dimensional isotropic harmonic oscillator) along which the closedness of the orbits and the isochronism of oscillations are preserved. Any deformation of orbits on these manifolds does not violate the hidden symmetry typically attributed to two-dimensional isotropic oscillator and the planar Kepler problem.

We will show that for $\sigma < 0$, there exists a one-parameter family of Hamiltonians of form (16.14) for which the trajectories of all finite orbits are closed and the third Kepler law holds: $T \propto (-E)^{-3/2}$. If the parameter of the family varies, the orbits are deformed. They can have the form of Kepler ellipses, circles, or Pascal's limacons. Orbits with self-intersection points can appear. When orbits are closed and the third Kepler law holds all representatives of the family have a hidden symmetry due to the existence of an additional integral of motion. For $\sigma > 0$, there exists a one-parameter family of Hamilton functions of form (16.14) that, along with the closedness of orbits, inherit from the two-dimensional isotropic harmonic oscillator the isochronism of nonlinear oscillations for all $E > 0$. All representatives of this family also have a hidden symmetry due to the existence of an additional integral of motion.

For $\sigma > 0$, there exists a one-parameter family of Hamilton functions of form (16.14) that, along with the closedness of orbits, inherit the isochronism of oscillations (already nonlinear) for all $E > 0$ from the two-dimensional isotropic harmonic oscillator. All representatives of this family also have a hidden symmetry (an additional integral of motion).

Moreover, in each of the cases $\sigma < 0$ and $\sigma > 0$, the corresponding one-parameter family is a particular case of a two-parameter family of Hamilton functions for each of which all finite motions are characterized by the same rotation number. The orbits of all finite motions are closed if this number is rational and are open otherwise.

To study Hamiltonian systems (16.14) we apply the approach developed in [160]. The approach is based on transforming an integrable Hamiltonian system with two degrees of freedom defined on the plane into a dynamic system defined on the sphere and inheriting the integrals of motion of the original system. Besides a few advantages from the stand point of analytical and numerical analysis a dynamic system on the sphere is interesting example of a dynamic system with one and a half degrees of freedom and with two first integrals of motion. Moreover, the first integrals of this system on the sphere allow one to obtain the scaling laws (with respect to one of the angular variables) related to linear transformations of one- and two-dimensional manifolds into themselves in the space of structural parameters $(\alpha_1, \beta_1, \alpha_2, \beta_2)$ of the original fractional classical mechanical systems with Hamilton function introduced by Eq. (16.14).

16.2 Dynamics on 2D sphere

The canonical equations for Hamiltonians of the form $H = H(p, r)$ are

$$\dot{\mathbf{r}} = \frac{\partial H}{\partial p}\frac{\mathbf{p}}{p} \qquad \dot{\mathbf{p}} = -\frac{\partial H}{\partial r}\frac{\mathbf{r}}{r}, \tag{16.15}$$

and have two integrals of motion (the energy $E = H$ and the moment $\mathbf{M} = [\mathbf{r}, \mathbf{p}]$). Expressing the vectors $\mathbf{p}$ and $\mathbf{r}$ in polar coordinates (p, ψ) and (r, φ), we obtain the noncanonical equations of motion

$$\dot{r} = \frac{\partial H}{\partial p}\cos(\psi - \varphi), \qquad \dot{p} = -\frac{\partial H}{\partial r}\cos(\psi - \varphi), \tag{16.16}$$

and

$$r\dot{\varphi} = \frac{\partial H}{\partial p}\sin(\psi - \varphi), \qquad p\dot{\psi} = -\frac{\partial H}{\partial r}\sin(\psi - \varphi). \tag{16.17}$$

Introducing new angular variables $\Psi = \psi + \varphi$ and $\Phi = \psi - \varphi$, we write the equations of motion as

$$\dot{r} = \frac{\partial H}{\partial p}\cos\Phi, \qquad \dot{p} = -\frac{\partial H}{\partial r}\cos\Phi, \tag{16.18}$$

and

$$\dot{\Phi} = \left(\frac{1}{p}\frac{\partial H}{\partial r} - \frac{1}{r}\frac{\partial H}{\partial p}\right)\sin\Phi, \qquad \dot{\Psi} = \left(\frac{1}{p}\frac{\partial H}{\partial r} + \frac{1}{r}\frac{\partial H}{\partial p}\right)\sin\Phi. \tag{16.19}$$

Hence, it is possible to reduce the equations to a dynamic system with one and a half degrees of freedom.

If Hamiltonian H has form (16.14), then p and r can be eliminated from Eqs. (16.18) and (16.19). Thus, we obtain the system of equations for the angular variables (Ψ, Φ)

$$\ddot{\Phi} = \cot\Phi\left\{2(\dot{\Phi})^2 + \frac{\theta_1 - \theta_2}{4D}(\varsigma_1\dot{\Phi} - \theta_1\dot{\Psi})(\varsigma_2\dot{\Phi} - \theta_2\dot{\Psi})\right\} \tag{16.20}$$

and

$$\ddot{\Psi} = \cot\Phi\left\{2\dot{\Phi}\dot{\Psi} + \frac{\varsigma_1 - \varsigma_2}{4D}(\varsigma_1\dot{\Phi} - \theta_1\dot{\Psi})(\varsigma_2\dot{\Phi} - \theta_2\dot{\Psi})\right\}, \tag{16.21}$$

where

$$\varsigma_i = \beta_i + \alpha_i, \qquad \theta_i = \beta_i - \alpha_i, \qquad i = 1,2, \tag{16.22}$$

$$D = \beta_2\alpha_1 - \beta_1\alpha_2 \neq 0.$$

Configuration space of dynamic system described by Eqs. (16.20) and (16.21) is identified as the surface of the unit sphere, Φ and Ψ are the polar and azimuthal angles. The state of the system is determined by the position of the point on the sphere and by the vector of angular velocity in the plane tangent to the sphere at a point (Φ, Ψ). It has to be noted, that the arbitrary real exponents α_i and β_i of Hamilton function given by Eq. (16.14) are contained only in the coefficients of Eqs. (16.20) and (16.21), which simplifies significantly an analysis of dynamic equations.

The values of the moduli of the vectors $\mathbf{r}$, and $\mathbf{p}$ are related to the angular variables (Φ, Ψ) by

$$r^{D-(\theta_1-\theta_2)} = \left(\frac{\theta_2 \dot{\Psi} - \varsigma_2 \dot{\Phi}}{2D \sin \Phi} \right)^{1-\alpha_2} \left(\frac{\theta_1 \dot{\Psi} - \varsigma_1 \dot{\Phi}}{-2\sigma D \sin \Phi} \right)^{\alpha_1 - 1}, \qquad (16.23)$$

and

$$p^{D-(\theta_1-\theta_2)} = \left(\frac{\theta_2 \dot{\Psi} - \varsigma_2 \dot{\Phi}}{2D \sin \Phi} \right)^{\beta_2 - 1} \left(\frac{\theta_1 \dot{\Psi} - \varsigma_1 \dot{\Phi}}{-2\sigma D \sin \Phi} \right)^{1-\beta_1}. \qquad (16.24)$$

On the plane of the angular velocities $(\dot{\Psi}, \dot{\Phi})$, the conditions $r > 0$ and $p > 0$ distinguish the admissible domains of their values

$$D(\theta_2 \dot{\Psi} - \varsigma_2 \dot{\Phi}) > 0, \qquad D\sigma(\theta_1 \dot{\Psi} - \varsigma_1 \dot{\Phi}) > 0. \qquad (16.25)$$

The dynamic system presented by Eqs. (16.20) and (16.21) has two integrals of motion [160]

$$I_1 = \frac{1}{D \sin^2 \Phi} [(\theta_2 - \theta_1)\dot{\Psi} - (\varsigma_2 - \varsigma_1)\dot{\Phi}] \left(\frac{\theta_1 \dot{\Psi} - \varsigma_1 \dot{\Phi}}{-2\sigma D \sin \Phi} \right)^{\alpha_1 - 1} \qquad (16.26)$$

and

$$I_2 = \frac{1}{D(\sin \Phi)^{D-2(\theta_2-\theta_1)}} \left(\frac{\theta_2 \dot{\Psi} - \varsigma_2 \dot{\Phi}}{2D \sin \Phi} \right)^{\theta_2} \left(\frac{\theta_1 \dot{\Psi} - \varsigma_1 \dot{\Phi}}{-2\sigma D \sin \Phi} \right)^{-\theta_1}, \qquad (16.27)$$

where the following notations have been introduced

$$I_1 = \frac{2E}{M}, \qquad I_2 = M^{D-2(\theta_2-\theta_1)}. \qquad (16.28)$$

here M is the momentum, $M = |\mathbf{M}|$.

Hereafter, we assume that $M \neq 0$, and take into account conditions given by Eq. (16.25), which require that no base of any power be negative. Moreover, we also assume that $\theta_1 \neq \theta_2$. Note that in the four-dimensional space of structural parameters, the condition $\theta_1 = \theta_2$ singles out a 3D manifold, which requires a separate analysis. Systems realizing the fall to the center correspond to this manifold. An example is dynamic system with Hamilton function $H = p^2 - 1/r^2$.

We eliminate the angular velocity $\dot{\Psi}$ from Eq. (16.26) and then rewrite Eqs. (16.26) and (16.27) as

$$1 = \left(\frac{\sin \Phi}{\sin \Phi_0}\right)^D \left(1 - \frac{1}{\nu_2}\frac{\Phi'}{\sin^2 \Phi}\right)^{\nu_2 D} \left(1 - \frac{1}{\nu_1}\frac{\Phi'}{\sin^2 \Phi}\right)^{-\nu_1 D}, \qquad (16.29)$$

and

$$\Psi' = \frac{2}{D}\frac{\sin^2 \Phi}{\nu_2 - \nu_1} + \frac{\varsigma_2 - \varsigma_1}{\nu_2 - \nu_1}\Phi', \qquad (16.30)$$

where the following notations have been introduced

$$\Phi' = \frac{d\Phi}{d\tau}, \qquad \tau = \frac{M}{ED}t, \qquad (16.31)$$

$$\nu_1 = \frac{\theta_1}{D}, \qquad \nu_2 = \frac{\theta_2}{D}, \qquad (16.32)$$

and for finite motion $\sin \Phi_0 < 1$ is defined by the relation

$$1 = \frac{E^{(\nu_2 - \nu_1)D}}{M^D}(\sin \Phi_0)^2 \left(\frac{\nu_2}{\nu_2 - \nu_1}\right)^{\nu_2 D} \left(\frac{\nu_1}{(-\sigma)(\nu_2 - \nu_1)}\right)^{-\nu_1 D}. \qquad (16.33)$$

Taking into account Eqs. (16.29) and (16.33) and follow [160] we write

$$\Phi(\tau) = \Phi(\tau; \nu_1\nu_2 | E, M) \qquad (16.34)$$

and

$$\Psi(\tau) = \Psi(\tau; \nu_1\nu_2, D, (\varsigma_2 - \varsigma_1) | E, M). \qquad (16.35)$$

The general motion of a representative point on the sphere is rotation with respect to the azimuthal angle Ψ accompanied by periodic oscillations with the period $\overline{T}$ with respect to the polar angle Φ: $\Phi(\tau + \overline{T}) = \Phi(\tau)$, where $\overline{T} = \overline{T}(\nu_1\nu_2, \sin \Phi_0)$. Moreover, we have $\Psi(\tau + \overline{T}) \neq \Psi(\tau)$ in the general case. Equation (16.29) can be solved for $\sin \Phi$, which can be treated as a function of Φ'. Performing the change $\Phi \to z = \cot \Phi$, we obtain the following expression for the period T ($T = \overline{T}E/MD$) of oscillations of the polar angle $\Phi(t)$

$$T = \frac{1/\nu_1 - 1/\nu_2}{\sin \Phi_0} \int\limits_{z_-}^{z_+} dz \frac{(1 - z/\nu_2)^{2\nu_2 - 1}(1 - z/\nu_1)^{-2\nu_1 - 1}}{\sqrt{(1 - z/\nu_2)^{2\nu_2}(1 - z/\nu_1)^{-2\nu_1} - \sin^2 \Phi_0}}. \quad (16.36)$$

We define the rotation number N of a trajectory on the sphere as the increment in the azimuthal angle Ψ for the period of the polar angle Φ, divided by 2π, and obtain

$$N = \frac{\sin \Phi_0}{\pi \nu_1 \nu_2 D} \int\limits_{z_-}^{z_+} dz \frac{(1 - z/\nu_2)^{-1}(1 - z/\nu_1)^{-1}}{\sqrt{(1 - z/\nu_2)^{2\nu_2}(1 - z/\nu_1)^{-2\nu_1} - \sin^2 \Phi_0}}, \quad (16.37)$$

where z_+ and z_- are roots of the equation

$$(1 - z/\nu_2)^{2\nu_2}(1 - z/\nu_1)^{-2\nu_1} = \sin^2 \Phi_0. \quad (16.38)$$

Let us note that although the function $\Psi(t)$ depends on the combination $\varsigma_2 - \varsigma$, the rotation number N is independent of this combination. This means that if we fix the values of the three combinations of the structural parameters in the four-dimensional space $(\alpha_1, \beta_1, \alpha_2, \beta_2)$,

$$\theta_1 = \beta_1 - \alpha_1, \qquad \theta_2 = \beta_2 - \alpha_2, \qquad D = \beta_2 \alpha_1 - \beta_1 \alpha_2, \quad (16.39)$$

then we come up with one-dimensional manifolds where the rotation number N dependency on the first integrals E and M is saved (see the next section).

16.3   Fractional Kepler problem and fractional harmonic oscillator problem

16.3.1   *One-dimensional manifold*

It follows from Eqs. (16.29), (16.31) and (16.34), (16.35) that the equations for the polar angle Φ depend only on the three combinations of the structural parameters α_1^0, β_1^0, α_2^0, β_2^0. They are constant on the straight lines

$$\alpha_1 = \alpha_1^0 + k_1 \gamma, \qquad \beta_1 = \beta_1^0 + k_1 \gamma, \quad (16.40)$$

$$\alpha_2 = \alpha_2^0 + k_2 \gamma, \qquad \beta_2 = \beta_2^0 + k_2 \gamma,$$

in the four-dimensional space of structural parameters (γ is the parameter of the straight line and $k_2/k_1 = (\beta_2 - \alpha_2)/(\beta_1 - \alpha_1) \equiv k_0$). On each of these straight lines, the dependence of the angle Φ and the rotation number N on the constant first integrals E and M is saved. Straight lines defined by Eq. (16.40) are assigned the Hamilton functions

$$H = p^{\alpha_1^0} r^{\beta_1^0} (rp)^{k_1\gamma} + \sigma p^{\alpha_2^0} r^{\beta_2^0} (rp)^{k_1\gamma}. \tag{16.41}$$

We illustrate this with examples of families of dynamic systems corresponding to two straight lines (16.40) passing through the points corresponding to the planar Kepler problem and to the isotropic oscillator problem.

16.3.2 *Fractional extension of the Kepler problem*

On the straight line (16.40) passing through the point ($\alpha_1^0 = 2, \beta_1^0 = \alpha_2^0 = 0, \beta_2^0 = -1$), which corresponds to the Kepler problem, we have $D = -2$, $\nu_1 = 1$, and $\nu_2 = 1/2$. Hence,

$$\alpha_1 = 2 + \gamma, \qquad \beta_1 = \gamma, \tag{16.42}$$

$$\alpha_2 = \frac{\gamma}{2}, \qquad \beta_2 = -1 + \frac{\gamma}{2}.$$

Equation (16.29) is solvable for Φ' on the line defined by Eq. (16.42)

$$\frac{M}{2E} \frac{\Phi'}{\sin^2 \Phi} = \frac{\sigma^2}{2M^3} \sqrt{\sin^2 \Phi - \sin^2 \Phi_0} \tag{16.43}$$

$$\times \left[\sqrt{\sin^2 \Phi - \sin^2 \Phi_0} \pm \sin \Phi \right],$$

where $\sin^2 \Phi_0 = -4EM^2/\sigma^2$. Integrating this equation yields two branches of the function $t = t(\Phi)$

$$\frac{\sigma^2 \sin^3 \Phi_0}{2M^3} (t - t_i) \tag{16.44}$$

$$= \left\{ \sin \Phi_0 \cot \Phi \mp \arctan \frac{\cot \Phi}{\sqrt{\cot^2 \Phi_0 - \cot^2 \Phi}} \right\} \Big|_{\Phi_i}^{\Phi}.$$

Integrating over the cycle $\Phi_0^- \Rightarrow \Phi_0^+ \Rightarrow \Phi_0^- \ (0 < \Phi_0^- \le \pi/2 \le \Phi_0^+ < \pi)$, we see that the period of oscillations of $\Phi(t)$ is subject to the dependence $T \propto |\sigma|(-E)^{-3/2}$, that is the third Kepler law holds for any γ. Treating Ψ as a function of Φ, we see that two branches of this function are determined by the equations

$$\frac{d\Psi}{d\Phi} + \gamma - 1 = \pm \frac{2\sin\Phi}{\sqrt{\sin^2\Phi - \sin^2\Phi_0}}. \tag{16.45}$$

Integrating over the cycle $\Phi_0^- \Rightarrow \Phi_0^+ \Rightarrow \Phi_0^-$, let us conclude that the increment in the azimuthal angle Ψ for the period T of the polar angle is equal to 2π independently of E, M, and the parameter γ. In the case when Eq. (16.42) holds, all finite orbits on the sphere are therefore closed for any γ. But the form of the orbits on the sphere depends on γ

$$\cos\Phi = \pm|\cos\Phi_0|\sin\left[\frac{1}{2}\Psi + \frac{1}{2}(1+\gamma)\Phi + \text{const}\right]. \tag{16.46}$$

Next, it can be shown that

$$r(\Phi,\gamma) = \left(\frac{M}{\sin\Phi}\right)^{1+\gamma/2} r^0(\Phi), \tag{16.47}$$

where $r^0(\Phi)$ is given by

$$\frac{1}{r^0(\Phi)} = \frac{|\sigma|}{2M}[\sin\Phi \pm \sqrt{\sin^2\Phi - \sin^2\Phi_0}]. \tag{16.48}$$

The evolution of the polar angle φ in the original system of polar coordinates on the plane (r, φ) is given by the expression

$$\varphi = (\Psi(\sin\Phi,\gamma) - \Phi)/2,$$

where $\Psi(\sin\Phi,\gamma)$ is a solution to Eq. (16.45). Therefore, the conditions that the orbits are closed on the plane (r, φ) are,

$$r(t + 2T,\gamma) = r(t,\gamma), \qquad \varphi(t + 2T,\gamma) = \varphi(t,\gamma). \tag{16.49}$$

These conditions hold for all values of γ.

Finally, we write the transformation formulas for orbits on the plane (r, φ) as the parameter γ varies

$$r = 2 \left(\frac{M}{\sin \Phi} \right)^{\gamma} r^0(\Phi), \qquad r' = 2 \left(\frac{M}{\sin \Phi} \right)^{\gamma'} r^0(\Phi), \qquad (16.50)$$

$$\Phi = \Phi_0^- - 2 \frac{\varphi' - \varphi}{\gamma' - \gamma}, \qquad (16.51)$$

where (r, φ) and(r', φ') are points of the corresponding orbits for dynamic systems with the Hamiltonians

$$H = p^2 (rp)^{\gamma} + \frac{\sigma}{r}(rp)^{\gamma/2}, \qquad H = p^2 (rp)^{\gamma'} + \frac{\sigma}{r}(rp)^{\gamma'/2}. \qquad (16.52)$$

These formulas permit obtaining an expression for the third integral of motion for $\gamma \neq 0$ from the known expression in the case $\gamma = 0$ (i.e., for the Kepler problem). This means that on straight line (16.42), all dynamic systems preserve the hidden symmetry of the Kepler problem. Figure 4 displays the numerically calculated deformation of orbits on the plane (r, φ) as the parameter γ varies. Singularities appear on the orbits because the conditions $\dot{r} = 0$ and $\dot{\varphi} = 0$ (i.e., $\Phi = \pi/2$, $\partial H/\partial p = 0$) are realized in some intervals of the values of γ. We note that the "reverse" motion along the orbit (the sign of $\dot{\varphi}$ is changed) is not related to the change in the direction of the vector of angular momentum: on the intervals of "inverse" motion, the velocity vector is directed opposite to the momentum vector.

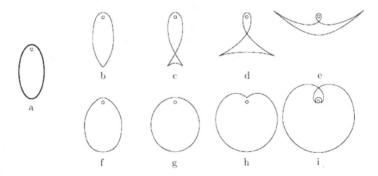

Fig. 4. Deformation of orbits on the plane (r, φ) as the parameter γ varies. (a) $\gamma = 0$ (the Kepler problem); (b) $\gamma = 0.3$; (c) $\gamma = 0.9$; (d) $\gamma = 2$; (e) $\gamma = 4$; (f) $\gamma = -0.9$; (g) $\gamma = -2$ (all orbits are circles whose centers are displaced from the origin); (h) $\gamma = -4$ (all orbits have the form of Pascal's limacons); (i) $\gamma = -8$ [160].

16.3.3   Fractional extension of the harmonic oscillator problem

Now we consider the straight line (16.40) passing through the point $(\alpha_1^0 = 2, \beta_1^0 = \alpha_2^0 = 0, \beta_2^0 = -1)$, which corresponds to the harmonic oscillator problem. On this straight line, we have $D = 4$, $\nu_1 = -1/2$ and $\nu_2 = 1/2$. Hence, we obtain

$$\alpha_1 = 2 + \gamma, \qquad \beta_1 = \gamma, \qquad \alpha_2 = \frac{\gamma}{2}, \qquad \beta_2 = -1 + \frac{\gamma}{2}. \qquad (16.53)$$

In this case, Eqs. (16.29) and (16.30) become

$$\dot{\Phi} = \pm 2\sigma^{1/2} \sin\Phi \sqrt{\frac{\sin^2\Phi}{\sin^2\Phi_0} \pm 1}, \quad \dot{\Psi} = \frac{2\mathcal{E}}{M}\sin^2\Phi - \gamma\dot{\Phi}. \qquad (16.54)$$

Here $\sin^2\Phi_0 = \sigma M^2/E^2$, and the differentiation is performed with respect to the original independent variable t. Integrating the first of the equations (with the lower sign under the radical sign), we find two branches of the function $t = t(\Phi)$

$$2\sigma^{1/2}(t - t_i) = \pm \arctan \frac{\cot\Phi}{\sqrt{\cot^2\Phi_0 - \cot^2\Phi}}\Big|_{\Phi_i}^{\Phi}. \qquad (16.55)$$

Integrating over the cycle $\Phi_0^- \Rightarrow \Phi_0^+ \Rightarrow \Phi_0^-$ $(0 < \Phi_0^- \leq \pi/2 \leq \Phi_0^+ < \pi)$, we see that the period of oscillations of $\Phi(t)$ is equal to $T = \sigma^{-1/2}$ and is independent of the constant first integrals E and M and the parameter γ. The azimuthal angle Ψ treated as a function of Φ is defined by the expression

$$\Psi - \Psi_i = \mp \arcsin \frac{\cos\Phi}{\cos\Phi_0}\Big|_{\Phi_i}^{\Phi} - \gamma(\Phi - \Phi_0). \qquad (16.56)$$

The increment in Ψ for the period T of the polar angle Φ is equal to π for all γ (the deformation of orbits does not violate their closedness). The values of r and φ are reconstructed from the formulas

$$r^2(t, \gamma) = \frac{E}{2\sigma}\left(\frac{M}{\sin\Phi}\right)^\gamma \left(1 \pm \sqrt{1 - \frac{\sin^2\Phi_0}{\sin^2\Phi_0} \pm 1}\right), \qquad (16.57)$$

and

$$2\varphi = \mp \arcsin \frac{\cos \Phi}{|\cos \Phi_0|} |_{\Phi_i}^{\Phi} - (1 + \gamma)(\Phi - \Phi_0^-). \tag{16.58}$$

In this case, we have $r(t + T, \gamma) = r(t, \gamma)$ and $\varphi(t + T, \gamma) = \varphi(t, \gamma) + \pi$ for all γ. All orbits of finite motions are therefore also closed in this case; moreover, all the corresponding dynamic systems with Hamiltonians $H = p^2(rp)^\gamma + \sigma r^2(rp)^{-2\gamma}$ are isochronous.

16.3.4 *Two-dimensional manifold*

We consider the set of straight lines (16.42) satisfying the conditions $\theta_1 = D$ and $\theta_1 = D/2$ or, explicitly, the conditions

$$\beta_1 - \alpha_1 = \beta_2\alpha_1 - \beta_1\alpha_2, \qquad \beta_2 - \alpha_2 = \frac{1}{2}(\beta_2\alpha_1 - \beta_1\alpha_2). \tag{16.59}$$

In particular, this set contains the points $(\alpha_2 = 2, \ \beta_1 = \alpha_2 = 0, \ \beta_2 = -1)$ and $(\alpha_1 = \beta_2 = 0, \ \beta_1 = 2, \ \alpha_2 = -1)$ corresponding to the Kepler problem in the r- and p-representations.

The equations for straight lines contained in this set are given by Eq. (16.42), where α_1^0, β_2^0, α_2^0, and β_2^0 satisfy conditions (16.59) and $k_2/k_1 = 1/2$. Solving, as an example, Eqs. (16.40) for α_1^0, and β_2^0 we obtain either

$$\alpha_2^0 = -1 + \frac{1}{2}\alpha_1^0, \qquad \beta_1^0 = 2(1 + \beta_2^0), \qquad D^0 = 2(1 + \beta_2^0) - \alpha_1^0 \neq 0, \tag{16.60}$$

or

$$\alpha_2^0 = \beta_2^0, \qquad \beta_1^0 = \alpha_1^0, \qquad D^0 = 0. \tag{16.61}$$

Setting $k_1 = 1$, $k_{21} = 1/2$, and $D^0 \neq 0$, we see that the desired set of straight lines is determined by the expressions

$$\alpha_1 = \alpha_1^0 + \gamma, \qquad \beta_1 = 2(1 + \beta_2^0) + \gamma, \tag{16.62}$$

$$\alpha_2 = -1 + \frac{1}{2}\alpha_1^0 + \frac{1}{2}\gamma, \qquad \beta_2 = \beta_2^0 + \frac{1}{2}\gamma, \tag{16.63}$$

or, in another form,

$$\alpha_1 - 2\alpha_2 = 2, \qquad \beta_1 - 2\beta_2 = 2, \tag{16.64}$$

$$\alpha_1 - \alpha_1^0 = 2(\beta_2 - \beta_2^0), \qquad \beta_1 - \beta_1^0 = 2(\alpha_2 - \alpha_2^0). \tag{16.65}$$

It follows from these expressions that all straight lines contained in this set belong to the two-dimensional manifold related to the condition $\nu_1/\nu_2 = 2$, and expressions (16.62) and (16.63) determine the position of the straight line passing through the point $(\alpha_1^0, \beta_1^0, \alpha_2^0, \beta_2^0)$ in the space of structural parameters $(\alpha_1, \beta_1, \alpha_2, \beta_2)$.

We note that for $\nu_1/\nu_2 = 2$, Eqs. (16.36) and (16.37) imply that $T \propto (-E)^{-3/2} 2D^{-1}$ and $N = 2/D$ on the manifold studied above.

The equation determining the orbit on the sphere becomes

$$\frac{d\Psi}{d\Phi} + \mathcal{A} = \pm \frac{D}{2} \frac{2\sin\Phi}{\sqrt{\sin^2\Phi - \sin^2\Phi_0}} \alpha_1 - \alpha_1^0 = 2(\beta_2 - \beta_2^0), \tag{16.66}$$

$$\beta_1 - \beta_1^0 = 2(\alpha_2 - \alpha_2^0),$$

where

$$\mathcal{A} = \frac{2}{D}(2 + \alpha_2^0 + \beta_2^0 - \alpha_1^0 - \beta_1^0 - \gamma). \tag{16.67}$$

Comparing this equation with Eq. (16.45), we see that the orbits of all finite motions are closed for rational values of D.

Therefore, a system with the Hamilton function

$$H = p^{-D}(rp)^\gamma - |\sigma| p^{-D/2}(rp)^{\gamma/2-1}, \tag{16.68}$$

all of whose finite orbits are characterized by the same rotation number $N = 2/|D|$, corresponds to a point with arbitrary values of $D^0 \neq 0$ and γ on two-dimensional manifold given by Eqs. (16.64) and (16.65). In Eqs. (16.64) and (16.65), we fix the values of the parameters ensuring the condition $D = D^0$, and as a result obtain a straight line of form (16.40) with the parameter γ, which lies on the above mentioned two-dimensional manifold. On this straight line, all systems are characterized by the same rotation number $N = 2/|D^0|$ (the same for all finite orbits). On the straight lines corresponding to rational values of D, all finite orbits are closed, while on the straight lines corresponding to irrational values of D, all finite orbits are open.

In particular, straight line (16.42) considered in Sec. 16.2 lies on the above mentioned two-dimensional manifold (this line corresponds to $D = -2$ and passes through the point ($\alpha_1^0 = 2$, $\beta_1^0 = \alpha_2^0 = 0$, $\beta_2^0 = -1$), as well as the straight line corresponding to $D = +2$ and passing through the point ($\alpha_1^0 = 0$, $\beta_1^0 = 2, \alpha_2^0 = -1$, $\beta_2^0 = 0$), which belongs to the Kepler problem in the momentum representation.

16.4 Asymptotic analysis of nearly circular orbits

16.4.1 *Circular orbits*

First considering circular orbits, let us note, that for any finite orbit, the angular variable Φ varies within the strip $0 \leq \pi/2 - \Delta\Phi \leq \Phi \leq \pi/2 + \Delta\Phi \leq \pi$, whose width is determined by the initial conditions. This strip can be contracted to the equatorial circle. Indeed, Eqs. (16.20) and (16.21) admit the solutions

$$\Phi = \Phi_0 \equiv \frac{\pi}{2}, \qquad \Psi = \Psi_0 + \Omega_0 t, \qquad (16.69)$$

which describe the uniform rotation of the representative point along the equator of the sphere with the angular velocity Ω_0; such solutions correspond to circular orbits on the plane (r, φ) for which $r = r_0 > 0$ and $p = p_0 > 0$. In the general case ($\theta_1 \neq \theta_2$),

$$r_0 = \left(\frac{E\theta_2}{\Delta\theta}\right)^{-\alpha_2/D} \left(-\frac{E\theta_1}{\sigma\Delta\theta}\right)^{\alpha_1/D}, \qquad (16.70)$$

$$p_0 = \left(\frac{E\theta_2}{\Delta\theta}\right)^{\beta_2/D} \left(-\frac{E\theta_1}{\sigma\Delta\theta}\right)^{-\beta_1/D},$$

where $\Delta\theta = \theta_2 - \theta_1$.

We note that on the circular orbits, the first integrals E and M depend on each other

$$M^D = \left|\frac{E}{\Delta\theta}E\right|^{\Delta\theta} |\sigma^{-1}\theta_1|^{-\theta_1}|\theta_2|^{\theta_2}. \qquad (16.71)$$

In the degenerate case $\theta_1 = \theta_2$, the circular orbits are assigned the levels $E = 0$ and $M = (-\sigma)^{\theta_1/D}$, to which the family of circular orbits such that $r_0 p_0 = M$ corresponds.

16.4.2 Asymptotic analysis

Let us note that Eqs. (16.20) and (16.21) permit constructing polynomial asymptotic solutions in a neighborhood of circular orbits. Such a solution can be described as rotation (i.e., a monotonic increase in the azimuthal angle Ψ) near the equator of the sphere, accompanied by small-amplitude oscillations with respect to the polar angle Φ.

We seek expansions of the functions $\Phi(t)$ and $\Upsilon(t) = \dot{\Psi}(t)$ in the form

$$\Phi = \frac{\pi}{2} + \sum_{n=1} \varepsilon^n \Phi_n(\tau), \qquad \Upsilon = C(\varepsilon) + \sum_{n=1} \varepsilon_n \Upsilon_n(\tau). \qquad (16.72)$$

The constant C is also expanded in a power series in ε

$$C = C_0 + \sum_{n=1} \varepsilon^n C_n. \qquad (16.73)$$

The equation for $\Phi_1(\tau)$ arising in the first order in ε has the form

$$\frac{d^2}{d\tau^2} \Phi_1(\tau) + \frac{1}{4} \frac{C_0^2 (\theta_1 - \theta_2)\theta_1\theta_2}{D} \Phi_1(\tau) = 0. \qquad (16.74)$$

Imposing the 2π-periodicity condition on the desired function, we obtain

$$C_0 = 2\sqrt{\frac{D}{(\theta_1 - \theta_2)\theta_1\theta_2}}. \qquad (16.75)$$

Then $\Phi_1(\tau) = \sin\tau$. Similarly, solving the equation for the function $\Upsilon_1(\tau)$, we obtain

$$\Upsilon_1(\tau) = \frac{\varsigma_1 - \varsigma_2}{\theta_1 - \theta_2} \cos\tau. \qquad (16.76)$$

Eliminating the secular term in the equation for $\Phi_2(\tau)$, we find that $C_1 = 0$ and

$$\Phi_2(\tau) = \frac{1}{6}\sqrt{\frac{D}{(\theta_1 - \theta_2)\theta_1\theta_2}}(\theta_1 + \theta_2)\sin 2\tau. \qquad (16.77)$$

The function $\Upsilon_2(\tau)$ is determined in a similar way.

Further, by eliminating the secular term in the equation for $\Phi_3(\tau)$, we have

$$C_2 = \frac{1}{12}\sqrt{\frac{D}{[(\theta_1 - \theta_2)\theta_1\theta_2]^3}}[D(\theta_1^2 - \theta_1\theta_2 + \theta_2^2) - 3\theta_1\theta_2(\theta_1 - \theta_2)]. \qquad (16.78)$$

For nearly circular orbits, we consider how their rotation numbers N depend on the small parameter ε. On the plane (r, φ), the rotation number is defined as the angular distance $\Delta\varphi$ between two successive apocenters (i.e. the points where $r = r_{\max}$) divided by 2π. It follows from the equations of motion that on the sphere, the apocenter points are assigned the points of intersection of the trajectory with the equator when passing from the "southern" hemisphere (with the pole $\Phi = 0$) to the "northern" hemisphere (with the pole $\Phi = \pi$). Therefore, the rotation number is given by the formula

$$N = \frac{1}{4\pi}\{\Psi(\tau)|_{\tau=2\pi} - \Psi(\tau)|_{\tau=0}\}. \tag{16.79}$$

Taking the above asymptotic expansion into account, we obtain $N = N(\varepsilon) \approx C(\varepsilon)/2$ for nearly circular orbits. For the circular orbit itself, the rotation number is not defined. We define it as the limit

$$N_0 = \lim_{\varepsilon \to 0} N(\varepsilon) = \frac{1}{2}C_0\sqrt{\frac{D}{(\theta_1 - \theta_2)\theta_1\theta_2}}, \tag{16.80}$$

which coincides with the value of rotation number (19) obtained earlier.

We note that for all finite orbits to have the same rotation number, it is necessary that $C(\varepsilon) = C_0 = const$, i.e. at least, the condition $C_2 = 0$ must hold.

A consequence of system (7) is the second-order equation for the angular variable Φ treated as a function of Ψ

$$\Phi'' = -\frac{1}{4D}\cot\Phi[(\theta_2 - \theta_1) - (\varsigma_2 - \varsigma_1)\Phi'](\varsigma_2\Phi' - \theta_2)(\varsigma_1\Phi' - \theta_1), \tag{16.81}$$

where $\Phi'' = d^2\Phi/d\Psi^2$ and $\Phi' = d\Phi/d\Psi$. It follows from the form of this equation that on the sphere, there exist three families of trajectories on which Φ and Ψ are linearly related. On the plane (Φ, Ψ), we consider the strip $0 \leq \Phi \leq \pi$, which is the evolvent of the sphere under study. In this case, the upper and lower boundaries of this strip correspond to the poles of the sphere, and the midline corresponds to its equator. On the evolvent of the sphere, the above three families of trajectories are then assigned three families of segments of straight lines bounded by the upper and lower boundaries of the strip and having the angular coefficients

$$k_1 = \frac{\theta_2 - \theta_1}{\varsigma_2 - \varsigma_1}, \qquad k_2 = \frac{\theta_2}{\varsigma_2}, \qquad k_3 = \frac{\theta_1}{\varsigma_1}. \tag{16.82}$$

Such segments are respectively called *segments of family* 1, 2, and 3. It has been shown in [160] that if the initial conditions are varied continuously such that the strip around the equator, which bounds the motion of the representative point, becomes wider, then the smooth finite orbits degenerate in the limit into a saw-tooth broken line whose components are alternating segments of three families, either 1 and 2, or 1 and 3, or 2 and 3, depending on the values of the structural parameters of the system. For shortness, such broken lines are called limit chains. Thus we have three types of limit chains: (1,2), (1,3), and (2,3). In the space of structural parameters, three domains related to a definite type of the limit chain are distinguished. In particular, the point corresponding to the Kepler case lies in one of these domains; another domain contains the point corresponding to the harmonic oscillator; and the third domain contains the point corresponding to the Kepler problem in the momentum representation. The prototype of the limit chain on the sphere can be either closed or open, depending on the values of structural parameters. We note that the segments composing the limit chain are assigned trajectories on the sphere, but they are not necessarily real trajectories on the original plane. For example, in the Kepler case, the limit chains consist of a segment of family 1 corresponding to the parabolic trajectory of the original problem and of a segment of family 2 that does not correspond to any trajectory of the original problem.

We also note that segments of two types composing the same limit chain correspond to trajectories on the sphere that lie at different integral levels. It is convenient to introduce limit chains because these chains can be assigned the rotation number N_∞ equal to the doubled length of a "tooth" of the chain divided by 2π, which can be easily derived from the values of the angular coefficients k_m. A numerical analysis provided by authors of work [160] confirms that this number is the limit of the rotation numbers of finite motions orbits degenerating (on the evolvent of the sphere) into the limit chain.

For all finite orbits to have the same rotation numbers, it is natural to require that the rotation numbers corresponding to the circular orbit (N_0) and to the limit chain (N_∞) be the same. It follows from the results of the asymptotic analysis that there is another condition that all coefficients except C_0 in the power expansion of $C(\varepsilon)$ are zero (in general, this condition is not *a priori* independent of the first condition). A numerical analysis leads to the conclusion that in the class of Hamiltonians (16.14), the joint realization of the two conditions

$$N_0 = N_\infty, \qquad C_2 = 0, \qquad (16.83)$$

is necessary and sufficient for the orbits of all finite motions in the system to have the same rotation numbers. If this number is rational, then all orbits are closed.

Let us list the sets of Hamiltonian systems of class (16.14) for which conditions given by Eq. (16.83) are satisfied. In the domain corresponding to chains of the form (1,2), conditions (16.83) are satisfied precisely on two-dimensional manifold defined by Eqs. (16.64) and (16.65). Analysis presented in [160] confirms that for any system with Hamilton function (16.68), the rotation numbers of all orbits are the same and are equal to $2/|D|$.

In the domain corresponding to chains of the form (2,3), conditions defined by Eq. (16.83) are satisfied on the manifold

$$\alpha_1 + \alpha_2 = 2, \qquad \beta_1 + \beta_2 = 2, \qquad (16.84)$$

containing the point that corresponds to the harmonic oscillator. This two-dimensional surface also splits into one-dimensional curves along which the rotation number is preserved. Namely, the rotation number is equal to $2/|D|$ for all orbits of any system with the Hamilton function

$$H = \frac{p^{1+D/2}}{r^{-1-D/2}}(rp)^\gamma + \sigma \frac{r^{1+D/2}}{p^{-1+D/2}}(rp)^{-\gamma}, \qquad \sigma > 0. \qquad (16.85)$$

Finally, in the third domain of the space of structural parameters (with chains of the form (1,3)), the conditions (16.83) are satisfied on the two-dimensional surface

$$\alpha_2 - 2\alpha_1 = 2, \qquad \beta_2 - 2\beta_1 = 2. \qquad (16.86)$$

The corresponding Hamilton function is obtained from Eq. (16.68) by the interchange $r \leftrightarrow p$.

Afterword

Fractional quantum mechanics emerged as a field over 15 years ago, attracting the attention of many researchers. The original idea behind fractional quantum mechanics was to develop a path integral over Lévy-like quantum paths instead of the well-known Feynman path integral over Brownian-like paths. The basic outcome of this implementation is an alternative path integral approach, which results in a new fundamental quantum equation – the fractional Schrödinger equation discovered by Laskin. In other words, if the Feynman path integral allows one to reproduce the Schrödinger equation, then the Laskin path integral over Lévy-like paths leads one to a fractional Schrödinger equation. Fractional quantum mechanics is a manifestation of a new non-Gaussian physical paradigm, based on deep relationships between the structure of fundamental physics equations and the fractal dimension of "underlying" quantum paths. The fractional Schrödinger equation includes the space derivative of fractional order α (α is the Lévy index) instead of the second order space derivative in the well-known Schrödinger equation. Thus, the fractional Schrödinger equation is the fractional differential equation in accordance with modern terminology. This is the main reason to coin the term, *fractional Schrödinger equation*, and the more general term - *fractional quantum mechanics*.

Today, there are two alternative fundamental approaches to fractional quantum mechanics - the Laskin path integral and the fractional Schrödinger equation. This book presents the first in depth systematic coverage of both theoretical approaches. The book pioneers quantum mechanical applications of the α-stable Lévy stochastic processes, the path integral over Lévy-like paths and Fox's H-function.

The H-function, never before used in quantum mechanics, is a well-suited mathematical tool to solve the fractional Schrödinger equation for quantum physical problems.

The book discovers and explores new fractional quantum mechanical equations. Remarkably, these equations are turned into the well-known equations of quantum mechanics in the particular case when the Lévy index $\alpha = 2$, and Lévy flight turns into Brownian motion. When $\alpha = 2$ Laskin's path integral becomes the well-known Feynman's path integral, and the fractional Schrödinger equation becomes the celebrated Schrödinger equation.

Fractional quantum mechanics brings us a new mathematical formulation of the Heisenberg Uncertainty Principle and a new fundamental physical concept of the quantum Lévy wave packet.

A new physical model - the quantum fractional oscillator has been introduced and studied. The symmetries of the quantum fractional oscillator and its non equidistant spectrum have been discovered and discussed. Exactly solvable models of fractional quantum mechanics have been explored in the book. They include a free particle solution to 1D and 3D fractional Schrödinger equations, quantum particle in the symmetric infinite potential well, bound state in δ-potential well, linear potential field and quantum kernel for a free particle in the box.

Time fractional quantum mechanics emerged soon after the discovery of fractional quantum mechanics. Time fractional quantum mechanics involves space fractality parameter α, which is the Lévy index, and time fractality parameter β, which is the order of the Caputo fractional time derivative. The manifestations of time fractional quantum mechanics are *time fractional Schrödinger equation* and *space-time fractional Schrödinger equation*, the invention of which was inspired by the fractional Schrödinger equation. *Space-time fractional quantum mechanics* introduced in this book involves two scale dimensional parameters, one of which can be considered as a time fractional generalization of the famous Planck's constant, while the other one can be interpreted as a time fractional generalization of the scale parameter of fractional quantum mechanics. The fractional generalization of Planck's constant is a fundamental dimensional parameter of time fractional quantum mechanics, while the time fractional generalization of Laskin's scale parameter plays a fundamental role in both time fractional quantum mechanics and time fractional classical mechanics. Time fractional quantum mechanical operators of coordinate, momentum and angular momentum have been introduced and their commutation relationship has been established. The pseudo-Hamilton quantum mechanical operator has been introduced and its Hermiticity has been proven in the framework of time fractional quantum mechanics. The general solution to the space-time fractional Schrödinger equation was found in the case when the pseudo-Hamilton operator does not depend on time. Energy of a quantum system in the framework of time fractional quantum mechanics was defined and calculated in terms of the Mittag-Leffler function. Two new functions associated with the Mittag-Leffler function have been launched and elaborated. These two new functions can be considered as a natural fractional generalization of the well-known trigonometric functions sine and cosine. A fractional generalization of the celebrated Euler's formula was discovered. A free particle space-time fractional quantum kernel was calculated in terms of Fox's H-function. It has been shown that a free particle space-time fractional quantum kernel can be alternatively expressed with the help of the Wright L-function.

The framework of time fractional quantum mechanics developed in this book, depending on the choices of fractality parameters α and β, covers the following fundamental quantum equations:

- The Schrödinger equation, $\alpha = 2$ and $\beta = 1$;
- Fractional Schrödinger equation (Laskin equation), $1 < \alpha \leq 2$ and $\beta = 1$;
- Time fractional Schrödinger equation (Naber equation), $\alpha = 2$ and $0 < \beta \leq 1$;
- Space-time fractional Schrödinger equation (Wang and Xu, Dong and Xu equation), $1 < \alpha \leq 2$ and $0 < \beta \leq 1$.

Fractional nonlinear quantum dynamics has been covered by the book. Fractional nonlinear Schrödinger equation, Nonlinear Hilbert–Schrödinger equation, fractional generalization of Zakharov system and fractional Ginzburg–Landau equation have been discovered and explored.

Fractional statistical mechanics has been introduced based on the path integral over Lévy fights. The path integral representations were obtained for density matrix and partition function of a statistical system.

The book introduces *fractional classical mechanics* as a classical counterpart of fractional quantum mechanics. The Lagrange, the Hamilton, the Hamilton–Jacobi, and the Poisson bracket approaches were developed and studied in the framework of fractional classical mechanics. A classical fractional oscillator model was introduced, and its exact analytical solution was found. A map between the energy dependence of the period of classical oscillations and the non-equidistant distribution of the energy levels for the quantum fractional oscillator has been established and discussed. The book presents *fractional Kepler's third law* which is a generalization of the well-known Kepler's third law. Fractional classical dynamics on two-dimensional sphere has been introduced and developed. It has been discovered that in the four-dimensional space of structural parameters, there exists a one-dimensional manifold, containing the case of the planar Kepler problem, along which the closedness of the orbits of all finite motions and the third Kepler law are saved. Similarly, it has been found that there exists a one-dimensional manifold, containing the case of the two-dimensional isotropic harmonic oscillator, along which the closedness of the orbits and the isochronism of nonlinear oscillations are saved.

The book serves as the first monograph and the first handbook on the topic, covering fundamentals and applications of fractional quantum mechanics, time fractional quantum mechanics, fractional statistical mechanics and fractional classical mechanics.

Appendix A

Fox H-function

A.1 Definition of the Fox H-function

Apart from the natural way in which Fox's H-function enters into fractional quantum mechanics and fractional statistical mechanics, its fractional derivatives and integrals are easily calculated by formally manipulating the parameters in the H-function. In other words, the fractional derivatives and integrals of Fox's H-function can be calculated within this general class of functions. This is the reason why Fox's H-function is a very important and efficient mathematical tool for fractional quantum mechanics and fractional statistical mechanics. Within the class of Fox's H-function various problems of fractional quantum and statistical mechanics can be expressed in a closed analytical way.

Some properties of the H-function in connection with Mellin-Barnes integrals were investigated by Barnes [161], Mellin [162], Dixon and Ferrar [163]. In an attempt to unify and extend the existing results on symmetric Fourier kernels, Fox [103] has defined the H-function in terms of a general Mellin-Barnes type integral. He has also investigated the most general Fourier kernel associated with the H-functions and obtained the asymptotic expansions of the kernel for large values of the argument. Asymptotic expansions and analytic continuations of the Fox H-function and its special cases were derived by Braaksma [164]. Many properties of the H-function are reported in the book [104] along with applications in statistics.

Fox's H-function is defined by the Mellin-Barnes integral [103], [164] (we follow the notations of the book [104])

$$
H_{p,q}^{m,n}(z) = H_{p,q}^{m,n}\left[z \left| \begin{matrix} (a_p, A_p) \\ (b_q, B_q) \end{matrix} \right. \right]
\tag{A.1}
$$

$$= H^{m,n}_{p,q} \left[z \left| \begin{matrix} (a_1, A_1), ..., (a_p, A_p) \\ (b_1, B_1), ..., (b_q, B_q) \end{matrix} \right. \right] = \frac{1}{2\pi i} \int_L ds \, z^s \, \chi(s),$$

where function $\chi(s)$ is given by

$$\chi(s) = \frac{\prod\limits_{j=1}^{m} \Gamma(b_j - B_j s) \prod\limits_{j=1}^{n} \Gamma(1 - a_j + A_j s)}{\prod\limits_{j=m+1}^{q} \Gamma(1 - b_j + B_j s) \prod\limits_{j=n+1}^{p} \Gamma(a_j - A_j s)} \tag{A.2}$$

and

$$z^s = \exp\{s \text{Log}|z| + i \arg z\},$$

here m, n, p and q are non-negative integers satisfying $0 \leq n \leq p$, $1 \leq m \leq q$; and the empty products are interpreted as unity. The parameters A_j $(j = 1, ..., p)$ and B_j $(j = 1, ..., q)$ are positive numbers; a_j $(j = 1, ..., p)$ and b_j $(j = 1, ..., p)$ are complex numbers such that

$$A_j(b_h + \nu) \neq B_h(a_j - \lambda - 1) \tag{A.3}$$

for $\nu, \lambda = 0, 1, ...$; $h = 1, ..., m$; $j = 1, ..., n$.

The L is a contour separating the points

$$s = \left(\frac{b_j + \nu}{B_j} \right), \quad (j = 1, ..., m; \quad \nu = 0, 1, ...),$$

which are the poles of $\Gamma(b_j - B_j s)$ $(j = 1, ..., m)$, from the points

$$s = \left(\frac{a_j - \nu - 1}{A_j} \right), \quad (j = 1, ..., n; \quad \nu = 0, 1, ...),$$

which are the poles of $\Gamma(1 - a_j + A_j s)$ $(j = 1, ..., n)$. The contour L exists on account of (A.3). These assumptions will be retained throughout.

In the contracted form the H-function in (A.1) will be denoted by one of the following notations:

$$H(z), \quad H^{m,n}_{p,q}(z), \quad H^{m,n}_{p,q} \left[z \left| \begin{matrix} (a_p, A_p) \\ (b_q, B_q) \end{matrix} \right. \right].$$

The Fox H-function is an analytic function of z which makes sense (i) for every $z \neq 0$ if $\mu > 0$ and (ii) for $0 < |z| < \beta^{-1}$ if $\mu = 0$, where

$$\mu = \sum_{j=1}^{q} B_j - \sum_{j=1}^{p} A_j \qquad (A.4)$$

and

$$\beta = \prod_{j=1}^{p} A_j^{A_j} \prod_{j=1}^{q} B_j^{-B_j}. \qquad (A.5)$$

Due to the occurrence of the factor z^s in (A.1), the H-function is in general multiple-valued, but is one-valued on the Riemann surface of $\log z$. The H-function is a generalization of Meijer's G-function [165], which is also defined by a Mellin-Barnes integral. The H-function reduces to the G-function if $A_j = 1$ and $B_k = 1$ for all $j = 1, 2, ..., p$ and $k = 1, 2, ..., q$,

$$G_{p,q}^{m,n}(z) = H_{p,q}^{m,n}\left[z \left| \begin{matrix} (a_1, 1), ..., (a_p, 1) \\ (b_1, 1), ..., (b_q, 1) \end{matrix} \right. \right].$$

If further $m = 1$ and $p \leq q$, then the H-function is expressible by

$$H_{p,q}^{1,n}\left[z \left| \begin{matrix} (a_1, 1)...(a_p, 1) \\ (b_1, 1)...(b_q, 1) \end{matrix} \right. \right] = \frac{\displaystyle\prod_{j=1}^{n} \Gamma(1 + b_1 - a_j) \, z^{b_1}}{\displaystyle\prod_{j=2}^{q} \Gamma(1 + b_1 - b_j) \prod_{j=n+1}^{p} \Gamma(a_j - b_1)}$$

$$\times {}_pF_{q-1}\left(\begin{matrix} 1 + b_1 - a_1, ..., 1 + b_1 - a_p \\ 1 + b_1 - b_2, ..., 1 + b_1 - b_q \end{matrix} ; (-1)^{p-n-1} z \right)$$

in terms of generalized hypergeometric functions ${}_pF_q$ [104].

Many well-known special functions, such as the error function, Bessel functions, Whittaker functions, Jacobi polynomials, and elliptic functions are included in the class of generalized hypergeometric functions. All of them can be expressed in terms of Fox's H-function. For example, the so-called Maitland's generalized hypergeometric function or the Wright function ${}_p\psi_q$ is

$${}_p\psi_q\left(\begin{matrix} (a_1, A_1), ..., (a_p, A_p) \\ (b_1, B_1), ..., (b_q, B_q) \end{matrix} ; -z \right) \qquad (A.6)$$

$$= H_{p,q+1}^{1,p} \left[z \left| \begin{array}{c} (1 - a_1, A_1), ..., (1 - a_p, A_p) \\ (0,1), (1 - b_q, B_q), ..., (1 - b_q, B_q) \end{array} \right. \right].$$

A special case of the Wright function is the generalized Mittag-Leffler $E_{\alpha,\beta}(z)$ function [149] given by

$$E_{\alpha,\beta}(z) =_1 \psi_1 \left(\begin{array}{c} (1,1) \\ (\beta, \alpha) \end{array} ; z \right) \tag{A.7}$$

$$= H_{1,2}^{1,1} \left[-z \left| \begin{array}{c} (0,1) \\ (0,1), (1 - \beta, \alpha) \end{array} \right. \right] = \sum_{k=0}^{\infty} \frac{z^k}{\Gamma(\alpha k + \beta)}.$$

The exponential function is expressed as

$$\frac{z^{b/B}}{B} \exp(-z^{1/B}) = H_{0,1}^{1,0} \left[z \left| \begin{array}{c} - \\ (b, B) \end{array} \right. \right]. \tag{A.8}$$

To represent an H-function in computable form let us consider the case when the poles $s = (b_j + \nu)/B_j$ $(j = 1, ..., m; \nu = 0, 1, ...)$ of $\prod_{j=1}^{m} {}' \Gamma(b_j - B_j s)$ are simple, that is, where

$$B_h(b_j + \lambda) \neq B_j(b_h + \nu), \qquad j \neq h,$$

$$h = 1, ..., m; \qquad \nu, \lambda = 0, 1, 2, ...,$$

and the prime means the product without the factor $j = h$.

Then we obtain the following expansion for the H-function

$$H_{p,q}^{m,n}(z)$$

$$= \sum_{h=1}^{m} \sum_{k=0}^{\infty} \frac{\prod_{j=1}^{n} \Gamma(1 - a_j + A_j s_{hk}) \prod_{j=1}^{m} {}' \Gamma(b_j - B_j s_{hk})}{\prod_{j=m+1}^{q} \Gamma(1 - b_j + B_j s_{hk}) \prod_{j=n+1}^{p} \Gamma(a_j - A_j s_{hk})} \frac{(-1)^k}{k!} \frac{z^{s_{hk}}}{B_h},$$

$$\tag{A.9}$$

$$s_{hk} = (b_h + k)/B_h$$

which exists for all $z \neq 0$ if $\mu > 0$ and for $0 < |z|, \beta^{-1}$ if $\mu = 0$, where μ and β are given by Eqs. (A.4) and (A.5). The prime means the product without the factor $j = h$.

The formula (A.9) can be used for the calculation of special values of the Fox function and to derive the asymptotic behavior for $z \to 0$.

The asymptotic expansion for $|z| \to \infty$ is treated in [164]. In particular, for $\mu > 0$ and $n \neq 0$

$$H_{p,q}^{m,n}(z) \sim \sum \operatorname{res}(\chi(s)\, z^s),$$

as $|z| \to \infty$ uniformly on every closed subsector of $|\arg z| \leq \frac{1}{2}\pi\lambda$. The residues have to be taken at the points $s = (a_j - 1 - \nu)/\alpha_j$ $(j = 1, ..., n;$ $\nu = 0, 1, ...)$ and λ is defined by

$$\lambda = \sum_{j=1}^{m} B_j + \sum_{j=1}^{n} A_j - \sum_{j=m+1}^{q} B_j - \sum_{j=n+1}^{p} A_j. \tag{A.10}$$

Symmetries in the parameters of the H-function are detected by regarding the definitions (A.1) and (A.2).

A.2 Fundamental properties of the H-function

The H-function introduced by Fox [103], possesses many interesting properties which can be used in time fractional quantum mechanics to calculate integrals and perform transformations to study limiting cases at particular choices of fractality parameters. We present the list of mainly used properties of the H-function. The results of this section follow readily from the definition of the H-function (A.1) and hence no proofs are given here.

Property 12.2.1 The H-function is symmetric in the pairs

$$(a_1, A_1), ..., (a_n, A_n) \quad \text{likewise} \quad (a_{n+1}, A_{n+1}), ..., (a_p, A_p),$$

and

$$(b_1, B_1), ..., (b_n, B_n) \quad \text{likewise} \quad (b_{m+1}, B_{m+1}), ..., (b_q, B_q).$$

Property 12.2.2 If one of the (a_j, A_j) $(j = 1, ..., n)$ is equal to one of the (b_j, B_j) $(j = m + 1, ..., q)$ or one of the (b_j, B_j) $(j = 1, ..., m)$ is equal to one of the (a_j, A_j) $(j = n + 1, ..., p)]$, then the H-function reduces to one of the lower order, and p, q and n (or) m decrease by unity.

Thus, we have the following reduction formula:

$$H_{p,q}^{m,n}\left[z\,\middle|\,\begin{array}{l}(a_1,A_1),...,(a_p,A_p)\\(b_1,B_1),...,(b_{q-1},B_{q-1}),(a_1,A_1)\end{array}\right] \qquad (A.11)$$

$$= H_{p-1,q-1}^{m,n-1}\left[z\,\middle|\,\begin{array}{l}(a_2,A_2),...,(a_p,A_p)\\(b_1,B_1),...,(b_{q-1},B_{q-1})\end{array}\right],$$

provided $n \geq 1$ and $q > m$.

Property 12.2.3

$$H_{p,q}^{m,n}\left[z\,\middle|\,\begin{array}{l}(a_1,A_1),...,(a_p,A_p)\\(b_1,B_1),...,(b_q,B_q)\end{array}\right] \qquad (A.12)$$

$$= H_{q,p}^{n,m}\left[\frac{1}{z}\,\middle|\,\begin{array}{l}(1-b_1,B_1),...,(1-b_q,B_q)\\(1-a_1,A_1),...,(1-a_p,A_p)\end{array}\right].$$

This is an important property of the H-function because it enables us to transform an H-function with $\mu = \sum_{j=1}^{m} B_j - \sum_{j=1}^{n} A_j > 0$ and $\arg x$ to one with $\mu < 0$ and $\arg(1/x)$ and vice versa.

Property 12.2.4

$$\frac{1}{k}H_{p,q}^{m,n}\left[z\,\middle|\,\begin{array}{l}(a_1,A_1),...,(a_p,A_p)\\(b_1,B_1),...,(b_q,B_q)\end{array}\right] \qquad (A.13)$$

$$= H_{p,q}^{m,n}\left[z^k\,\middle|\,\begin{array}{l}(a_1,kA_1),...,(a_p,kA_p)\\(b_1,kB_1),...,(b_q,kB_q)\end{array}\right],$$

where $k > 0$.

Property 12.2.5

$$z^\sigma H_{p,q}^{m,n}\left[z\,\middle|\,\begin{array}{l}(a_1,A_1),...,(a_p,A_p)\\(b_1,B_1),...,(b_q,B_q)\end{array}\right] \qquad (A.14)$$

$$= H_{p,q}^{m,n}\left[z\,\middle|\,\begin{array}{l}(a_1+\sigma A_1,A_1),...,(a_p+\sigma A_p,A_p)\\(b_1+\sigma B_1,B_1),...,(b_q+\sigma B_q,B_q)\end{array}\right].$$

Property 12.2.6

$$H_{p+1,q+1}^{m,n+1}\left[z\,\middle|\,\begin{array}{l}(0,\gamma),...,(a_p,A_p)\\(b_q,B_q),...,(r,\gamma)\end{array}\right] \qquad (A.15)$$

$$= (-1)^r H_{p+1,q+1}^{m+1,n} \left[z \left| \begin{array}{c} (a_p, A_p), ..., (0, \gamma) \\ (r, \gamma), ..., (b_q, B_q) \end{array} \right. \right],$$

where $p \leq q$.

Property 12.2.7

$$H_{p+1,q+1}^{m+1,n} \left[z \left| \begin{array}{c} (a_p, A_p), ..., (1 - r, \gamma) \\ (1, \gamma), ..., (b_q, B_q) \end{array} \right. \right] \tag{A.16}$$

$$= (-1)^r H_{p+1,q+1}^{m+1,n} \left[z \left| \begin{array}{c} (1 - r, \gamma), ..., (a_p, A_p) \\ (b_q, B_q), ..., (1, \gamma) \end{array} \right. \right],$$

where $p \leq q$.

In the above properties given by Eqs. (A.11), (A.15) and (A.16) the branches of the H-function are suitably chosen.

Property 12.2.8

$$\int_0^\infty dx x^\rho e^{-ux} H_{p,q}^{m,n} \left[\omega x^{-r} \left| \begin{array}{c} (a_1, A_1), ..., (a_p, A_p) \\ (b_1, B_1), ..., (b_q, B_q) \end{array} \right. \right] \tag{A.17}$$

$$= u^{-\rho-1} H_{p,q+1}^{m+1,n} \left[\omega u^r \left| \begin{array}{c} (a_1, A_1), ..., (a_p, A_p) \\ (1 + \rho, r), (b_1, B_1), ..., (b_q, B_q) \end{array} \right. \right],$$

where $m \cdot n \neq 0$; $r, p \geq 0$; $a, a^* > 0$; $|\arg \omega| < a^* \pi/2$; $\mathrm{Re}\,\alpha + r$, $\min_{1 \leq j \leq m} \mathrm{Re}(b_j/B_j) > 0$.

Some important identities, needed for fractional quantum and statistical mechanics, are (see [104], page 13)

$$\frac{d^r}{dz^r} \left\{ z^\lambda H_{p,q}^{m,n} \left[\beta z^\delta \left| \begin{array}{c} (a_p, A_p) \\ (b_q, B_q) \end{array} \right. \right] \right\} \tag{A.18}$$

$$= z^{\lambda-r} H_{p+1,q+1}^{m,n+1} \left[\beta z^\delta \left| \begin{array}{c} (-\lambda, \delta), (a_p, A_p) \\ (b_q, B_q), (r - \lambda, \delta) \end{array} \right. \right],$$

and

$${}_0 D_z^\nu \left\{ z^\alpha H_{p,q}^{m,n} \left[(az)^\beta \left| \begin{array}{c} (a_j, A_j) \\ (b_j, B_j) \end{array} \right. \right] \right\} \tag{A.19}$$

$$= z^{\alpha-\nu} H^{m,n+1}_{p+1,q+1} \left[(az)^\beta \; \middle| \; \begin{matrix} (-\alpha,\beta),(a_j,A_j) \\ (b_j,B_j),(\nu-\alpha,\beta) \end{matrix} \right],$$

which hold for arbitrary ν, for a, $b > 0$ and $\alpha + \beta \min(b_j/B_j) > -1$ ($1 \le j \le m$). The $_0 D_z^\nu$ notes fractional derivative of order ν (see for definition [30], [37], [38]).

Further interesting and important properties of Fox's H-function and the expressions for elementary and special functions by the H-function are listed in [104].

A.2.1   Cosine transform of the H-function

The cosine transform of H-function $H^{m,n}_{p,q}\left[at^\alpha \middle| \begin{smallmatrix}(a_p,A_p)\\(b_q,B_q)\end{smallmatrix} \right]$ is defined by [166]

$$\int_0^\infty dt\, t^{\nu-1} \cos(xt) H^{m,n}_{p,q}\left[at^\alpha \; \middle| \; \begin{matrix}(a_p,A_p)\\(b_q,B_q)\end{matrix} \right] \tag{A.20}$$

$$= \frac{\pi}{x^\nu} H^{n+1,m}_{q+1,p+2}\left[\frac{x^\alpha}{a} \; \middle| \; \begin{matrix} (1-b_q,B_q),(\frac{\nu+1}{2},\frac{\alpha}{2}) \\ (\nu,\alpha),(1-a_p,A_p),(\frac{\nu+1}{2},\frac{\alpha}{2}) \end{matrix} \right],$$

where $\mathrm{Re}[\nu + \alpha \min_{1\le j \le m}(\frac{b_j}{B_j})] > 0$, $\mathrm{Re}[\nu + \alpha \max_{1 \le j \le n}(\frac{a_j-1}{A_j})] < 0$, $|\arg a| < \pi\lambda/2$, with λ defined by Eq. (A.10).

A.2.2   Some functions expressed in terms of H-function

Exponential function is

$$\exp\left\{ -i\frac{D_\alpha |p|^\alpha t}{\hbar} \right\} = H^{1,0}_{0,1}\left[i\frac{D_\alpha t}{\hbar}|p|^\alpha \; \middle| \; \begin{matrix} - \\ (0,1) \end{matrix} \right]. \tag{A.21}$$

In the notations of [150], the Wright function $\phi(a,\beta;z)$ is expressed as

$$\phi(a,\beta;z) = \sum_{k=0}^\infty \frac{z^k}{k!\Gamma(ak+\beta))}, \qquad -1 < a < 0, \qquad \beta \in \mathbb{C}, \tag{A.22}$$

where $\mathbb{C}$ stands for the field of complex numbers.

In terms of Fox's H-function the function $\phi(a,\beta;z)$ has the following representation

$$\phi(a,\beta;z) = H^{1,0}_{0,2}\left[-z \; \middle| \; \begin{matrix} - \\ (0,1),(1-\beta,\alpha) \end{matrix} \right], \tag{A.23}$$

$$-1 < a < 0, \qquad \beta \ge 0.$$

A.2.3   Lévy α-stable distribution in terms of H-function

As an example of application of the cosine transform of H-function let us present the $K^{(0)}(x, \tau)$ defined by Eq. (7.3) in terms of $H_{2,2}^{1,1}$-function. With the help of Eq. (A.21) we have

$$K^{(0)}(x, \tau) = \frac{1}{2\pi\hbar} \int_{-\infty}^{\infty} dp \exp\left\{ i\frac{px}{\hbar} - i\frac{D_\alpha |p|^\alpha \tau}{\hbar} \right\} \qquad (A.24)$$

$$= \frac{1}{\pi\hbar} \int_{0}^{\infty} dp \cos(i\frac{px}{\hbar}) H_{0,1}^{1,0} \left[i\frac{D_\alpha t}{\hbar} p^\alpha \left| \begin{matrix} - \\ (0,1) \end{matrix} \right. \right].$$

Then using Eq. (A.20) gives us

$$K^{(0)}(x, \tau) = \frac{1}{\pi\hbar} \int_{0}^{\infty} dp \cos(i\frac{px}{\hbar}) H_{0,1}^{1,0} \left[i\frac{D_\alpha t}{\hbar} p^\alpha \left| \begin{matrix} - \\ (0,1) \end{matrix} \right. \right] \qquad (A.25)$$

$$= \frac{1}{x} H_{2,2}^{1,1} \left[\frac{1}{\hbar^\alpha} \left(\frac{\hbar}{iD_\alpha \tau} \right) x^\alpha \left| \begin{matrix} (1,1),(1,\alpha/2) \\ (1,\alpha),(1,\alpha/2) \end{matrix} \right. \right], \qquad x \geq 0,$$

or, for any x

$$K^{(0)}(x, \tau) = \frac{1}{|x|} H_{2,2}^{1,1} \left[\frac{1}{\hbar^\alpha} \left(\frac{\hbar}{iD_\alpha \tau} \right) |x|^\alpha \left| \begin{matrix} (1,1),(1,\alpha/2) \\ (1,\alpha),(1,\alpha/2) \end{matrix} \right. \right]. \qquad (A.26)$$

Applying Fox H-function Property 12.2.4 given by Eq. (A.13), we can write $K^{(0)}(x, \tau)$ as

$$K^{(0)}(x, \tau) = \frac{1}{\alpha|x|} H_{2,2}^{1,1} \left[\frac{1}{\hbar} \left(\frac{\hbar}{iD_\alpha \tau} \right)^{1/\alpha} |x| \left| \begin{matrix} (1,1/\alpha),(1,1/2) \\ (1,1),(1,1/2) \end{matrix} \right. \right],$$

which is a free particle quantum kernel representation given by Eq. (7.23).

In terms of Lévy probability distribution function $L_\alpha(z)$ defined by Eq. (6.15) a free particle quantum kernel $K^{(0)}(x, \tau)$ is expressed as (see Eq. (6.26))

$$K^{(0)}(x, \tau) = \frac{1}{2\pi\hbar} \int_{-\infty}^{\infty} dp \exp\left\{ i\frac{px}{\hbar} - i\frac{D_\alpha |p|^\alpha \tau}{\hbar} \right\} \qquad (A.27)$$

$$= \frac{1}{\hbar} \left(\frac{iD_\alpha \tau}{\hbar} \right)^{-1/\alpha} L_\alpha \left\{ \frac{1}{\hbar} \left(\frac{\hbar}{iD_\alpha \tau} \right)^{1/\alpha} |x| \right\}.$$

By comparing Eqs. (A.26) and (A.27) we obtain the expression of Lévy probability distribution function L_α in terms of Fox's $H_{2,2}^{1,1}$-function

$$\frac{1}{\hbar} \left(\frac{iD_\alpha \tau}{\hbar} \right)^{-1/\alpha} L_\alpha \left\{ \frac{1}{\hbar} \left(\frac{\hbar}{iD_\alpha \tau} \right)^{1/\alpha} |x| \right\} \qquad (A.28)$$

$$= \frac{1}{\alpha |x|} H_{2,2}^{1,1} \left[\frac{1}{\hbar} \left(\frac{\hbar}{iD_\alpha \tau} \right)^{1/\alpha} |x| \, \Bigg| \, \begin{matrix} (1, 1/\alpha), (1, 1/2) \\ (1, 1), (1, 1/2) \end{matrix} \right].$$

Appendix B

Fractional Calculus

B.1 Brief introductory remarks

Most of mathematical theory applicable to the study of fractional calculus was developed prior to the turn of the 20th century. However, in the past 100 years numerous applications and physical manifestations of fractional calculus have been found. In some cases, the mathematics has had to change to meet the requirements of physical reality.

The concept for fractional integrals and derivatives was a natural outgrowth of integer order integrals and derivatives in the same way as the fractional exponent follows from the more traditional integer order exponent. While one cannot imagine the multiplication of a quantity a fractional number of times, there seems no practical restriction to placing a non-integer into the exponential position. Similarly, the common formulation for the fractional integral can be derived directly from a traditional expression of the repeated integration of a function.

B.1.1 Definition of the Riemann–Liouville fractional derivative

To introduce fractional derivative we begin with the n-fold (n is an integer) iterated integral of a function $\psi(x)$

$$_cI_x^n\psi(x) = \frac{1}{(n-1)!}\int\limits_c^x dy(x-y)^{n-1}\psi(y)$$

or

$$_cI_x^n\psi(x) = \frac{1}{\Gamma(n)}\int\limits_c^x dy(x-y)^{n-1}\psi(y), \tag{B.1}$$

313

if we take into account that the Gamma function $\Gamma(n)$ for integer n is $\Gamma(n) = (n-1)!$. We will call $_cI_x^n f(x)$ defined by Eq. (B.1) as integral of order n.

The definition of the Riemann–Liouville fractional integral can be derived as generalization of the above equations to any fractional order $\alpha > 0$

$$_cI_x^\alpha \psi(x) = \frac{1}{\Gamma(\alpha)} \int_c^x dy (x-y)^{\alpha-1} \psi(y),$$

$$_xI_c^\alpha \psi(x) = \frac{1}{\Gamma(\alpha)} \int_x^c dy (y-x)^{\alpha-1} \psi(y),$$

where the right-hand sided and left-hand sided fractional integrals have been introduced.

A fractional derivative $_cD_x^\alpha$ of a function $f(x)$ of order α can be defined as the n-th ordinary derivative of the Riemann–Liouville fractional integral of order $n - \alpha$, where $n = [\alpha] + 1$ and $[\alpha]$ is the integer part of α. That is

$$_cD_x^\alpha \psi(x) = \frac{d^n}{dx^n} (_cI_x^\alpha \psi(x)) = \frac{1}{\Gamma(n-\alpha)} \frac{d^n}{dx^n} \int_c^x dy (x-y)^{n-\alpha-1} \psi(y), \quad \text{(B.2)}$$

and

$$_xD_c^\alpha \psi(x) = \frac{d^n}{dx^n} (_xI_c^\alpha \psi(x)) = \frac{1}{\Gamma(n-\alpha)} \frac{d^n}{dx^n} \int_x^c dy (y-x)^{n-\alpha-1} \psi(y). \quad \text{(B.3)}$$

Further, we introduce fractional derivatives $_{-\infty}D_x^\alpha$ and $_xD_\infty^\alpha$ by choosing $c = -\infty$ and $b = \infty$.

$$_{-\infty}D_x^\alpha \psi(x) = \frac{1}{\Gamma(n-\alpha)} \frac{d^n}{dx^n} \int_{-\infty}^x dy (x-y)^{n-\alpha-1} \psi(y), \quad \text{(B.4)}$$

and

$$_xD_\infty^\alpha \psi(x) = \frac{1}{\Gamma(n-\alpha)} \frac{d^n}{dx^n} \int_x^\infty dy (y-x)^{n-\alpha-1} \psi(y), \quad \text{(B.5)}$$

with $n - 1 \leq \alpha < n$.

B.1.2 Definition of Caputo fractional derivative

The Caputo fractional derivative ∂_t^β is defined by

$$\partial_t^\beta f(t) = \frac{1}{\Gamma(n-\beta)} \int\limits_0^t d\tau \, \frac{f^{(n)}(\tau)}{(t-\tau)^{\beta+1-n}}, \qquad n-1 < \beta < n, \qquad (B.6)$$

where $f^{(n)}(\tau)$ is defined as

$$f^{(n)}(\tau) = \frac{d^n}{d\tau^n} f(\tau), \qquad \beta = n. \qquad (B.7)$$

It was invented by Caputo in [143], see also [39].

B.1.3 Definition of quantum Riesz fractional derivative

To understand the quantum Riesz fractional derivative given by Eq. (3.9) let us consider the one-dimensional case. We define operator $(\hbar\nabla)^\alpha$ in terms of introduced by Eqs. (B.4) and (B.5) operators $_{-\infty}D_x^\alpha$ and $_xD_\infty^\alpha$ as [37]

$$(\hbar\nabla)^\alpha = -\frac{\hbar^\alpha}{2\cos(\pi\alpha/2)} \{_{-\infty}D_x^\alpha +_x D_\infty^\alpha\}, \qquad 1 < \alpha \le 2. \qquad (B.8)$$

Substituting Eqs. (B.4) and (B.5) with $n = 2$ into Eq. (B.8) yields

$$(\hbar\nabla)^\alpha \psi(x,t) = -\frac{\hbar^\alpha}{2\cos(\pi\alpha/2)} \frac{1}{\Gamma(2-\alpha)} \left\{ \frac{d^2}{dx^2} \int\limits_{-\infty}^x dy(x-y)^{1-\alpha}\psi(y) \right.$$

$$\left. + \frac{d^2}{dx^2} \int\limits_x^\infty dy(y-x)^{1-\alpha}\psi(y) \right\}, \qquad 1 < \alpha \le 2, \qquad (B.9)$$

where $\Gamma(2-\alpha)$ is the Gamma function.

It follows from definition given by Eq. (B.9) that the quantum Riesz operator $(\hbar\nabla_x)^\alpha$ has the convenient property

$$\int\limits_{-\infty}^{\infty} dx e^{-i\frac{px}{\hbar}} (\hbar\nabla)^\alpha \psi(x,t) = -|p|^\alpha \varphi(p,t) \qquad (B.10)$$

or

$$(\hbar\nabla)^\alpha \psi(x,t) = -\frac{1}{2\pi\hbar} \int\limits_{-\infty}^{\infty} dp e^{i\frac{px}{\hbar}} |p|^\alpha \varphi(p,t), \qquad (B.11)$$

where wave function in coordinate space $\psi(x,t)$ and wave function in the momentum representation $\varphi(p,t)$ are related to each other by the Fourier transforms

$$\psi(x,t) = \frac{1}{2\pi\hbar} \int\limits_{-\infty}^{\infty} dp e^{i\frac{px}{\hbar}} \varphi(p,t) \qquad (B.12)$$

and

$$\varphi(p,t) = \int\limits_{-\infty}^{\infty} dx e^{-i\frac{px}{\hbar}} \psi(x,t). \qquad (B.13)$$

In other words, the Fourier transform of quantum Riesz fractional derivative of order α is equivalent to a multiplication of the wave function in momentum representation $\varphi(p,t)$ by $|p|^\alpha$. We use Eq. (B.11) as the definition of the quantum Riesz fractional derivative of order α.

To generalize the Riesz fractional derivative of order α to arbitrary space dimension D we use Eqs. (B.10) and (B.11). Let $\psi(\mathbf{r},t)$ be the wave function on R^D, $\mathbf{r} \in R^D$ and $\varphi(\mathbf{p},t)$ its Fourier transform on P^D, $\mathbf{p} \in P^D$. The function $\varphi(\mathbf{p},t)$ is the wave function in momentum representation. The operator $(\Delta)^{\alpha/2}$ is defined by

$$(\Delta)^{\alpha/2} = (\partial^2/\partial r_1^2 + \partial^2/\partial r_2^2 + \cdots + \partial^2/\partial r_D^2)^{\alpha/2}. \qquad (B.14)$$

Then D-dimensional quantum Riesz fractional derivative is

$$\int d^D r e^{-i\frac{\mathbf{p}\mathbf{r}}{\hbar}} (-\hbar^2\Delta)^{\alpha/2} \psi(\mathbf{r},t) = |\mathbf{p}|^\alpha \varphi(\mathbf{p},t) \qquad (B.15)$$

or

$$(-\hbar^2\Delta)^{\alpha/2}\psi(\mathbf{r},t) = \frac{1}{(2\pi\hbar)^D} \int d^D p\, e^{i\frac{\mathbf{pr}}{\hbar}} |\mathbf{p}|^\alpha \varphi(\mathbf{p},t), \qquad (\text{B.16})$$

where wave function in coordinate D-dimensional space $\psi(\mathbf{r},t)$ and wave function in D-dimensional momentum representation $\varphi(\mathbf{p},t)$ are related to each other by the Fourier transforms

$$\psi(\mathbf{r},t) = \frac{1}{(2\pi\hbar)^D} \int d^D p\, e^{i\frac{\mathbf{pr}}{\hbar}} \varphi(\mathbf{p},t) \qquad (\text{B.17})$$

and

$$\varphi(\mathbf{p},t) = \int d^D r\, e^{-i\frac{\mathbf{pr}}{\hbar}} \psi(\mathbf{r},t). \qquad (\text{B.18})$$

Appendix C

Calculation of the Integral

To calculate the integral in Eq. (10.74) we follow by [116] and write

$$\int_{-\infty}^{\infty} dp \frac{e^{ipx/\hbar}}{|p|^\alpha + \lambda^\alpha} = 2\pi\lambda^{1-\alpha} I(\lambda x/\hbar), \qquad \lambda > 0, \qquad \text{(C.1)}$$

where the following notation has been introduced

$$I(w) = \frac{1}{\pi} \int_0^{\infty} dy \frac{\cos wy}{|y|^\alpha + 1}. \qquad \text{(C.2)}$$

To calculate the integral $I(w)$ we apply the calculation techniques based on the Mellin transform as it was first done by de Oliveira $et\ al.$ [116].

The Mellin $\mathcal{M}[f(x)](z)$ and inverse Mellin transforms $\mathcal{M}^{-1}[f(x)]$ are defined by

$$\mathcal{M}[f(x)](z) = \int_0^{\infty} dx x^{z-1} f(x), \qquad \text{(C.3)}$$

$$\mathcal{M}^{-1}[F(x)](z) = \frac{1}{2\pi i} \int_{c-i\infty}^{c+i\infty} dz x^{-z} F(z). \qquad \text{(C.4)}$$

Since the Mellin transform of $I(w)$ takes only those positive values of w, and since $I(-w) = I(w)$, we only need to replace w by $|w|$ at the end of the calculation for the result to be valid for all w. To obtain the Mellin transform of $I(w)$ one can use the equation [116], [167]

$$\mathcal{M}_w[\cos(wy)](z) = y^{-z}\Gamma(z)\cos\frac{\pi z}{2}. \tag{C.5}$$

Then the Mellin transform of $I(w)$ reads

$$\mathcal{M}_w[I(w)](z) = \frac{1}{\pi}\Gamma(z)\cos\frac{\pi z}{2}\int\limits_0^\infty dy\frac{y^{-z}}{y^\alpha+1}. \tag{C.6}$$

To calculate the integral $\int\limits_{-\infty}^\infty dy y^{-z}/(y^\alpha+1)$ we use formula (see, formula 3.241.2, p. 322 in [86])

$$\int\limits_0^\infty \frac{x^{\mu-1}dx}{1+x^\nu} = \frac{\pi}{\nu}\operatorname{cosec}\frac{\pi\mu}{\nu} = \frac{1}{\nu}B(\frac{\mu}{\nu},\frac{\nu-\mu}{\nu}), \qquad [\operatorname{Re}\nu > \operatorname{Re}\mu > 0], \tag{C.7}$$

where $B(\mu/\nu,(\nu-\mu)/\nu)$ is Beta function[1].
Then we have

$$\int\limits_0^\infty dy\frac{y^{-z}}{y^\alpha+1} = \frac{1}{\alpha}B(\frac{1-z}{\alpha},1-\frac{1-z}{\alpha}). \tag{C.8}$$

Using this result and the identity

$$\cos(\frac{\pi z}{2}) = \sin(\frac{\pi(1-z)}{2}) = \frac{\pi}{\Gamma(\frac{1-z}{2})\Gamma(1-\frac{1-z}{2})}, \tag{C.9}$$

we express $\mathcal{M}_w[I(w)](z)$ given by Eq. (C.6) in the form

$$\mathcal{M}_w[I(w)](z) = \frac{1}{\alpha}\frac{\Gamma(\frac{1-z}{\alpha})\Gamma(1-\frac{1-z}{\alpha})}{\Gamma(\frac{1-z}{2})\Gamma(1-\frac{1-z}{2})}. \tag{C.10}$$

Thus, $I(w)$ is given by the inverse Mellin transform

$$I(w) = \frac{1}{2\pi i\alpha}\int\limits_{c-i\infty}^{c+i\infty} dz w^{-z}\frac{\Gamma(\frac{1-z}{\alpha})\Gamma(1-\frac{1-z}{\alpha})}{\Gamma(\frac{1-z}{2})\Gamma(1-\frac{1-z}{2})}. \tag{C.11}$$

---

[1]The Beta function has familiar representation in terms of the Gamma function

$$B(a,b) = \frac{\Gamma(a)\Gamma(b)}{\Gamma(a+b)}, \qquad \operatorname{Re}a > 0, \quad \operatorname{Re}b > 0.$$

By comparing this expression with Eqs. (A.1) and (A.2) we see that $I(w)$ is expressed in terms of Fox's $H_{2,3}^{2,1}$-function in the following way,

$$I(w) = \frac{1}{\alpha} H_{2,3}^{2,1} \left[w \left| \begin{array}{l} (1 - 1/\alpha, 1/\alpha), (1/2, 1/2) \\ (0, 1), (1 - 1/\alpha, 1/\alpha), (1/2, 1/2) \end{array} \right. \right]. \qquad (C.12)$$

By applying the H-function Properties 12.2.4 and 12.2.5 given by Eqs. (A.13) and (A.14) respectively, we can rewrite the above equation as

$$I(w) = \frac{1}{|w|} H_{2,3}^{2,1} \left[|w|^{\alpha} \left| \begin{array}{l} (1, 1), (1, \alpha/2) \\ (1, \alpha), (1, 1), (1, \alpha/2) \end{array} \right. \right], \qquad (C.13)$$

where we replaced w by $|w|$ in order for this expression to hold also for the negative values of w since $I(-w) = I(w)$.

Having $I(w)$ given by Eq. (C.13), we can present Eq. (C.1) in the form [116]

$$\int_{-\infty}^{\infty} dp \frac{e^{ipx/\hbar}}{|p|^{\alpha} + \lambda^{\alpha}} = \frac{2\pi\hbar}{\lambda^{\alpha}|x|} H_{2,3}^{2,1} \left[(\frac{\lambda}{\hbar})^{\alpha}|x|^{\alpha} \left| \begin{array}{l} (1, 1), (1, \alpha/2) \\ (1, \alpha), (1, 1), (1, \alpha/2) \end{array} \right. \right], \qquad (C.14)$$

with $\lambda > 0$.

Let us note that when $\alpha = 2$ Eq. (C.13) becomes

$$I(w)|_{\alpha=2} = \frac{1}{|w|} H_{2,3}^{2,1} \left[|w|^{2} \left| \begin{array}{l} (1, 1), (1, 1) \\ (1, 2), (1, 1), (1, 1) \end{array} \right. \right] \qquad (C.15)$$

$$= \frac{1}{|w|} H_{0,1}^{1,0} \left[|w|^{2} \left| \begin{array}{l} - \\ (1, 2) \end{array} \right. \right] = \frac{|w|}{2} \exp(-|w|),$$

due to the H-function Property 12.2.2 given by Eq. (A.11). Therefore, it follows from Eqs. (C.14) and (C.15) that

$$\int_{-\infty}^{\infty} dp \frac{e^{ipx/\hbar}}{|p|^{2} + \lambda^{2}} = \frac{\pi}{\lambda} \exp(-\lambda|x|/\hbar). \qquad (C.16)$$

Appendix D

Polylogarithm

D.1 Polylogarithm as a power series

The polylogarithm $\mathrm{Li}_s(z)$ is defined by [168] - [170]

$$\mathrm{Li}_s(z) = \sum_{n=1}^{\infty} \frac{z^n}{n^s} = \frac{z}{\Gamma(s)} \int_0^{\infty} dt \frac{t^{s-1}}{e^t - z}, \qquad (\mathrm{D.1})$$

here s is real parameter and z is the complex argument. The name *polylogarithm* comes from the fact that the function $\mathrm{Li}_s(z)$ may be introduced as the repeated integral of itself,

$$\mathrm{Li}_{s+1}(z) = \int_0^z dt \frac{\mathrm{Li}_s(t)}{t}. \qquad (\mathrm{D.2})$$

It is easy to see that for $z = 1$ the polylogarithm $\mathrm{Li}_s(1)$ reduces to the well-known Riemann zeta function

$$\mathrm{Li}_s(1) = \zeta(s) = \sum_{n=1}^{\infty} \frac{1}{n^s}, \qquad \mathrm{Re}(s) > 1. \qquad (\mathrm{D.3})$$

The quantum statistical mechanics is the best known field where the polylogarithm arises in natural way. Indeed, if we note that the Bose–Einstein distribution function $BE(t)$ is given by

$$BE(t) = \frac{t^{s-1}}{e^{t-\mu} - 1}, \qquad (\mathrm{D.4})$$

and the Fermi–Dirac distribution function $FD(t)$ is given by

$$FD(t) = \frac{t^{s-1}}{e^{t-\mu} + 1},\tag{D.5}$$

then the integrals of the Bose–Einstein distribution function and the Fermi-Dirac distribution function are respectively

$$\int\limits_0^\infty dt\, BE(t) = \int\limits_0^\infty dt\, \frac{t^{s-1}}{e^{t-\mu} - 1} = \Gamma(s)\mathrm{Li}_s(e^\mu)\tag{D.6}$$

and

$$\int\limits_0^\infty dt\, FD(t) = \int\limits_0^\infty dt\, \frac{t^{s-1}}{e^{t-\mu} + 1} = -\Gamma(s)\mathrm{Li}_s(-e^\mu),\tag{D.7}$$

where $\Gamma(s)$ is the Gamma function.

For our purposes we are looking for power series representation (about $\mu = 0$) of the polylogarithm $\mathrm{Li}_s(e^{-u})$

$$\mathrm{Li}_s(e^{-\mu}) = \frac{1}{\Gamma(s)} \int\limits_0^\infty dt\, \frac{t^{s-1}}{e^{t+\mu} - 1}.\tag{D.8}$$

The power series representation of the polylogarithm $\mathrm{Li}_s(e^{-\mu})$ can be found by using the Mellin transform, see, [171],

$$M_s(r) = \int\limits_0^\infty du\, u^{r-1}\mathrm{Li}_s(e^{-u}) = \int\limits_0^\infty du\, u^{r-1} \sum_{n=1}^\infty \frac{e^{-nu}}{n^s} = \Gamma(r)\zeta(s+r),\tag{D.9}$$

where $\zeta(s+r)$ is the Riemann zeta function defined by Eq. (D.3).

The inverse Mellin transform then gives

$$\mathrm{Li}_s(e^{-u}) = \frac{1}{2\pi i} \int\limits_{c-i\infty}^{c+i\infty} dr\, u^{-r} M_s(r)\tag{D.10}$$

$$= \frac{1}{2\pi i} \int\limits_{c-i\infty}^{c+i\infty} dr\, u^{-r} \Gamma(r)\zeta(s+r), \qquad r > 0,$$

where c is a constant to the right of the poles of the integrand. The path of integration may be converted into a closed contour, and the simple poles of the integrand are those of of the Riemann zeta function $\zeta(s + r)$ at $r = 1 - s$ with residue $+1$ and the Gamma function $\Gamma(r)$ at $r = -l$ with residues $(-1)^l/l!$, here $l = 0, -1, -2, ...$

Summing the residues yields, for $|\mu| < 2\pi$ and $s \neq 1, 2, 3, ...$

$$\text{Li}_s(e^{-\mu}) = \Gamma(1 - s)\mu^{s-1} + \sum_{l=0}^{\infty} \frac{\zeta(s - l)}{l!}(-\mu)^l. \qquad (D.11)$$

This equation gives us the power series representation of the polylogarithm $\text{Li}_s(e^{-\mu})$ (about $\mu = 0$).

If the parameter s is a positive integer n, both the Gamma function $\Gamma(1 - s)$ and the $l = n - 1$ term become infinite, although their sum remains finite [171].

For integer $l > 0$ we have

$$\lim_{s \to l+1} \left[\Gamma(1 - s)(\mu)^{s-1} + \frac{\zeta(s - l)}{l!}(-\mu)^l \right] \qquad (D.12)$$

$$= \frac{(-\mu)^l}{l!} \left(\sum_{m=1}^{l} \frac{1}{m} - \ln(\mu) \right),$$

and for $l = 0$

$$\lim_{s \to 1} \left[\Gamma(1 - s)(\mu)^{s-1} + \zeta(s) \right] = -\ln(\mu). \qquad (D.13)$$

D.2 Properties of function $\mathcal{V}(k)$

It is easy to see that $\mathcal{J}(k)$ given by Eq. (11.41) with $\mathcal{J}_n$ defined by Eq. (11.7) can be expressed in terms of the polylogarithm

$$\mathcal{J}(k) = 2\mathcal{J} \sum_{n=1}^{\infty} \frac{\cos(kna)}{n^s} = 2\mathcal{J} \, \text{Re}\{\text{Li}_s(e^{-ika})\}. \qquad (D.14)$$

Then, the function $\mathcal{V}(k)$ defined by Eq. (11.40) is written as

$$\mathcal{V}(k) = 2\mathcal{J}\zeta(s) \, \text{Re}\left\{ 1 - \frac{\text{Li}_s(e^{-ika})}{\zeta(s)} \right\}, \qquad (D.15)$$

where $\zeta(s)$ is the Riemann zeta function given by Eq. (D.3).

Taking into account power series representation given by Eq. (D.11), we find from Eq. (D.15) that in the limit $a \to 0$ function $\mathcal{V}(k)$ has the following behavior, depending on value of the parameter s, [133]

$$\mathcal{V}(k) \sim D_s|ka|^{s-1}, \qquad 2 \leq s < 3, \qquad (\text{D.16})$$

$$\mathcal{V}(k) \sim -J(ka)^2 \ln ka, \qquad s = 3, \qquad (\text{D.17})$$

$$\mathcal{V}(k) \sim \frac{J\zeta(s-2)}{2}(ka)^2, \qquad s > 3, \qquad (\text{D.18})$$

where $\zeta(s)$ is the Riemann zeta function and the coefficient D_s is defined by

$$D_s = \frac{\pi J}{\Gamma(s)\sin(\pi(s-1)/2)}, \qquad (\text{D.19})$$

with $\Gamma(s)$ being the Gamma function.

Appendix E

Fractional Generalization of Trigonometric Functions cos z and sin z

Following [146] we introduce two new functions $Ec_\beta(z)$ and $Es_\beta(z)$ defined by the series

$$Ec_\beta(z) = \sum_{m=0}^{\infty} \frac{(-1)^{m\beta} z^{2m}}{\Gamma(2\beta m + 1)} \tag{E.1}$$

and

$$Es_\beta(z) = \sum_{m=0}^{\infty} \frac{(-1)^{m\beta} z^{(2m+1)}}{\Gamma(\beta(2m + 1) + 1)}. \tag{E.2}$$

It is easy to see that the following two identities hold for functions $Ec_\beta(z)$ and $Es_\beta(z)$

$$Ec_\beta(z) = \frac{i^\beta E_\beta(-i^\beta z) - (-i)^\beta E_\beta(i^\beta z)}{i^\beta - (-i)^\beta}, \tag{E.3}$$

and

$$Es_\beta(z) = \frac{E_\beta(i^\beta z) - E_\beta(-i^\beta z)}{i^\beta - (-i)^\beta}, \tag{E.4}$$

where $E_\beta(z)$ is the Mittag-Leffler function given by Eq. (12.61) and i is imaginary unit, $i = \sqrt{-1}$.

Hence, functions $Ec_\beta(z)$ and $Es_\beta(z)$ can be considered as a natural fractional generalization of the well-known trigonometric functions $\cos(z)$, and $\sin(z)$ respectively. Indeed, when $\beta = 1$ $Ec_\beta(z)$ and $Es_\beta(z)$ become

$$Ec_\beta(z)|_{\beta=1} = Ec_1(z) = \sum_{m=0}^{\infty} \frac{(-1)^m z^{2m}}{\Gamma(2m + 1)} \tag{E.5}$$

$$= \sum_{m=0}^{\infty} \frac{(-1)^m z^{2m}}{(2m)!} = \cos z,$$

and

$$Es_{\beta}(z)|_{\beta=1} = Es_1(z) = \sum_{m=0}^{\infty} \frac{(-1)^m z^{(2m+1)}}{\Gamma((2m+1)+1)} \tag{E.6}$$

$$= \sum_{m=0}^{\infty} \frac{(-1)^m z^{(2m+1)}}{(2m+1)!} = \sin z.$$

The new expression for the Mittag-Leffler function $E_{\beta}(i^{\beta}z)$ given by Eq. (12.66) can be considered as a fractional generalization [146] of the celebrated Euler's formula, which is recovered from Eq. (12.66) in the limit case $\beta = 1$,

$$e^{iz} = \cos z + i \sin z. \tag{E.7}$$

Two new expressions (E.3) and (E.4) are a fractional generalization of the well-known equations

$$\cos z = \frac{1}{2}(e^{iz} + e^{-iz}), \tag{E.8}$$

and

$$\sin z = \frac{1}{2i}(e^{iz} - e^{-iz}). \tag{E.9}$$

Let us note that Eq. (12.66) can be further generalized to

$$E_{\beta,\gamma}(i^{\sigma}z) = Ec_{\beta,\gamma}^{\sigma}(z) + i^{\sigma}Es_{\beta,\gamma}^{\sigma}(z), \tag{E.10}$$

$$0 < \beta \le 1, \qquad 0 < \gamma \le 1, \qquad 0 < \sigma \le 1,$$

where two indices Mittag-Leffler function $E_{\beta,\gamma}(z)$ is defined by

$$E_{\beta,\gamma}(z) = \sum_{m=0}^{\infty} \frac{z^m}{\Gamma(\beta m + \gamma)}, \tag{E.11}$$

and two new functions $Ec^\sigma_{\beta,\gamma}(z)$ and $Es^\sigma_{\beta,\gamma}(z)$ are introduced as

$$Ec^\sigma_{\beta,\gamma}(z) = \sum_{m=0}^{\infty} \frac{(-1)^{m\sigma} z^{2m}}{\Gamma(2\beta m + \gamma)} \tag{E.12}$$

and

$$Es^\sigma_{\beta,\gamma}(z) = \sum_{m=0}^{\infty} \frac{(-1)^{m\sigma} z^{(2m+1)}}{\Gamma(\beta(2m+1) + \gamma)}, \tag{E.13}$$

respectively.

In terms of new generalized functions $Ec^\sigma_{\beta,\gamma}(z)$ and $Es^\sigma_{\beta,\gamma}(z)$, the functions $Ec_\beta(z)$ and $Es_\beta(z)$ defined by Eqs. (E.1) and (E.2) are given by

$$Ec_\beta(z) = Ec^\sigma_{\beta,\gamma}(z)|_{\sigma=\beta,\gamma=1}, \qquad Es_\beta(z) = Es^\sigma_{\beta,\gamma}(z)|_{\sigma=\beta,\gamma=1}. \tag{E.14}$$

Bibliography

1. W. Heisenberg, *Z. Phys.* **33**, 879 (1925).
2. M. Born, W. Heisenberg and P. Jordan, *Z. Phys.* **36**, 557 (1926).
3. W. Heisenberg, *The Physical Principles of Quantum Theory* (University of Chicago Press, 1930).
4. E. Schrödinger, *Ann. Phys.* NY, **79**, 361 (1926).
5. E. Schrödinger, *Abhandlungen zur Wellenmechanik* (Johann Ambrosius Barth Verlag, 1926).
6. E. Schrödinger, *Collected Papers on Wave Mechanics* (AMS Chelsea Publishing, 2003).
7. R. P. Feynman, *Rev. Mod. Phys.* **20**, 367 (1948); *Phys. Rev.* **84**, 108 (1951).
8. L. S. Schulman, *Techniques and Applications of Path Integration* (Wiley-Interscience, 1981).
9. H. Kleinert, *Path Integrals in Quantum Mechanics, Statistics and Polymer Physics* (World Scientific, 1990).
10. P. A. M. Dirac, *Physikalische Zeitschrift der Sowjetunion* Band 3, Heft 1, 64 (1933).
11. R. P. Feynman, *The development of the space-time view of quantum electrodynamics*, Nobel Lecture, December 11, 1965 (http://www.feynmanlectures.info/other/Feynmans_Nobel_Lecture.pdf).
12. R. P. Feynman and A. R. Hibbs, *Quantum Mechanics and Path Integrals* (McGraw-Hill, 1965).
13. A. A. Slavnov and L. D. Faddeev, *Introduction to Quantum Gauge Field Theory* (Nauka, 1978).
14. J. Glimm and A. Jaffe, *Quantum Physics: A Functional Integral Point of View*, 2nd ed. (Springer, 1987).
15. A. M. Polyakov, *Gauge Field and Strings* (Harwood, 1987).
16. C. Itzykson and J.-B. Zuber, *Quantum Field Theory* (McGraw-Hill, 1985).
17. F. A. Berezin, *The Method of Second Quantization* (Academic Press, 1966).
18. A. Beraudo, J. P. Blaizot, P. Faccioli and G. Garberoglio, *Nucl. Phys. A* **846**, 104 (2010).
19. R. P. Feynman, *Statistical Mechanics* (Benjamin, 1972).

20. H. Kleinert, *Path Integrals in Quantum Mechanics, Statistics, Polymer Physics, and Financial Markets* (World Scientific, 2006), Chap. 2.
21. J. R. Klauder and E. C. G. Sudarshan, *Fundamentals of Quantum Optics* (Benjamin, 1968).
22. M. Lax, *Fluctuation and Coherence Phenomena in Classical and Quantum Physics* (Gordon and Breach, 1968).
23. M. Kac, *On Some Connection between Probability Theory and Differential and Integral Equations*, Proc. *2nd Berkeley Sympos. Math. Stat. and Prob.* (University of California Press, 1951), pp. 189-215. Reprinted in Mark Kac: *Probability, Number Theory, and Statistical Physics – Selected Papers*, Eds. K. Baclawski and M.D. Donsker, (The MIT Press, 1979).
24. M. Kac, *Probability and Related Topics in Physical Sciences* (Interscience, 1959), Chap. IV.
25. F. W. Wiegel, *Introduction to Path-Integral Methods in Physics and Polymer Science* (World Scientific, 1986).
26. L. S. Schulman, *Techniques and Applications of Path Integration* (John Wiley & Sons, 1981).
27. B. B. Mandelbrot, *The Fractal Geometry of Nature* (W. H. Freeman, 1982).
28. J. Feder, *Fractals* (Plenum Press, 1988).
29. E. W. Montroll and M. F. Shlezinger, *The Wonderful World of Random Walks in Nonequalibrium Phenomena II. From Stochastics to Hydrodynamics* (North-Holland, 1984).
30. K. B. Oldham and J. Spanier, *The Fractional Calculus* (Academic Press, 1974).
31. R. Metzler and J. Klafter, *Phys. Rep.* **339**, 1 (2000).
32. J. Klafter, S. C. Lim and R. Metzler, *Fractional Dynamics* (World Scientific, 2011).
33. G. W. Leibniz, *Letter from Hanover, Germany to G.F.A. L'Hospital, September 30, 1695, Leibnizens Mathematische Schriften*, **2**, 301 (Olms Verlag, 1962), first published in 1849.
34. G. W. Leibniz, *Letter from Hanover, Germany to Johann Bernoulli, December 28, 1695, Leibnizens Mathematische Schriften*, **3**, 226 (Olms Verlag, 1962), first published in 1849.
35. G. M. Zaslavsky, *Phys. Rep.* **371**, 461 (2002).
36. E. Di Nezza, G. Patalluci and E. Valdinoci, *Bull. Sci. Math.* **136**, 521 (2012) (https://arxiv.org/pdf/1104.4345.pdf).
37. S. G. Samko, A. A. Kilbas and O. I. Marichev, *Fractional Integrals and Derivatives, Theory and Applications* (Gordon and Breach, 1993).
38. K. S. Miller and B. Ross, *An Introduction to the Fractional Calculus and Fractional Differential Equations* (Wiley, 1993).
39. I. Podlubny, *Fractional Differential Equations*, vol. 198 of Mathematics in Science and Engineering (Academic Press, 1999).
40. R. Herrmann, *Fractional Calculus: An Introduction for Physicists* (World Scientific, 2011).
41. R. Hilfer, *Applications of Fractional Calculus in Physics* (World Scientific, 2000).

42. A. A. Kilbas H. M. Srivastava and J. J. Trujillo, *Theory and Applications of Fractional Differential Equations* (Elsevier Science, 2006).

43. V. Tarasov, *Quantum Mechanics of non-Hamiltonian and Dissipative Systems,* Vol. 7 *of Monograph Series on Nonlinear Science and Complexity* (Elsevier Science, 2008).

44. M. F. Shlesinger, B. J. West and J. Klafter, *Nature* **363**, 31 (1993).

45. J. Klafter, A. Blumen and M. F. Shlesinger, *Phys. Rev. A* **35**, 3081 (1987).

46. M. F. Shlesinger, B. J. West and J. Klafter, *Phys. Rev. Lett.* **57**, 1100 (1987).

47. G. M. Zaslavsky, *Physica D* **76**, 110 (1994).

48. A. I. Saichev and G. M. Zaslavsky, *Chaos* **7**, 753 (1997).

49. G. Zimbardo, P. Veltri, G. Basile and S. Principato, *Phys. Plasmas* **2**, 2653 (1995).

50. R. N. Mantegna and H. E. Stanley, *Nature* **376**, 46 (2002).

51. G. M. Viswanathan, V. Afanasyev, S. V. Buldyrev, E. J. Murphy, P. A. Prince and H. E. Stanley, *Nature* **381**, 413 (1996).

52. B. J. West and W. Deering, *Phys. Rep.* **246**, 1 (1994).

53. F. Bardou, J. P. Bouchaud, O. Emile, A. Aspect, and C. Cohen-Tannoudji, *Phys. Rev. Lett.* **72**, 203 (1994).

54. B. Saubamea, M. Leduc, and C. Cohen-Tannoudji, *Phys. Rev. Lett.* **83**, 3796 (1999).

55. S. Schaufler, W. P. Schleich and V. P. Yakovlev, *Phys. Rev. Lett.* **83**, 3162 (1999).

56. H. Katori, S. Schlipf and H. Walter, *Phys. Rev. Lett.* **79**, 2221 (1997).

57. N. Mercadier, W. Guerin, M. Chevrollier and R. Kaiser, *Lévy flights of photons in hot atomic vapours,* arXiv:0904.2454v1.

58. B. G. Klappauf, W. H. Oskay, D. A. Steck and M. G. Raizen, *Phys. Rev. Lett.* **81**, 4044 (1998).

59. B. Sundaram and G. Zaslavsky, *Phys. Rev. E* **59**, 7231 (1999).

60. A. Iomin and G. Zaslavsky, *Chaos* **10**, 147 (2000).

61. P. Lévy, *Théorie de l'Addition des Variables Aléatoires,* 2nd ed. (Gauthier-Villars, Paris, 1954).

62. A. Y. Khintchine and P. Lévy, *C.R. Acad. Sci.* (Paris) **202**, 374 (1936).

63. W. Feller, *An Introduction to Probability Theory and its Applications* (John Wiley & Sons, 1966).

64. B. V. Gnedenko and A. N. Kolmogorov, *Limit Distributions for Sums of Random Variables,* (Addison-Wesley 1954).

65. V. M. Zolotarev, *One-dimensional Stable Distributions* (American Mathematical Society,1986).

66. V. V. Uchaikin and V. M. Zolotarev, *Chance and Stability: Stable Distributions and their Applications* (VSP, 1999).

67. N. Laskin, *Phys. Lett. A* **268**, 298 (2000).

68. N. Wiener, *Proc. London Math. Soc.* **22**, 454 (1924).

69. K. Nishimoto, *Fractional Calculus* (University of New Haven Press, 1989).

70. P. L. Butzer and U. Westphal, *An Introduction to Fractional Calculus,* Ch. 1 in *Applications of Fractional Calculus in Physics* (Ed. R. Hilfer) (World Scientific, 2000) pp. 1-85.

71. B. A. Stickler, *Phys. Rev. E* **88**, 012120 (2013).
72. F. Pinsker, W. Bao, Y. Zhang, H. Ohadi, A. Dreismann and J. Baumberg, *Phys. Rev. B* **92**, 195310 (2015) (https://arxiv.org/pdf/1508.03621.pdf).
73. C. Weisbuch et al., *Phys. Rev. Lett.* **69**, 3314 (1992).
74. D. A. Tayurskii and Y. V. Lysogorskiy, *Chinese Science Bulletin* **56**, 3617 (2011) (https://arxiv.org/pdf/1012.2949.pdf).
75. R. Herrmann, *J. Phys. G* **34**, 607 (2007) (https://arxiv.org/abs/nucl-th/0610091).
76. R. Herrmann, *Physica A* **389**, 4613 (2010) (https://arxiv.org/abs/1007.1084).
77. S. Longhi, *Opt. Lett.* **40**, 1117 (2015) (https://arxiv.org/pdf/1501.02061.pdf).
78. J. C. Gutiérrez-Vega, *Opt. Lett.* **32**, 1521 (2007).
79. S. C. Lim, *Physica A* **363**, 269 (2006).
80. G. Parisi and Y.-S. Wu, *Sci. Sinica* **24**, 483 (1981).
81. S. C. Lim and S. V. Muniandy, *Phys. Lett. A* **324**, 396 (2004).
82. N. Laskin, *Physica A* **287**, 482 (2000).
83. J. Zinn-Justin, *Quantum Field Theory and Critical Phenomena* (Clarendon Press, 1996).
84. A. Einstein, *Ann. Phys.* (Leipzig) **17**, 549 (1905). *English translation: Investigations on the Theory of Brownian Movement* (Dover, 1956).
85. C. W. Gardiner, *Handbook of Stochastic Methods: For Physics, Chemistry and the Natural Sciences*, Springer Series in Synergetics, (Springer, 1996).
86. I. S. Gradshteyn and I. M. Ryzhik, *Table of Integrals, Series, and Products*, 7th ed., Editors A. Jeffrey and D. Zwillinger (Academic Press, 2007).
87. http://dlmf.nist.gov/10.22.51
88. M. Riesz, *Acta Math.* **81**, 1 (1949).
89. V. V. Yanovsky, A. V. Chechkin, D. Schertzer, A. V. Tur, *Physica A* **282**, 13 (2000).
90. J. Holtsmark, *Ann. Phys.* **58**, 577 (1917).
91. S. Chandrasekhar, *Rev. Mod. Phys.* **15**, 1 (1943).
92. N. Laskin, *Phys. Rev. E* **62**, 3135 (2000).
93. N. Laskin, *Commun. Nonlinear Sci. Numer. Simul.* **12**, 2 (2007).
94. L. D. Landau and E. M. Lifshitz, *Quantum Mechanics (Non-relativistic Theory)*, Vol. 3, 3rd ed., *Course of Theoretical Physics* (Butterworth-Heinemann, 2003).
95. N. Laskin, *Chaos* **10**, 780 (2000).
96. N. Laskin, *Phys. Rev. E* **66**, 056108 (2002).
97. D. A. Tayurskii and Y. V. Lysogorskiy, *J. Phys. Conf. Ser.* **394**, 012004 (2012).
98. N. Laskin, *Phys Rev. E* **93**, 066104 (2016).
99. J. Dong and M. Xu, *J. Math. Phys.* **49**, 052105 (2008).
100. F. Bloch, *Z. für Phys.* **74**, 295 (1932).
101. W. Heisenberg, *Z. für Phys.* **43**, 172 (1927), (English translation in J.A. Wheeler and W.H. Zurek, (eds.) *Quantum Theory and Measurements* (Princeton University Press, 1983), pp. 62-84).

102. E. H. Kennard, *Z. für Phys.* **44**, 326 (1927).
103. C. Fox, *Trans. Am. Math. Soc.* **98**, 395 (1961).
104. A. M. Mathai and R.. K. Saxena, *The H-function with Applications in Statistics and Other Disciplines* (Wiley Eastern, 1978).
105. M. Mathai, R. K. Saxena and H. J. Haubold, *The H-Function* (Springer, 2009).
106. H. M. Srivastava, K. C. Gupta and S.P. Goyal, *The H-function of One and Two Variables with Applications* (South Asian Publishers, 1982).
107. H. Kleinert, *Path Integrals in Quantum Mechanics, Statistics, Polymer Physics, and Financial Markets*, 4th ed. (World Scientific, 2006), Chap. 1, (1.382).
108. H. Kleinert, *Path Integrals in Quantum Mechanics, Statistics, Polymer Physics, and Financial Markets* (World Scientific; 1990), Chap. 1, (1.390).
109. X. Y. Guo and M. Y. Xu, *J. Math. Phys.* **47**, 082104 (2006).
110. K. Berkelman, *Rep. Prog. Phys.* **49**, 1 (1986).
111. D. Ruelle, *Statistical Mechanics* (W.A. Benjamin, Inc., 1969).
112. S. Ş. Bayin, *J. Math. Phys.* **53**, 042105 (2012).
113. S. Ş. Bayin, *J. Math. Phys.* **53**, 084101 (2012).
114. S. S. Bayin, *Mathematical Methods in Science and Engineering* (*Supplements of Chap.* **14**, Wiley, 2006).
115. J. Dong and M. Xu, *J. Math. Phys.* **48**, 072105 (2007).
116. E. C. de Oliveira, F. S. Costa and J. Vaz Jr., *J. Math. Phys.* **51**, 123517 (2010).
117. D. J. Griffiths, *Introduction to Quantum Mechanics*, 2nd ed. (Pearson Education, 2005), Sec. 2.5.2.
118. A. Lin, X. Jiang and F. Miao, *J. of Shandong Univ.* (*Engineering Science*) **40**, 139 (2010).
119. J. L. Heilbron and T. S. Kuhn, *The Genesis of the Bohr Atom* in Historical Studies in the Physical Sciences 1, 2, ed. R. McCormmach (Princeton University Press, 1969), pp. 211-290.
120. N. Bohr, *Phil. Mag.* **26**, 1, 476, 857 (1913).
121. N. Bohr, *Collected Works*, Vol. 4. ed. J. Rud Nielsen (North-Holland, 1977).
122. L. D. Landau and E. M. Lifshitz, *Mechanics*, Vol. 1, 3rd ed., *Course of Theoretical Physics* (Pergamon Press, 1976).
123. P. Felmer, A. Quaas and J. Tan, *Proc. of the Royal Society of Edinburgh A* **142**, 1237 (2012).
124. S. Secchi, *J. Math. Phys.* **54**, 031501 (2013).
125. S. Secchi, *Topol. Methods Nonlinear Anal.* **47**, 19 (2016)
126. G. Chen, Y. Zheng, *Fractional nonlinear Schrödinger equations with singular potential in R^n*, *arXiv*:1511.09124.
127. M. Cheng, *J. Math. Phys.* **53**, 043507 (2012).
128. M. M. Fall, F. Mahmoudi and E. Valdinoci, *Nonlinearity* **28**, 1937 (2015).
129. M. M. Fall and E. Valdinoci, *Commun. Math. Phys.* **329**, 383 (2014).
130. A. S. Davydov, *Journal of Theoretical Biology* **38**, 559 (1973).
131. A. S. Davydov and N. I. Kislukha, *Physica Status Solidi B* **59**, 465 (1973).
132. G. M. Zaslavsky, A. A. Stanislavsky and M. Edelman, *Chaos* **16**, 013102 (2006) (https://arxiv.org/abs/nlin/0508018).
133. N. Laskin and G. M. Zaslavsky, *Physica A* **368**, 38 (2006).

134. N. Laskin, *Exciton-Phonon Dynamics with Long-Range Interaction*, in *Dynamical Systems and Methods*, Eds. A. C. J. Luo, J. A. T. Machado, D. Baleanu, (Springer, 2012), pp. 311-322 (https://arxiv.org/ftp/arxiv/papers/1104/1104.1310.pdf).

135. A. S. Davydov, *Solitons in Molecular Systems* (Reidel, 1985).

136. A. S. Davydov, *Sov. Phys. Usp.* **25**, 898 (1982).

137. A. C. Scott, *Phys. Rep.* **217**, 1 (1992).

138. G. M. Zaslavsky, *Fractional Kinetics of Hamiltonian Chaotic Systems*, Ch. V in *Applications of Fractional Calculus in Physics* (Ed. R. Hilfer) (World Scientific, 2000), pp. 203–239.

139. V. E. Zakharov, *Sov. J. of Exp. Theor. Phys.* **35**, 908 (1972).

140. http://dlmf.nist.gov/1.17

141. M. Naber, *J. Math. Phys.* **45**, 3339 (2004).

142. G. C. Wick, *Phys. Rev.* **96**, 1124 (1954).

143. M. Caputo, *Geophys. J. R. Astr. Soc.* **13**, 529 (1967).

144. S. Wang and M. Xu, *J. Math. Phys.* **48**, 043502 (2007).

145. J. Dong and M. Xu, *J. Math. Anal. Appl.* **344**, 1005 (2008).

146. N. Laskin, *Chaos, Solitons and Fractals*, **102**, 16 (2017).

147. S. Ş. Bayin, *J. Math. Phys.* **54**, 012103 (2013).

148. M. G. Mittag-Leffler, *C. R. Acad. Sci. Paris* **137**, 554 (1903).

149. H. J. Haubold, A. M. Mathai and R. K. Saxena, *Mittag-Leffler functions and their applications* (https://arXiv:0909.0230v2 [math.CA], 2009).

150. H. Bateman, *Higher transcendental functions*, A. Erdélyi, W. Magnus, F. Oberhettinger and F. G. Tricomi Eds., Vol. I (McGraw-Hill, 1953), Sec. 1.6, formula (2).

151. H. Bateman, *Higher Transcendental Functions*, A. Erdélyi, W. Magnus, F. Oberhettinger and F. G. Tricomi Eds., Vol. III (McGraw-Hill, 1955), Sec. 18, formula (27).

152. A. M. Mathai, *A Handbook of Generalized Special Functions for Statistical and Physical Sciences* (Clarendon, 1993).

153. Z. Korichi and M. Meftah, *Theor. Math. Phys.* **186**, 433 (2016).

154. L. D. Landau and E. M. Lifshitz, *Statistical Physics*, Second revised and enlarged edition, Vol. 5 of *Course of Theoretical Physics* (Pergamon Press, 1969).

155. N. Laskin, *The European Physical Journal Special Topics* **222**, 1929 (2013).

156. http://dlmf.nist.gov/5.12.1

157. http://dlmf.nist.gov/8.17.1

158. http://dlmf.nist.gov/8.17.7

159. V. M. Eleonsky, V. G. Korolev and N. E. Kulagin, *Physica D* **110**, 223 (1997); *Chaos* **7**, 710 (1997).

160. V. M. Eleonskii, V. G. Korolev and N. E. Kulagin, *Theor. Math. Phys.* **136**, 1131 (2003).

161. E. W. Barnes, *Proc. London Math. Soc.* **6**, 141 (1908).

162. H. J. Mellin, *Math. Ann.* **68**, 395 (1910).

163. A. L. Dixon and W. L. Ferrat, *Q. J. Math. Oxford Ser.* **7**, 81 (1936).

164. B. L. J. Braaksma, *Compos. Math.* **15**, 239 (1964).

165. C. S. Meijer, *Proc. Nederl. Akad. Wetensch* **49**, 227, 344, 457, 632, 765, 936, 1062, 1165 (1946).
166. A. P. Prudnikov, Y. A. Brychkov and O. I. Marichev, *Integrals and Series* (Gordon and Breach, 1990).
167. F. Oberhettinger, *Tables of Mellin Transforms* (Springer-Verlag, 1974).
168. L. Lewin, *Dilogarithms and Associated Functions* (MacDonald, 1958).
169. H. Bateman, *Higher Transcendental Functions*, A. Erdélyi, W. Magnus, F. Oberhettinger and F. G. Tricomi Eds., Vol. I (McGraw-Hill, 1953), Sec. 1.11, formula (14).
170. http://dlmf.nist.gov/25.12
171. J. E. Robinson, *Phys. Rev.* **83**, 678 (1951).

Index

Printed in the United States
By Bookmasters